Integration Theory

T0272010

CHAPMAN & HALL MATHEMATICS SERIES

Editors:

Professor Keith Devlin	Professor Derek Goldrei	Dr James Montaldi
St Mary's College	Open University	Université de Lille
USA	UK	France

OTHER TITLES IN THE SERIES INCLUDE

Dynamical Systems
Differential equations, maps and chaotic behaviour
D.K. Arrowsmith and C.M. Place

Network Optimization
V.K. Balakrishnan

Algebraic Numbers and Algebraic Functions
P.M. Cohn

Elements of Linear Algebra
P.M. Cohn

Control and Optimization
B.D. Craven

Sets, Functions and Logic
A foundation course in mathematics
Second edition
K. Devlin

Functions of Two Variables
S. Dineen

The Dynamic Cosmos
M.S. Madsen

Full information on the complete range of Chapman & Hall mathematics books is available from the publishers.

Integration Theory

W. Filter
Professor of Analysis
University of Palermo
Italy

and

K. Weber
Professor of Mathematics
Technikum Winterthur
Switzerland

CHAPMAN & HALL
London · Weinheim · New York · Tokyo · Melbourne · Madras

Published by Chapman & Hall, 2–6 Boundary Row, London SE1 8HN, UK

Chapman & Hall, 2–6 Boundary Row, London SE1 8HN, UK

Chapman & Hall GmbH, Pappelallee 3, 69469 Weinheim, Germany

Chapman & Hall USA, 115 Fifth Avenue, New York, NY 10003, USA

Chapman & Hall Japan, ITP-Japan, Kyowa Building, 3F, 2-2-1 Hirakawacho, Chiyoda-ku, Tokyo 102, Japan

Chapman & Hall Australia, 102 Dodds Street, South Melbourne, Victoria 3205, Australia

Chapman & Hall India, R. Seshadri, 32 Second Main Road, CIT East, Madras 600 035, India

First edition 1997

© 1997 W. Filter and K. Weber

Printed in Great Britain by St Edmundsbury Press Ltd, Bury St Edmunds, Suffolk

ISBN 0 412 57680 5

A catalogue record for this book is available from the British Library

∞ Printed on permanent acid-free text paper, manufactured in accordance with ANSI/NISO Z39.48-1992 and ANSI/NISO Z39.48-1984 (Permanence of Paper).

Contents

Preface

This book contains the material from an introductory course on integration theory taught at ETH (the Swiss Federal Institute of Technology) in Zurich. Students taking the course are in their third or fourth year of tertiary studies and therefore have had substantial prior exposure to mathematics. The course assumes some familiarity with the concepts presented in the preceding courses.

Since this book is addressed to a wider audience and since different institutes have different programmes, the same assumptions cannot be made here. As explaining everything in detail would have resulted in a book of daunting dimensions, whose very size would discourage all but those of epic heroism and dedication, we have chosen a compromise: we explain in detail in the text itself only those ideas which are essential to the development of the subject matter and we have appended a separate glossary of all definitions used, adding explanations and examples as needed. The reader is, however, expected to be familiar with the basic properties of the Riemann integral as well as with basic facts from point-set topology; the latter are especially needed for Chapter 5, 'Measures on Hausdorff Spaces'.

We have chosen this course in order to preserve the character of an introduction at an intermediate level, which should nevertheless be accessible to those with limited prior knowledge, who are willing to postpone questions on matters not central to the development of the theory.

Years of experience have convinced us of the importance to the student of *active* learning, especially when confronted by new concepts from a still unfamiliar theory. We have therefore included a large number of exercises. These place less emphasis on originality and the majority of them have been kept elementary, so that even an average student should be able to complete them successfully.

The concept of the integral developed in this book is 'the optimal' one in the sense explained in *Integration Theory I* by C. Constantinescu and K. Weber (Wiley-Interscience, New York, 1985). (We shall occasionally point readers who are interested in further pursuing a deep analysis to this book and refer to it as [CW].) This concept of the integral was proposed by I.E. Segal and R.A. Kunze in their book *Integrals and Operators* (McGraw-Hill, New York, 1968), but they did not develop the theory. It results in a larger class of integrable functions than the 'usual' integrals in general – it coincides with them in the σ-finite case – and unifies abstract integration theory with the theory of integration on Hausdorff spaces. In fact, in the context of this theory, the latter simply becomes a special case of the abstract theory. We also mention that in the topological case, we arrive at Bourbaki's 'essential integral'.

There is methodical emphasis on the structural perspective. The fundamental notion for our approach is that of a vector lattice. Naturally, the basic properties of vector lattices which are needed in this book are discussed in detail. For the sake of perspicuity, we first discuss vector lattices of real-valued functions, which is completely adequate for the first chapters. Our discussion later moves to the abstract framework needed for an adequate account of, for example, L^p-spaces and the Radon–Nikodym theorem.

We wish to express our deeply felt gratitude to our teacher, Professor Corneliu Constantinescu. His enthusiasm for the theory of measure and integration and his encompassing knowledge have inspired us and left their indelible stamp upon us. We hope that we have succeeded in writing this book in his spirit.

Our sincere thanks also go to Imre Bokor for his excellent translation of the original German text and to Helmut Köditz for his expert help in producing the LaTeX files.

Finally, we would like to thank our publishers, Chapman & Hall, for their ever congenial cooperation.

Introduction

Many mathematical problems have contributed to the development of measure and integration theory. One of the most significant of these is known as the **measure problem** in $\mathbb{R}^n$.

In essence, the problem is to assign to each element A of a sufficiently broad set of subsets of $\mathbb{R}^n$ a positive number – the **content** or **measure** of A – in such a way that several obvious conditions are satisfied:

(i) The content of the union of two disjoint sets is the sum of their individual contents.

(ii) If A is a subset of B, then the content of A does not exceed that of B.

(iii) Congruent sets have the same content.

Systems of sets for which such a content can be defined are not difficult to find. Fix an orthogonal coordinate system in $\mathbb{R}^n$. Then the set $\mathfrak{R}$ of all n-dimensional rectangular prisms whose faces are parallel to the coordinate axes has this property. Recall that such a rectangular prism is the Cartesian product of closed intervals, that is, it is of the form $\prod_{k=1}^{n} [\alpha_k, \beta_k]$, where $\alpha_k \leq \beta_k$ for every $k \in \{1, \ldots, n\}$. Note that because of our requirements, for each content λ on $\mathfrak{R}$ there must be a number $\gamma \geq 0$ such that

$$\lambda\left(\prod_{k=1}^{n} [\alpha_k, \beta_k] \right) = \gamma \prod_{k=1}^{n} (\beta_k - \alpha_k),$$

and conversely, given any number $\gamma \geq 0$, this formula defines a content on $\mathfrak{R}$. Moreover, $\gamma > 0$ is the only interesting case, and if we insist that the unit cube $\prod_{k=1}^{n} [0, 1]$ have content 1, then γ must be 1. We shall use this normalization henceforth.

More generally, we may consider sets of $\mathbb{R}^n$ which are unions of finitely many sets of $\mathfrak{R}$. Each such set A may be partitioned into a finite number of rectangular prisms of $\mathfrak{R}$ which have at most boundary points in common.

If $A = \bigcup_{k=1}^{m} A_k$ and $A = \bigcup_{i=1}^{r} B_i$ are two such decompositions, then

$$\sum_{k=1}^{m} \lambda(A_k) = \sum_{i=1}^{r} \lambda(B_i).$$

If we define this number – which is independent of the decomposition of A – to be the content of A, then we obtain a content λ on the set $\mathfrak{I}$ of all finite unions of rectangular prisms of $\mathfrak{R}$, which agrees on $\mathfrak{R}$ with the previously defined content. We say that we have extended $(\mathfrak{R}, \lambda)$ to $(\mathfrak{I}, \lambda)$. The method used here for the extension is called the **partition** or **dissection method**.

Clearly the system $(\mathfrak{I}, \lambda)$ does not meet our practical needs. We cannot yet assign a content to rectangular prisms whose edges are not parallel to the coordinate axes. We may, however, attempt a further extension.

Let $\mathfrak{J}$ be the set of all subsets A of $\mathbb{R}^n$ with the property that for every $\varepsilon > 0$ there are sets $B \in \mathfrak{I}$ and $C \in \mathfrak{I}$ such that $B \subset A \subset C$ and $\lambda(C) - \lambda(B) < \varepsilon$. For each such set A the set of all intervals $[\lambda(B), \lambda(C)]$ with $B, C \in \mathfrak{I}$ and $B \subset A \subset C$ is a nesting of intervals. This nesting determines a unique real number, which we define to be the content of A and denote by $\lambda(A)$. It can be shown that λ satisfies conditions (i)–(iii). The method just described is called the **method of exhaustion**.

$\mathfrak{J}$ is quite a remarkable system of sets. The elements of $\mathfrak{J}$ are called Jordan measurable. While the reader is no doubt familiar with the system $\mathfrak{J}$ from elementary analysis, we list its properties:

(a) $\emptyset \in \mathfrak{J}$.

(b) $A \cup B \in \mathfrak{J}$, $A \cap B \in \mathfrak{J}$, and $A \setminus B \in \mathfrak{J}$ for all $A, B \in \mathfrak{J}$.

Systems of sets having these properties are called rings of sets. Thus $\mathfrak{J}$ admits certain algebraic operations. Since it can also be shown that a set which is congruent to a set in $\mathfrak{J}$ is itself in $\mathfrak{J}$, we have arrived at a system which is, in some sense, closed.

Is $(\mathfrak{J}, \lambda)$ sufficient for our purposes? Unfortunately not. In fact, there are even compact sets which are not contained in $\mathfrak{J}$. For applications of analysis, however, it is essential that it be possible to assign a content to at least the compact sets, which are of particular significance. Our task is not yet complete! A further extension is necessary. But clearly neither dissection nor the method of exhaustion offers further assistance: neither of these classical methods gives rise to a proper extension of $(\mathfrak{J}, \lambda)$. We have thus reached the precise point at which the methods of modern measure theory – which we present in detail in this book – begin to play a part.

What measure and integration theory preserves is the idea of extension to obtain more complex objects from simpler basic forms. The methods differ from the elementary ones, however, in that they make greater use of convergence; that is, they incorporate topological methods.

We are faced with a similar problem when considering the Riemann integral in $\mathbb{R}^n$. The space $\mathcal{R}$ of Riemann integrable functions also has insufficient scope for practical applications because, for example, the convergence theorems for $\mathcal{R}$ are not strong enough; for instance, the (pointwise) limit of an increasing sequence of Riemann integrable functions from $[0,1]$ to $[0,1]$ need not be Riemann integrable. We must therefore extend the Riemann integral on $\mathcal{R}$ to a larger class of functions possessing better properties, in particular stronger convergence properties. The two extension problems are so closely related that the solution of one leads to that of the other, as we shall see.

We can approach the problem from another angle. As the reader knows, the Riemann integral can be used to define a notion of distance on the space $\mathcal{R}$ of Riemann integrable functions by letting

$$d(f,g) := \int |f(x) - g(x)|\, dx.$$

Similarly, we may define the distance

$$d(A, B) := \lambda(A \triangle B)$$

for sets $A, B \in \mathfrak{J}$, where $A \triangle B := (A \setminus B) \cup (B \setminus A)$ denotes the symmetric difference of A and B. In this way, $\mathcal{R}$ and $\mathfrak{J}$ become pseudometric spaces. Such spaces may be 'completed'. We may ask whether the problem of extension can be solved by such a completion. We shall see that there is in fact a close connection between the objects which arise from the methods of measure and integration theory and these topological completions. One difficulty cannot be overlooked, however. Topologically completing $(\mathcal{R}, d)$ or $(\mathfrak{J}, d)$ results in spaces of abstract objects which can no longer necessarily be interpreted as functions on $\mathbb{R}^n$ or as subsets of $\mathbb{R}^n$. The original question concerning these concrete objects cannot be answered through the process of completion. Thus, we must find another approach. In this book, we follow the course which was first presented by the American mathematician P.J. Daniell in 1918. Daniell solved the problem of extending the integral in the sense of the examples described here. This allowed the measure problem to be solved easily. The objects obtained also fit in with the abstract process of completion mentioned above.

In Daniell's approach to integration theory, order properties rear their head everywhere. The linear functionals ℓ to be extended to integrals are *positive* and satisfy the condition that $\inf_{n \in \mathbb{N}} \ell(f_n) = 0$ for every *decreasing* sequence $(f_n)_{n \in \mathbb{N}}$ whose pointwise *infimum* is 0. (This is called 'null-continuity'). The natural domains of these functionals are *vector lattices* of functions, and so on. Chapter 2 therefore contains a broad discussion of all these notions, which are of importance later. Then, in Chapter 3, we carry out the extension procedure for null-continuous positive linear functionals and define the integral somewhat more generally than in the classical Daniell approach. This is achieved by taking the concept of a null set further than in other approaches. Of course, we also pay attention to the powerful convergence theorems like the monotone convergence theorem and the Lebesgue convergence theorem. It is not surprising that the problem of extending a positive real-valued measure μ and defining the integral with respect to μ is just a special case of the general theory: consider the functional ℓ_μ which is obtained starting from the 'obvious' definition $\ell_\mu(e_A) := \mu(A)$ for characteristic functions e_A. This important special case is studied in detail in Chapter 4. In particular, it is shown that the Daniell integrals derived from measures are precisely the ones which have the Stone property. Locally integrable functions as well as measurable sets and functions are treated in this chapter, including Egoroff's theorem and the concept of convergence in measure. Moreover, product measures and Fubini's theorem are discussed in detail. Needless to say, our development is illustrated by numerous examples, Lebesgue measure being possibly the most important of them.

We then proceed with the more advanced theory. Because of our general definition of the integral, the theory of measure and integration in Hausdorff spaces, presented in part in Chapter 5, turns out to be another special case of our earlier considerations. Locally compact and metric spaces receive special attention. Lusin's theorem and the Riesz representation theorem are presented and we use Vitali's covering theorem to show that n-dimensional Lebesgue measure solves the measure problem in $\mathbb{R}^n$ discussed above.

Chapter 6 is devoted to the basics of the $\mathcal{L}^p$-spaces, including a discussion of convergence properties, and to the concept of uniform integrability which allows us to formulate conditions which are both necessary and sufficient for the conclusion of the Lebesgue convergence theorem to hold. The properties of these spaces are considerably improved when we pass to the L^p-spaces where functions in $\mathcal{L}^p$ are identified whenever they are equal almost everywhere. But this transition finally leads us out of the realm of vector lattices of functions. Chapter 7 therefore contains a discussion of abstract vector lattices before studying some aspects of L^p-spaces.

Abstract vector lattices also provide the appropriate tool for investigating spaces of measures (Chapter 8). We discuss the structure of these spaces, touching upon the Hahn decomposition and absolute continuity (which is formulated also as a vector lattice-theoretic property), and we present (one version of) the Radon–Nikodym theorem. Finally, Chapter 9 contains a brief introduction to the theory of functions of locally finite variation and absolutely continuous functions on $\mathbb{R}$. Vector lattice techniques are applied here too.

1

Preliminaries

___

This chapter contains notation, definitions and results needed later, without aspiring to be a complete list. The reader is assumed to be familiar with elementary calculus, including the Riemann integral, a concise introduction to which can be found at the beginning of Chapter 3. Some acquaintance with the notions from point-set topology presented below is also assumed.

As usual, $\mathbb{N}$ denotes $\{1, 2, 3, \ldots\}$, the set of natural numbers. We write $\mathbb{Z}, \mathbb{Q}, \mathbb{R}$ for the sets of integers, of rational numbers and of real numbers, respectively.

We write $x := y$ when we mean that x is defined to be equal to y. For example, we could have written $\mathbb{N} := \{1, 2, 3, \ldots\}$ above.

We write $P \Rightarrow Q$, with P and Q propositions, to denote that P **implies** Q. We say that P and Q are **equivalent** or that P holds if and only if Q holds, if P implies Q and Q implies P, and we write $P \Leftrightarrow Q$ in this case. We write $P :\Leftrightarrow Q$ when P is defined to be equivalent to Q. For example, we would write: 'The integer n is divisible by 3 $:\Leftrightarrow n = 3k$ for some integer k.'

We occasionally use the symbol $\exists$ as abbreviation for 'there is (are) ...', and $\forall$ for 'for all ...'.

Given a set X, we write $x \in X$ to denote that x is an element of X. The expression $x \notin X$ signifies that x is not an element of X. If P is a property referring to the elements of X, we write

$$\{x \in X \mid P(x)\}$$

for the set of all $x \in X$ with the property P. The symbol $\emptyset$ denotes the **empty set**, i.e. the set having no elements.

If A and B are sets, then A is said to be a **subset** of B (in symbols: $A \subset B$) if every element of A is an element of B. We write $\mathfrak{P}(X)$ for the **power set** of the set X, i.e. $\mathfrak{P}(X)$ denotes the set of all subsets of X. If P is a property of subsets of a set X, we write

$$\{A \subset X \mid P(A)\}$$

for the set of all subsets of X with the property P.

Given sets A, B we write $A \cup B$ for the **union** of A and B, and $A \cap B$ for the **intersection** of A and B, i.e.

$$A \cup B := \{x \mid x \in A \text{ or } x \in B\} \quad \text{and} \quad A \cap B := \{x \mid x \in A \text{ and } x \in B\}.$$

This can be extended easily to more than two terms. For example

$$A_1 \cup \cdots \cup A_n := \{x \mid x \in A_j \text{ for some } j \in \{1, \ldots, n\}\}.$$

The sets A and B are said to be **disjoint** if $A \cap B = \emptyset$. The **difference** $A \setminus B$ of A and B is the set of all $x \in A$ not belonging to B.

Given a set $\mathfrak{R}$ of sets, we write

$$\bigcup \{A \mid A \in \mathfrak{R}\} \quad \text{or} \quad \bigcup_{A \in \mathfrak{R}} A$$

for the **union** of the sets belonging to $\mathfrak{R}$. In other words, it is the set of all those x which belong to A for some $A \in \mathfrak{R}$. If I is a set and if A_ι is also a set whenever $\iota \in I$, then we write

$$\bigcup \{A_\iota \mid \iota \in I\} \quad \text{or} \quad \bigcup_{\iota \in I} A_\iota$$

for the union of all the sets A_ι with ι running through I. We write $\bigcup_{k=1}^n A_k$ for $\bigcup \{A_k \mid k \in \{1, \ldots, n\}\}$. Note that $\bigcup_{A \in \emptyset} A$ and $\bigcup_{\iota \in \emptyset} A_\iota$ are empty.

The corresponding notation for **intersections** is defined analogously. However, we always assume in the case of intersection that $\mathfrak{R}$ and I are non-empty.

Recall **De Morgan's laws**. Given an arbitrary set A and a non-empty set of sets $\mathfrak{R}$:

$$A \setminus \left(\bigcup \{B \mid B \in \mathfrak{R}\} \right) = \bigcap \{A \setminus B \mid B \in \mathfrak{R}\},$$
$$A \setminus \left(\bigcap \{B \mid B \in \mathfrak{R}\} \right) = \bigcup \{A \setminus B \mid B \in \mathfrak{R}\}.$$

Let X and Y be sets. The **Cartesian product** of X and Y is the set

$$X \times Y := \{(x, y) \mid x \in X, y \in Y\}$$

of all ordered pairs (x, y), with $x \in X$ and $y \in Y$.

A **map** or **mapping** or **function** from a set X to a set Y is a subset φ of $X \times Y$ such that for each $x \in X$ there is a unique $y \in Y$ with $(x, y) \in \varphi$. We write $y = \varphi(x)$ for $(x, y) \in \varphi$ in such a case. The set X is called the **domain** of φ, and Y is called the **codomain** of φ. The **range** of φ is the set of all $y \in Y$ such that there is an $x \in X$ with $(x, y) \in \varphi$. We usually write mappings φ from X to Y as

$$\varphi : X \longrightarrow Y, \quad x \longmapsto \varphi(x)$$

or $\varphi : X \to Y$. We write Y^X for the set of all mappings from X to Y.

Let X, Y, Z be sets and $\varphi : X \to Y$ and $\psi : Y \to Z$ mappings. Then the **composition** $\psi \circ \varphi$ is defined by

$$\psi \circ \varphi : X \longrightarrow Z, \quad x \longmapsto \psi(\varphi(x)).$$

Let X, Y be sets. Then a mapping $\varphi : X \to Y$ is called

- **injective** if $\varphi(x_1) = \varphi(x_2)$ implies that $x_1 = x_2$, for every $x_1, x_2 \in X$;
- **surjective** if for every $y \in Y$ there is an $x \in X$ with $y = \varphi(x)$;
- **bijective** if it is both injective and surjective.

Given a bijective mapping $\varphi : X \to Y$ there is a unique bijective **inverse mapping**, denoted by φ^{-1}, from Y to X satisfying $\varphi \circ \varphi^{-1} = id_Y$ and $\varphi^{-1} \circ \varphi = id_X$, where for the set A

$$id_A : A \longrightarrow A, \quad x \longmapsto x$$

is the **identity map** on A.

Let $\varphi : X \to Y$ be a map and take $A \subset X$, $B \subset Y$. Then

$$\varphi(A) := \{\varphi(x) \mid x \in A\} \quad \text{and} \quad \varphi^{-1}(B) := \{x \in X \mid \varphi(x) \in B\}$$

are called respectively the **image** of A and the **pre-image** of B under φ. The mapping

$$\varphi|_A : A \longrightarrow Y, \quad x \longmapsto \varphi(x)$$

is called the **restriction** of φ to A.

Let $I \neq \emptyset$ be a set and for each $\iota \in I$, let $A_\iota \neq \emptyset$ be a set. The **axiom of choice** asserts that there is a function $\varphi : I \to \bigcup_{\iota \in I} A_\iota$ assigning to each $\iota \in I$ an element $x_\iota \in A_\iota$. (Note that this is trivially true if I is finite.)

A set X is **finite** if for some $n \in \mathbb{N}$ there is an injective mapping $\varphi : X \to \{1, \ldots, n\}$. The set X is called **countable** if there is an injective mapping $\varphi : X \to \mathbb{N}$. Every finite set is therefore countable. The set X is called **uncountable** if it is not countable.

A **sequence** in a set X is a mapping $\varphi : \mathbb{N} \to X$. In general, we write $(x_n)_{n \in \mathbb{N}}$ instead of φ, where $x_n := \varphi(n)$ for $n \in \mathbb{N}$.

Let I be a finite set, and for each $\iota \in I$, let X_ι be a set. We define the **Cartesian product** $\prod_{\iota \in I} X_\iota$ to be the set of all mappings

$$\varphi : I \to \bigcup_{\iota \in I} X_\iota$$

such that $\varphi(\iota) \in X_\iota$ for each $\iota \in I$. We denote the elements of $\prod_{\iota \in I} X_\iota$ suggestively by $(x_\iota)_{\iota \in I}$, with $x_\iota \in X_\iota$ for $\iota \in I$. Let J, K be disjoint subsets of I with $I = J \cup K$. Then the map

$$\prod_{\iota \in I} X_\iota \longrightarrow \left(\prod_{\iota \in J} X_\iota\right) \times \left(\prod_{\iota \in K} X_\iota\right), \quad (x_\iota)_{\iota \in I} \longmapsto ((x_\iota)_{\iota \in J}, (x_\iota)_{\iota \in K})$$

is bijective. Thus we may identify $\prod_{\iota \in I} X_\iota$ with $\left(\prod_{\iota \in J} X_\iota\right) \times \left(\prod_{\iota \in K} X_\iota\right)$ in a natural way.

For n sets $X_1, \ldots, X_n$, we define $\prod_{k=1}^{n} X_k$, or $X_1 \times \cdots \times X_n$, to be the set of all ordered n-tuples $(x_1, \ldots, x_n)$, with $x_k \in X_k$ for every $k \in \{1, \ldots, n\}$. We identify the **Cartesian product** $\prod_{k=1}^{n} X_k$ in a natural way with $\prod_{k \in \{1, \ldots, n\}} X_k$ via the map which assigns to each $(x_1, \ldots, x_n) \in \prod_{k=1}^{n} X_k$ the element $(x_k)_{k \in \{1, \ldots, n\}}$ of $\prod_{k \in \{1, \ldots, n\}} X_k$.

Let X be a set. A **relation** R on X is a subset R of $X \times X$. We write xRy if $(x, y) \in R$. The relation $\sim$ is called an **equivalence relation** on X if for all $x, y, z \in X$:

$$x \sim x; \qquad x \sim y \implies y \sim x; \qquad x \sim y \text{ and } y \sim z \implies x \sim z.$$

In this case, $\{y \in X \mid x \sim y\}$ is called the **equivalence class** of x.

We next recall the notions and facts from topology we shall need below. Let X be a set. A set $\mathfrak{T}$ of subsets of X is called a **topology** on X if

$$\emptyset \in \mathfrak{T}, \; X \in \mathfrak{T};$$
$$T_1, T_2 \in \mathfrak{T} \implies T_1 \cap T_2 \in \mathfrak{T};$$
$$\mathfrak{S} \subset \mathfrak{T} \implies \bigcup \{S \mid S \in \mathfrak{S}\} \in \mathfrak{T}.$$

The pair $(X, \mathfrak{T})$ is called a **topological space** and the elements of $\mathfrak{T}$ are called the **open** subsets of X. (When there is no danger of confusion we speak simply of 'the topological space X'.) If the complement $X \setminus F$ of $F \subset X$ is open, then F is called **closed**. Every subset A of X contains a largest open subset $\overset{\circ}{A}$ of X called the **interior** of A, and is contained in a smallest closed subset $\overline{A}$ of X called the **closure** of A. We have

$$\overset{\circ}{A} = \bigcup \{T \in \mathfrak{T} \mid T \subset A\} \quad \text{and} \quad \overline{A} = \bigcap \{F \mid F \supset A, \, F \text{ closed}\}.$$

The topology $\mathfrak{P}(X)$ on X is the **discrete topology** on X. Thus every subset of X is both open and closed in the discrete topology on X.

A subset $\mathfrak{B}$ of the topology $\mathfrak{T}$ is called a **base** for $\mathfrak{T}$ if every open T can be written as the union of elements of $\mathfrak{B}$.

Let $(X, \mathfrak{T})$ be a topological space, and let $A \subset X$. Then

$$\{A \cap T \mid T \in \mathfrak{T}\}$$

is known as the **relative topology** on A and A (with this topology) is called a **subspace** of X. If $\mathfrak{B}$ is a base for $\mathfrak{T}$, then $\{A \cap B \mid B \in \mathfrak{B}\}$ is a base for the relative topology on A. The subset A of X is called **dense** in X if $\overline{A} = X$, and it is called **nowhere dense** in X if $\overline{A}$ has empty interior.

X is called **separable** if it possesses a countable dense subset. If X possesses a countable base, then it is separable.

Let x be a point in the topological space X. Then $U \subset X$ is a **neighbourhood** of x if there is some open subset T of X such that $x \in T \subset U$. A **Hausdorff space** is a topological space in which any pair of distinct points possesses two disjoint neighbourhoods.

Let A be a subset of the topological space X and take $x \in X$. Then x is an **interior point** of A if $x \in \overset{\circ}{A}$, i.e. if there is a neighbourhood of x contained in A. We have $x \in \overline{A}$ if and only if $U \cap A \neq \emptyset$ for every neighbourhood U of x. The point x is an **accumulation point** of A if $x \in \overline{A \setminus \{x\}}$, i.e. if given any neighbourhood U of x, $(U \setminus \{x\}) \cap A \neq \emptyset$.

The topological space $(X, \mathfrak{T})$ is called **compact** if it is Hausdorff and satisfies the following condition: given any subset $\mathfrak{G}$ of $\mathfrak{T}$ with $X = \bigcup_{S \in \mathfrak{G}} S$ (such a set $\mathfrak{G}$ is called an **open cover** of X), there is a finite subset $\mathfrak{G}_1$ of $\mathfrak{G}$ with $X = \bigcup_{S \in \mathfrak{G}_1} S$. This condition is equivalent to the finite intersection property: if $\mathfrak{F}$ is a subset of the set of closed sets of X such that $\bigcap_{F \in \mathfrak{F}} F = \emptyset$, then there is a finite subset $\mathfrak{F}_1$ of $\mathfrak{F}$ with $\bigcap_{F \in \mathfrak{F}_1} F = \emptyset$. A subset A of the topological space X is called **compact** if it is compact in the relative topology. Every compact subset of the Hausdorff space X is closed, and every closed subset of a compact set is compact. Moreover, finite unions and arbitrary intersections of compact sets are compact. The subset A of X is **relatively compact** if $\overline{A}$ is compact.

A Hausdorff space X is called **locally compact** if every point of X possesses a compact neighbourhood. In this case, every neighbourhood of the point $x \in X$ contains a compact neighbourhood of x.

For $k = 1, \ldots, n$ let X_k be a topological space. Then the **product topology** on the Cartesian product $X := X_1 \times \cdots \times X_n$ is the topology for which all sets of the form $T_1 \times \cdots \times T_n$ (T_k open in X_k for every $k \in \{1, \ldots, n\}$) constitute a base. If K_k is a compact subset of X_k for each $k \in \{1, \ldots, n\}$, then $K_1 \times \cdots \times K_n$ is a compact subset of X.

Let $(X, \mathfrak{T})$ and $(Y, \mathfrak{G})$ be two topological spaces and $f : X \to Y$ a mapping. Then f is said to be **continuous at** $x \in X$ if given any neighbourhood V of $f(x)$, the set $f^{-1}(V)$ is a neighbourhood of x. The map f is called **continuous** if it is continuous at every $x \in X$, or equivalently, if $f^{-1}(S) \in \mathfrak{T}$ for every $S \in \mathfrak{G}$, or equivalently, if the pre-image of every closed subset of Y is closed in X. If f is continuous, then the image under f of every compact subset of X is compact in Y. A bijective continuous map $f : X \to Y$ for which f^{-1} is also continuous, is a **homeomorphism** of X onto Y.

We shall need the following version of **Urysohn's theorem**. Let K be a compact subset of the locally compact space X and T an open set containing K. Then there is a continuous real-valued function f on X such that $f(X) \subset [0, 1]$, $f|_K = 1$, and $f = 0$ outside some compact set L with $K \subset L \subset T$.

The **Tietze extension theorem** holds under the same hypotheses: given a continuous function $f : K \to \mathbb{R}$, there is a continuous function $g : X \to \mathbb{R}$ such that $g|_K = f$ and g vanishes outside some compact set L satisfying $K \subset L \subset T$.

Let X be a set. A mapping $d : X \times X \to \mathbb{R}_+$ is a **pseudometric** on X and (X, d) is called a **pseudometric space**, if for all $x, y, z \in X$:

$$d(x, x) = 0;$$
$$d(x, y) = d(y, x);$$
$$d(x, z) \leq d(x, y) + d(y, z) \qquad \text{(triangle inequality)}.$$

If in addition the pseudometric d satisfies

$$d(x, y) = 0 \implies x = y,$$

then it is called a **metric** on X; in this case (X, d) is a **metric space**. Every pseudometric d on X generates a natural topology on X. This **topology generated by d** is given by defining $T \subset X$ to be d-open if and only if for each $x \in T$ there is an $\varepsilon > 0$ with

$$B(x, \varepsilon) := \{ y \in X \mid d(x, y) < \varepsilon \}$$

a subset of T. Topological notions and assertions in a pseudometric space always refer to the topology generated by the pseudometric. If d is a metric, then X is Hausdorff.

A topological space $(X, \mathfrak{T})$ is **(pseudo)metrizable** if there is a (pseudo)metric d on X which generates the topology $\mathfrak{T}$. Every locally compact space with a countable base is metrizable.

A sequence $(x_n)_{n \in \mathbb{N}}$ in the pseudometric space (X, d) is said to **converge** to $x \in X$ if for every $\varepsilon > 0$ there is an $m \in \mathbb{N}$ such that

$$d(x_n, x) < \varepsilon \qquad \text{for all } n \geq m.$$

A subset A of X is dense in X if for every $x \in X$ there is a sequence $(x_n)_{n \in \mathbb{N}}$ in A converging to x.

A pseudometric space is separable if and only if it has a countable base. Hence every subspace of a separable pseudometric space is separable.

Let (X, d) and (Y, d') be pseudometric spaces. Then the map $f : X \to Y$ is continuous at $x \in X$ if and only if for each $\varepsilon > 0$ there is a $\delta > 0$ with

$$d'(f(y), f(x)) < \varepsilon \qquad \text{whenever } d(y, x) < \delta.$$

The map f is called **uniformly continuous** if for each $\varepsilon > 0$ there is a $\delta > 0$ such that

$$d'(f(y), f(x)) < \varepsilon \qquad \text{for all } x, y \in X \text{ with } d(y, x) < \delta.$$

Finally, f is an **isometry** if it is bijective and

$$d'(f(y), f(x)) = d(y, x) \qquad \text{for all } x, y \in X.$$

A sequence $(x_n)_{n \in \mathbb{N}}$ in the pseudometric space (X, d) is a **Cauchy sequence** if for every $\varepsilon > 0$ there is an $m \in \mathbb{N}$ such that

$$d(x_k, x_l) < \varepsilon \qquad \text{for all } k, l \geq m.$$

(X, d) is called **complete** if every Cauchy sequence in X converges to an element of X. A complete (pseudo)metric space (Y, d') is called a **completion** of the (pseudo)metric space (X, d) if there is an isometry of (X, d) onto a dense subspace of (Y, d').

Now let X be a real vector space. A **norm** on X is a mapping $\|\cdot\| : X \to \mathbb{R}$ such that for all $x, y \in X$ and every $\alpha \in \mathbb{R}$:

$$\|x\| \geq 0; \qquad \|x\| = 0 \iff x = 0;$$
$$\|\alpha x\| = |\alpha| \, \|x\|;$$
$$\|x + y\| \leq \|x\| + \|y\| \qquad \text{(triangle inequality)}.$$

If $\|\cdot\|$ is a norm on X, then $(X, \|\cdot\|)$ is called a **normed space**. A normed space is always a metric space with respect to the **induced metric**

$$d(x, y) := \|x - y\| \qquad \text{for all } x, y \in X.$$

If $(X, \|\cdot\|)$ is norm complete (i.e. complete with respect to the induced metric), then it is called a **Banach space**.

If X is a compact metric space, then $\mathcal{C}(X)$, the space of all continuous real-valued functions on X, is a separable Banach space with respect to the **supremum norm**

$$\|f\| := \sup\{|f(x)| \,|\, x \in X\}.$$

(This theorem is a consequence of the Stone–Weierstrass theorem and may be found in most books on Banach spaces.)

Let X be a real vector space with an inner product $< \cdot \,|\, \cdot >$. Then the **induced norm** on X is defined by

$$\|x\| := \sqrt{< x|x >} \qquad \text{for all } x \in X.$$

The space X is a **Hilbert space** if it is a Banach space with respect to the induced norm.

For every $n \in \mathbb{N}$, the map

$$(x, y) \longmapsto \sum_{k=1}^{n} x_k y_k \qquad \text{(where } x = (x_1, \dots, x_n), \, y = (y_1, \dots, y_n))$$

is an inner product on $\mathbb{R}^n$. With respect to the induced **Euclidean norm**

$$\|x\| := \left(\sum_{k=1}^{n} x_k^2 \right)^{1/2},$$

or, equivalently, the induced **Euclidean metric**

$$d(x, y) := \left(\sum_{k=1}^{n} (x_k - y_k)^2 \right)^{1/2},$$

the space $\mathbb{R}^n$ is a complete locally compact space with a countable base.

Every open subset of $\mathbb{R}$ can be written as a union of countably many pairwise disjoint open intervals. We denote **open intervals** by $]a, b[$, i.e.

$$]a, b[:= \{x \in \mathbb{R} \,|\, a < x < b\},$$

while the **closed interval** $[a, b]$ is defined as

$$[a, b] := \{x \in \mathbb{R} \,|\, a \le x \le b\}.$$

A sequence $(f_n)_{n \in \mathbb{N}}$ of real-valued functions on a set X is called **uniformly convergent** to a real-valued function f on X if for every $\varepsilon > 0$ there is an $m \in \mathbb{N}$ such that

$$|f_n(x) - f(x)| < \varepsilon \qquad \text{for all } n \ge m \text{ and all } x \in X.$$

In such a case, if each f_n is continuous, then f is also continuous.

2

Function spaces and functionals

2.1 Ordered sets and lattices

There are several paths leading to the notion of the integral. As mentioned in the Introduction, this book employs the method developed by P. J. Daniell. Daniell's construction relies heavily upon the natural order structure of function spaces.

The order structure proves to be fundamental even in relation to the more subtle points of integration theory. This fact alone makes some observations about ordered sets indispensable.

An **ordered set** is a set X together with a relation $\leq$ on X satisfying

(i) $x \leq x$ for every $x \in X$;

(ii) $x \leq y$ and $y \leq x \implies x = y$, for all $x, y \in X$;

(iii) $x \leq y$ and $y \leq z \implies x \leq z$, for all $x, y, z \in X$.

Such a relation is called a **partial ordering** on X. Note that there may very well be elements of X which are not comparable by such a relation: i.e. there may be $x, y \in X$ for which neither $x \leq y$ nor $y \leq x$. An ordered set is **totally ordered** if given any distinct elements x and y of X, either $x \leq y$ or $y \leq x$.

Given elements x, y of the ordered set X, we write $x \geq y$ if and only if $y \leq x$. We write $x < y$ to indicate that $x \leq y$ but $x \neq y$, and finally we write $x > y$ if and only if $y < x$.

Let X be an ordered set and $A \subset X$. An element $x \in X$ is

- an **upper bound** for A $(A \leq x)$ if $y \leq x$ for every $y \in A$;

- a **lower bound** for A $(x \leq A)$ if $x \leq y$ for every $y \in A$;

- a **largest element** or **maximum** of A $(x = \max A)$ if $A \leq x$ and $x \in A$;

- a **smallest element** or **minimum** of A $(x = \min A)$ if $x \leq A$ and $x \in A$.

A largest or smallest element of the set $A \subset X$ is always uniquely determined. If, for example, x and y are largest elements of A, then according to the definition of largest elements of A, $x \leq y$ and $y \leq x$. Thus (ii) ensures that $x = y$.

$A \subset X$ is said to be

- **bounded above** if $A \leq x$ for some $x \in X$;
- **bounded below** if $x \leq A$ for some $x \in X$;
- **bounded** if $x \leq A \leq y$ for some $x, y \in X$.

The empty set is bounded whenever $X \neq \emptyset$.

$x \in X$ is called the

- **supremum** of A $(x = \bigvee A)$ if x is the smallest upper bound of A;
- **infimum** of A $(x = \bigwedge A)$ if x is the largest lower bound of A.

Given $x, y \in X$, define

$$x \vee y := \bigvee \{x, y\} \qquad \text{and} \qquad x \wedge y := \bigwedge \{x, y\}$$

whenever these elements exist.

The existence of suprema and infima in an ordered set is by no means always assured. Their existence distinguishes a large class of ordered sets, namely lattices. An ordered set X is called a **lattice** if $x \vee y$ and $x \wedge y$ exist for all $x, y \in X$. By induction, it is clear that suprema and infima exist for arbitrary finite, non-empty subsets of a lattice.

In a lattice X, the maps

$$X \times X \longrightarrow X, \quad (x, y) \longmapsto x \vee y$$
$$X \times X \longrightarrow X, \quad (x, y) \longmapsto x \wedge y$$

may be thought of as binary algebraic operations (**lattice operations**). The following properties follow directly from the definitions:

$$x \vee y = y \vee x, \quad x \wedge y = y \wedge x \qquad (\textbf{commutativity})$$
$$x \vee (y \vee z) = (x \vee y) \vee z, \quad x \wedge (y \wedge z) = (x \wedge y) \wedge z \qquad (\textbf{associativity})$$
$$x \wedge (x \vee y) = x, \quad x \vee (x \wedge y) = x$$

These simple laws illustrate an important property of lattice operations, namely **duality**. Each law corresponds to a dual one which is obtained by interchanging all occurrences of $\wedge$ and $\vee$. This duality will prove to be an important 'regularity factor' when the lattice structure is relevant. We often use it implicitly, in that we deal with only one of a pair of dual statements, leaving it to the reader to formulate and prove its companion.

Given ordered sets X and Y, the map $\varphi : X \to Y$ is said to be **increasing** if $\varphi(x) \leq \varphi(y)$ whenever $x \leq y$. X and Y are said to be **order isomorphic** if there is a bijective map $\varphi : X \to Y$ with both φ and φ^{-1} increasing. Such a map φ is called an **order isomorphism**.

We next present several examples whose importance will become apparent later.

(a) Every totally ordered set X is a lattice. Given $x, y \in X$, $x \vee y = \max\{x, y\}$ and $x \wedge y = \min\{x, y\}$.

(b) $\mathbb{Q}$ and $\mathbb{R}$ are totally ordered sets and therefore lattices by example (a). $\mathbb{Q}$ and $\mathbb{R}$ have neither a smallest nor a largest element.

Let $A := \{\alpha \in \mathbb{Q} | -\sqrt{2} < \alpha < \sqrt{2}\}$. A is bounded in $\mathbb{Q}$ and $\mathbb{R}$. The number 2 is an upper bound, -2 is a lower bound. In $\mathbb{R}$, $\bigvee A = \sqrt{2}$ and $\bigwedge A = -\sqrt{2}$. Neither $\bigvee A$ nor $\bigwedge A$ exists in $\mathbb{Q}$. The existence of these elements is characteristic of $\mathbb{R}$. We have:

Theorem 2.1 *Every non-empty bounded subset A of $\mathbb{R}$ has both a supremum and an infimum.*

Proof. Let $B := \{\gamma \in \mathbb{R} | A \leq \gamma\}$ and $C := \mathbb{R} \setminus B$. (C, B) is a Dedekind cut in $\mathbb{R}$, i.e. $C \cup B = \mathbb{R}$, $C \cap B = \emptyset$ and $c \leq b$ for every $c \in C$ and $b \in B$. Hence, by Dedekind's axiom, there is a unique number $\gamma \in \mathbb{R}$ such that $C \leq \gamma$ and $\gamma \leq B$. Thus $\gamma = \bigvee A$. Similarly, $\bigwedge A$ exists. □

Theorem 2.1 reveals the most important single order property of $\mathbb{R}$. We will continually appeal to this property.

A lattice X with the property that $\bigvee A$ and $\bigwedge A$ exist for every non-empty bounded subset A of X is said to be **conditionally complete**. $\mathbb{R}$ is thus a conditionally complete lattice.

(c) The needs of integration theory make it convenient to extend the ordered set $\mathbb{R}$ by adjoining two elements to act as the largest and smallest elements respectively. We set

$$\overline{\mathbb{R}} := \mathbb{R} \cup \{-\infty, \infty\}$$

and extend the ordering $\leq$ from $\mathbb{R}$ to $\overline{\mathbb{R}}$ by requiring that

$$-\infty < \alpha \text{ and } \alpha < \infty \text{ for all } \alpha \in \mathbb{R}, \qquad -\infty < \infty.$$

$\overline{\mathbb{R}}$ is order isomorphic to the closed interval $[-1, 1]$. The map

$$\varphi : \overline{\mathbb{R}} \longrightarrow [-1, 1], \quad \alpha \longmapsto \begin{cases} \frac{\alpha}{1+|\alpha|} & \text{for } \alpha \in \mathbb{R} \\ 1 & \text{for } \alpha = \infty \\ -1 & \text{for } \alpha = -\infty \end{cases}$$

is an order isomorphism. $\overline{\mathbb{R}}$ is therefore totally ordered. This extension strengthens the completeness properties of $\mathbb{R}$. In fact, we have the following theorem, whose proof is obvious.

Theorem 2.2 *Every subset of $\overline{\mathbb{R}}$ has both a supremum and an infimum. If $A \subset \overline{\mathbb{R}}$ is bounded in $\mathbb{R}$, then these coincide with the supremum and infimum in $\mathbb{R}$. If A is not bounded above in $\mathbb{R}$, then $\bigvee A = \infty$. If A is not bounded below in $\mathbb{R}$, then $\bigwedge A = -\infty$.*

A lattice in which every subset has both a supremum and an infimum is called **complete**. $\overline{\mathbb{R}}$ is a complete lattice.

The elements ∞ and $-\infty$ should not be regarded as new numbers. The arithmetic of real numbers cannot be extended to $\overline{\mathbb{R}}$ without relinquishing some of its properties. Given $\alpha \in \mathbb{R}$, we define

$$\infty + \infty := \infty, \qquad\qquad\qquad (-\infty) + (-\infty) := -\infty,$$
$$\infty + \alpha := \infty, \qquad\qquad\qquad (-\infty) + \alpha := -\infty,$$

$$\infty\alpha := \alpha\infty := \begin{cases} \infty & \text{for } \alpha > 0 \\ -\infty & \text{for } \alpha < 0, \end{cases}$$

$$(-\infty)\alpha := \alpha(-\infty) := \begin{cases} -\infty & \text{for } \alpha > 0 \\ \infty & \text{for } \alpha < 0, \end{cases}$$

$$\infty\infty := (-\infty)(-\infty) := \infty, \qquad\qquad \infty(-\infty) := (-\infty)\infty := -\infty,$$
$$\infty 0 := 0\infty := (-\infty)0 := 0(-\infty) := 0.$$

The last formula is convenient without incurring difficulties. On the other hand, $\infty + (-\infty)$ and $(-\infty) + \infty$ cannot be defined sensibly. This fact is the source of certain technical difficulties when we later introduce the integral. Finally, we define $|\infty| := |-\infty| := \infty$.

(d) We use inclusion, denoted $\subset$, to order the power set $\mathfrak{P}(X)$ of a set X. Unlike $\mathbb{R}$, $\overline{\mathbb{R}}$ and $\mathbb{Q}$, $\mathfrak{P}(X)$ is not totally ordered. However:

Theorem 2.3 $\mathfrak{P}(X)$ *is a complete lattice with respect to* $\subset$. *Given any non-empty subset* $\mathfrak{R}$ *of* $\mathfrak{P}(X)$,

$$\bigvee \mathfrak{R} = \bigcup_{A \in \mathfrak{R}} A \qquad and \qquad \bigwedge \mathfrak{R} = \bigcap_{A \in \mathfrak{R}} A.$$

Furthermore $\bigvee \emptyset = \emptyset$ *and* $\bigwedge \emptyset = X$. $\emptyset$ *is the smallest element of* $\mathfrak{P}(X)$, X *the largest.*

After these examples we turn again to more general considerations.

Let X be a set. A **family** in X **indexed by the set** I is a map $\varphi : I \to X$. The elements of I are called the **indices** of the family, and I the **indexing set**. We often write $(x_\iota)_{\iota \in I}$ to denote such a family. An important special case is $I = \mathbb{N}$. In this case, we speak of a **sequence** $(x_n)_{n \in \mathbb{N}}$ in X.

A family $(A_\iota)_{\iota \in I}$ in $\mathfrak{P}(X)$ is called **disjoint** if the sets A_ι are pairwise disjoint, i.e. if

$$A_{\iota_1} \cap A_{\iota_2} = \emptyset \qquad \text{whenever } \iota_1 \neq \iota_2.$$

Let X be an ordered set. A family $(x_\iota)_{\iota \in I}$ in X is

- **directed up** or **upward directed** if for each $\iota_1, \iota_2 \in I$ there is a λ in I with $x_{\iota_1} \leq x_\lambda$ and $x_{\iota_2} \leq x_\lambda$;
- **directed down** or **downward directed** if for each $\iota_1, \iota_2 \in I$ there is a λ in I with $x_\lambda \leq x_{\iota_1}$ and $x_\lambda \leq x_{\iota_2}$.

A sequence $(x_n)_{n \in \mathbb{N}}$ in X is

- **increasing** if $x_n \leq x_{n+1}$ for every $n \in \mathbb{N}$;
- **decreasing** if $x_{n+1} \leq x_n$ for every $n \in \mathbb{N}$;
- **strictly increasing** if $x_n < x_{n+1}$ for every $n \in \mathbb{N}$;
- **strictly decreasing** if $x_{n+1} < x_n$ for every $n \in \mathbb{N}$;
- **monotone** if it is either increasing or decreasing.

Thus increasing and decreasing sequences are important special cases of upward and downward directed families respectively.

Given a family $(x_\iota)_{\iota \in I}$ in X, define

$$\bigvee_{\iota \in I} x_\iota := \bigvee \{x_\iota \mid \iota \in I\} \quad \text{and} \quad \bigwedge_{\iota \in I} x_\iota := \bigwedge \{x_\iota \mid \iota \in I\}$$

provided the supremum (or infimum) exists.

If $(x_n)_{n \in \mathbb{N}}$ is an increasing sequence in the ordered set X and $\bigvee_{n \in \mathbb{N}} x_n$ exists, then we write $\uparrow x_n := \bigvee_{n \in \mathbb{N}} x_n$. If $(x_n)_{n \in \mathbb{N}}$ is decreasing and $\bigwedge_{n \in \mathbb{N}} x_n$ exists, then we write $\downarrow x_n := \bigwedge_{n \in \mathbb{N}} x_n$. Note that the notation $\uparrow x_n$ conveys two pieces of information:

1. $(x_n)_{n \in \mathbb{N}}$ is increasing;
2. $\bigvee_{n \in \mathbb{N}} x_n = \uparrow x_n$ exists.

The corresponding comments hold for $\downarrow x_n$. Since we only use the symbols $\uparrow$ and $\downarrow$ for sequences, we omit the indexing set and use the simplified notation instead of writing $\uparrow_{n \in \mathbb{N}}$ or $\downarrow_{n \in \mathbb{N}}$. The symbols $\uparrow$ and $\downarrow$ always refer unambiguously to sequences. This convention will spare us from one piece of cluttered notation.

Finally, we write simply $x_n \uparrow x$ $(x_n \downarrow x)$ when the sequence $(x_n)_{n \in \mathbb{N}}$ is increasing (decreasing) and $x = \uparrow x_n$ $(x = \downarrow x_n)$.

Let X be a complete lattice. Given a sequence $(x_n)_{n \in \mathbb{N}}$ in X, let

$$\limsup_{n \to \infty} x_n := \bigwedge_{n \in \mathbb{N}} \left(\bigvee_{m \geq n} x_m \right) \quad \text{(\textbf{limes superior})},$$

$$\liminf_{n \to \infty} x_n := \bigvee_{n \in \mathbb{N}} \left(\bigwedge_{m \geq n} x_m \right) \quad \text{(\textbf{limes inferior})}.$$

Note that

$$\liminf_{n \to \infty} x_n \leq \limsup_{n \to \infty} x_n$$

always holds.

The limes superior and the limes inferior define a concept of convergence for complete lattices. A sequence $(x_n)_{n \in \mathbb{N}}$ in X is said to be **order convergent** to $x \in X$ if

$$\liminf_{n \to \infty} x_n = \limsup_{n \to \infty} x_n = x.$$

x is called the **order limit** of $(x_n)_{n \in \mathbb{N}}$ or simply the **limit** of $(x_n)_{n \in \mathbb{N}}$. We write

$$x = \lim_{n \to \infty} x_n.$$

These definitions provide a natural notion of convergence for $\overline{\mathbb{R}}$ and $\mathfrak{P}(X)$ in particular. The following theorems show that this agrees with the usual notions in $\overline{\mathbb{R}}$.

Theorem 2.4 *Let $(\alpha_n)_{n \in \mathbb{N}}$ be a sequence in $\overline{\mathbb{R}}$ and α an element of $\overline{\mathbb{R}}$. Then:*

(a) If the α_n and α are elements of $\mathbb{R}$, then the following are equivalent:

(a1) $(\alpha_n)_{n \in \mathbb{N}}$ is order convergent to α.

(a2) For each $\varepsilon > 0$ there is an $m \in \mathbb{N}$ with $|\alpha - \alpha_n| < \varepsilon$ whenever $n \geq m$.

(b) If $\alpha = \infty$, then the following are equivalent:

(b1) $(\alpha_n)_{n \in \mathbb{N}}$ is order convergent to α.

(b2) For each $\gamma \in \mathbb{R}$ there is an $m \in \mathbb{N}$ with $\alpha_n \geq \gamma$ whenever $n \geq m$.

(c) If $\alpha = -\infty$, then the following are equivalent:

(c1) $(\alpha_n)_{n \in \mathbb{N}}$ is order convergent to α.

(c2) For each $\gamma \in \mathbb{R}$ there is an $m \in \mathbb{N}$ with $\alpha_n \leq \gamma$ whenever $n \geq m$.

Proof. (a1)$\Rightarrow$(a2). By hypothesis,

$$\alpha = \bigvee_{n \in \mathbb{N}} \bigwedge_{m \geq n} \alpha_m = \bigwedge_{n \in \mathbb{N}} \bigvee_{m \geq n} \alpha_m.$$

Take $\varepsilon > 0$. Then there is an $n \in \mathbb{N}$ with

$$\bigwedge_{m \geq n} \alpha_m \in \,]\alpha - \varepsilon, \alpha + \varepsilon[\quad \text{and} \quad \bigvee_{m \geq n} \alpha_m \in \,]\alpha - \varepsilon, \alpha + \varepsilon[.$$

Thus $\alpha_m \in \,]\alpha - \varepsilon, \alpha + \varepsilon[$ and $|\alpha - \alpha_m| < \varepsilon$ for every $m \geq n$.

(a2)$\Rightarrow$(a1). The hypothesis implies that for each $\varepsilon > 0$ there is an $n \in \mathbb{N}$ such that

$$\alpha - \varepsilon < \alpha_m < \alpha + \varepsilon$$

for every $m \geq n$. Thus

$$\alpha - \varepsilon \leq \bigwedge_{m \geq n} \alpha_m \leq \bigvee_{m \geq n} \alpha_m \leq \alpha + \varepsilon.$$

Since ε is arbitrary,

$$\alpha = \liminf_{n \to \infty} \alpha_n = \limsup_{n \to \infty} \alpha_n.$$

(b) and (c) can be proven analogously. $\square$

Since the Daniell integral is constructed essentially by a process of passing to monotone limits, i.e. taking limits of monotone sequences, the following result is of particular importance.

Theorem 2.5 *Let X be a complete lattice, $(x_n)_{n \in \mathbb{N}}$ a sequence in X and x an element of X. Then*

(a) $x = \uparrow x_n$ *implies that* $x = \lim_{n \to \infty} x_n$;

(b) $x = \downarrow x_n$ *implies that* $x = \lim_{n \to \infty} x_n$.

Proof. (a) Suppose that $x = \uparrow x_n$. Then $(x_n)_{n \in \mathbb{N}}$ is increasing and hence $x_n = \bigwedge_{m \geq n} x_m$ for every $n \in \mathbb{N}$. Thus

$$x = \bigvee_{n \in \mathbb{N}} \bigwedge_{m \geq n} x_m = \limsup_{n \to \infty} x_n.$$

On the other hand, $x = \bigvee_{m \geq n} x_m$ for every $n \in \mathbb{N}$, and thus

$$x = \bigwedge_{n \in \mathbb{N}} \bigvee_{m \geq n} x_m = \liminf_{n \to \infty} x_n.$$

The proof of (b) is similar. $\qquad\qquad\qquad\qquad\qquad\qquad\qquad\qquad\square$

Notation: In the rest of this book we frequently use lattice operations simultaneously in $\overline{\mathbb{R}}$ and in lattices of functions, which are introduced in the next section. In order to clarify the distinction, we use the symbols sup and inf in $\overline{\mathbb{R}}$ instead of $\bigvee$ and $\bigwedge$ in the case of suprema or infima of sequences or sets.

For the lattice operations on the two elements α and β of $\overline{\mathbb{R}}$, however, we retain the notation $\alpha \vee \beta$ and $\alpha \wedge \beta$.

Exercises

1. Are the following relations order relations?

   (a) $X :=$ the set of all oxen, $a \preccurlyeq b$ if and only if a is at least as stupid as b.

   (b) $X :=$ the set of all men, $a \preccurlyeq b$ if and only if a is a relative of b.

2. Do there exist special cases in which $\mathfrak{P}(X)$ is totally ordered?

3. Consider the relation $\preccurlyeq$ on $\mathbb{N}$, defined by:

$$m \preccurlyeq n \iff n = km \text{ for some } k \in \mathbb{N}.$$

   Show that:

   (a) $\preccurlyeq$ is an order relation on $\mathbb{N}$, and if $m \preccurlyeq n$, then $m \leq n$.

   (b) $\mathbb{N}$ is a conditionally complete lattice with respect to $\preccurlyeq$. Given a nonempty family $(n_\iota)_{\iota \in I}$ in $\mathbb{N}$, $\bigwedge_{\iota \in I} n_\iota$ is the greatest common divisor of all n_ι. If $(n_\iota)_{\iota \in I}$ is bounded above, $\bigvee_{\iota \in I} n_\iota$ is the least common multiple of all the n_ι.

(c) Given $m, n, p \in \mathbb{N}$, $mp \preccurlyeq np$ whenever $m \preccurlyeq n$.

(d) $1 \preccurlyeq 2$ but it is not the case that $2 \preccurlyeq 3$. Hence it does not follow from $m \preccurlyeq n$ that $m + p \preccurlyeq n + p$. This shows that the order relation $\preccurlyeq$ does not bear the same relationship to the arithmetic operations on $\mathbb{N}$ as does $\leq$.

4. Determine which of the following pairs are order isomorphic:

$$[-1, 1] \text{ and } \mathbb{R} \text{ (or } \overline{\mathbb{R}}\text{),}$$
$$]-1, 1[\text{ and } \mathbb{R} \text{ (or } \overline{\mathbb{R}}\text{),}$$
$$]-1, 1[\text{ and } [-1, 1].$$

5. Define a relation on $\mathbb{R}^2$ by

$$(\alpha, \beta) \preccurlyeq (\gamma, \delta) \iff \alpha < \gamma \text{ or } (\alpha = \gamma \text{ and } \beta \leq \delta).$$

Show that $\mathbb{R}^2$ is a lattice with respect to $\preccurlyeq$, which is not conditionally complete. (Consider the sequence $\big((1/n, 0)\big)_{n \in \mathbb{N}}$.)

6. This exercise investigates the relation $\preccurlyeq$ on $\mathbb{R}$ defined by

$$\alpha \preccurlyeq \beta \iff \text{for some } \gamma \in \mathbb{R}, \ \gamma \geq 1, \ \beta = \alpha\gamma.$$

Verify the following properties of this relation.

(a) $\preccurlyeq$ is an order relation.

(b) Given $\alpha, \beta > 0$, $\alpha \preccurlyeq \beta \iff \alpha \leq \beta$.

(c) Given $\alpha, \beta < 0$, $\alpha \preccurlyeq \beta \iff \alpha \geq \beta$.

(d) If $\alpha \neq 0$, then neither $\alpha \preccurlyeq 0$ nor $0 \preccurlyeq \alpha$ holds.

(e) $\mathbb{R}$ is not a lattice with respect to $\preccurlyeq$.

(f) The sequence $(\alpha_n)_{n \in \mathbb{N}}$ in $\mathbb{R}$ converges to $\alpha \neq 0$ with respect to $\preccurlyeq$ if and only if $(\alpha_n)_{n \in \mathbb{N}}$ converges to α in the usual sense.

(g) $(0)_{n \in \mathbb{N}}$ is the only sequence which converges to 0 with respect to $\preccurlyeq$.

7. Determine whether the sequences $(A_n)_{n \in \mathbb{N}}$, $(B_n)_{n \in \mathbb{N}}$, $(C_n)_{n \in \mathbb{N}}$, $(D_n)_{n \in \mathbb{N}}$ in $\mathfrak{P}(\mathbb{R})$ are order convergent and determine each limit which exists.

$$A_n := \{1, n\}, \qquad B_n := [1, n],$$

$$C_n := [n-1, n], \qquad D_n := \begin{cases} \{0\} & \text{if } n \text{ is even} \\ \{1\} & \text{if } n \text{ is odd.} \end{cases}$$

8. Let X be a topological space and $\mathfrak{T}$ the set of open sets of X. We set $\mathfrak{S} := \{U \in \mathfrak{T} \mid \overset{\circ}{\overline{U}} = U\}$. Show that $\mathfrak{T}$ and $\mathfrak{S}$ are complete lattices with respect to inclusion. Determine $U \wedge V$, $U \vee V$, $\bigwedge_{\iota \in I} U_\iota$, $\bigvee_{\iota \in I} U_\iota$ for $U, V \in \mathfrak{T}$ ($U, V \in \mathfrak{S}$) and a non-empty family $(U_\iota)_{\iota \in I}$ in $\mathfrak{T}$ (in $\mathfrak{S}$).

2.2 The spaces $\mathbf{R}^X$ and $\overline{\mathbf{R}}^X$

Given a set X, let

- $\mathbb{R}^X$ denote the set of **real-valued functions** on X, thus

$$f \in \mathbb{R}^X : \Longleftrightarrow f : X \longrightarrow \mathbb{R};$$

- $\overline{\mathbb{R}}^X$ denote the set of **extended real-valued functions** on X, thus

$$f \in \overline{\mathbb{R}}^X : \Longleftrightarrow f : X \longrightarrow \overline{\mathbb{R}}.$$

The set $\mathbb{R}^X$ admits naturally defined algebraic operations. For $f, g \in \mathbb{R}^X$ and $\alpha \in \mathbb{R}$ define

$$
\begin{align}
(f + g)(x) &:= f(x) + g(x) \quad \text{for every } x \in X, \tag{1} \\
(\alpha f)(x) &:= \alpha f(x) \qquad\qquad \text{for every } x \in X, \tag{2} \\
(fg)(x) &:= f(x)g(x) \qquad\; \text{for every } x \in X. \tag{3}
\end{align}
$$

$\mathbb{R}^X$ is a real vector space with respect to operations (1) and (2) above, and a commutative algebra with unity with respect to (1)–(3). (A real vector space E is called an algebra if it is equipped with a bilinear mapping (a 'product') $E \times E \to E$, $(x, y) \mapsto xy$ satisfying $x(yz) = (xy)z$ for all $x, y, z \in E$. The algebra is called commutative if $xy = yx$ for all $x, y \in E$. Moreover, the element $e \in E$ is called a unity if $ex = xe = x$ for all $x \in E$.) As we have no further use for the consequences of the algebra structure of $\mathbb{R}^X$, we do not explore it further.

The attempt to extend operation (1) to $\overline{\mathbb{R}}^X$ poses a problem, since addition is not well defined on all of $\overline{\mathbb{R}}$. We consider $f + g$ ($f, g \in \overline{\mathbb{R}}^X$) to be defined by (1) if $f(x) + g(x)$ is defined for every $x \in X$. On the other hand, αf ($\alpha \in \overline{\mathbb{R}}$) and fg can be defined by (2) and (3) without difficulty, bearing in mind the definitions in Section 2.1.

Given $f \in \overline{\mathbb{R}}^X$ we define $-f := (-1)f$, and for $f, g \in \overline{\mathbb{R}}^X$ we define $f - g := f + (-g)$ provided the sum on the right is meaningful.

Given $A \subset X$ we denote the **characteristic function** of A by e_A^X or simply e_A. Then

$$
e_A : X \longrightarrow \mathbb{R}, \quad x \longmapsto
\begin{cases}
1 & \text{if } x \in A \\
0 & \text{if } x \in X \setminus A.
\end{cases}
$$

Functions of the form αe_X are called **constant functions** on X. We often write simply α instead of αe_X.

Of particular importance is the order relation on $\overline{\mathbb{R}}^X$ defined as follows. Given $f, g \in \overline{\mathbb{R}}^X$,

$$f \leq g : \Longleftrightarrow f(x) \leq g(x) \text{ for every } x \in X.$$

$f \in \overline{\mathbb{R}}^X$ is called **positive** if $f \geq 0$.

The definitions of the relation $\leq$ and of the algebraic operations have something in common. They are induced 'pointwise' by the corresponding relation and operations in $\overline{\mathbb{R}}$. Consequently the properties of $\overline{\mathbb{R}}^X$ follow easily from those of $\overline{\mathbb{R}}$. We begin with the properties of $\leq$.

Theorem 2.6 $\leq$ *is an order relation on* $\overline{\mathbb{R}}^X$ *with respect to which* $\overline{\mathbb{R}}^X$ *is a complete lattice. Given* $\mathcal{F} \subset \overline{\mathbb{R}}^X$,

$$\left(\bigvee \mathcal{F}\right)(x) = \sup\{f(x) \mid f \in \mathcal{F}\} \quad \text{for every } x \in X,$$
$$\left(\bigwedge \mathcal{F}\right)(x) = \inf\{f(x) \mid f \in \mathcal{F}\} \quad \text{for every } x \in X.$$

Given a sequence $(f_n)_{n \in \mathbb{N}}$ *in* $\overline{\mathbb{R}}^X$,

$$\left(\limsup_{n \to \infty} f_n\right)(x) = \limsup_{n \to \infty} f_n(x) \quad \text{for every } x \in X,$$
$$\left(\liminf_{n \to \infty} f_n\right)(x) = \liminf_{n \to \infty} f_n(x) \quad \text{for every } x \in X,$$

and if $(f_n)_{n \in \mathbb{N}}$ *converges, then*

$$\left(\lim_{n \to \infty} f_n\right)(x) = \lim_{n \to \infty} f_n(x) \quad \text{for every } x \in X.$$

Finally,

$$f_n \uparrow f \iff f_n(x) \uparrow f(x) \quad \text{for every } x \in X,$$
$$f_n \downarrow f \iff f_n(x) \downarrow f(x) \quad \text{for every } x \in X.$$

The proofs are straightforward, since all the operations are defined pointwise.

For $f \in \overline{\mathbb{R}}^X$ define

$$f^+ := f \vee 0 \qquad \text{(the \textbf{positive part} of } f),$$
$$f^- := (-f) \vee 0 \qquad \text{(the \textbf{negative part} of } f),$$

and

$$|f| : X \longrightarrow \overline{\mathbb{R}}, \quad x \longmapsto |f(x)| \qquad \text{(the \textbf{absolute value} of } f).$$

We draw attention to the fact that f^- is positive. The reason for choosing this definition will become apparent in our study of vector lattices.

The algebraic and order structures of $\overline{\mathbb{R}}^X$ are not completely independent! There are numerous useful connections linking them, making these structures 'compatible'. Our next theorem lists some of the most important of them.

Theorem 2.7 *Given* $f, g, h \in \overline{\mathbb{R}}^X$ *and* $\alpha \in \mathbb{R}$, *we have the following whenever the relevant operations are defined.*

(a) *If* $f \leq g$, *then* $f + h \leq g + h$ *and* $-f \geq -g$.

(b) *If* $0 \leq f$ *and* $0 \leq g$, *then* $0 \leq fg$.

(c) $f + g = f \vee g + f \wedge g$.

(d) $f = f^+ - f^-$, $|f| = f^+ + f^-$.

(e) $|f - g| = f \vee g - f \wedge g$.

(f) $f \vee g = \frac{1}{2}(f + g + |f - g|)$, $f \wedge g = \frac{1}{2}(f + g - |f - g|)$.

(g) $|f \vee g| \leq |f| + |g|$, $|f \wedge g| \leq |f| + |g|$.

(h) $|\alpha f| = |\alpha| \, |f|$.

Finally, given $f_1, f_2, g_1, g_2 \in \overline{\mathbb{R}}^X$,

(i) $|f_1 \vee f_2 - g_1 \vee g_2| \leq |f_1 - g_1| + |f_2 - g_2|$ *and*
$|f_1 \wedge f_2 - g_1 \wedge g_2| \leq |f_1 - g_1| + |f_2 - g_2|$

provided the relevant operations are defined.

Proof. The proofs follow a similar pattern. We prove two assertions as examples and leave the rest for the reader as exercises.

(c) If $f(x) = \infty$, then $g(x) \neq -\infty$, and both sides of the equation are equal to ∞.

If $f(x) = -\infty$, then $g(x) \neq \infty$, and both sides of the equation are equal to $-\infty$.

The cases $g(x) = \infty$ and $g(x) = -\infty$ are analogous.

Suppose that $f(x) \in \mathbb{R}$ and $g(x) \in \mathbb{R}$.

Case 1: $f(x) \leq g(x)$. Then $f(x) \vee g(x) = g(x)$ and $f(x) \wedge g(x) = f(x)$.

Case 2: $f(x) \geq g(x)$. Then $f(x) \vee g(x) = f(x)$ and $f(x) \wedge g(x) = g(x)$.

Thus in both cases

$$f(x) \vee g(x) + f(x) \wedge g(x) = f(x) + g(x).$$

(f) (We deal with only the first relation.) If $f(x) = \infty$, then $g(x) \in \mathbb{R}$. Therefore both sides of the equation are equal to ∞.

If $f(x) = -\infty$, then $g(x) \in \mathbb{R}$. Thus $|f(x) - g(x)| = \infty$, and the right-hand side is not defined. This case is excluded.

The cases $g(x) = \infty$ and $g(x) = -\infty$ can be dealt with analogously.

Suppose that $f(x) \in \mathbb{R}$ and $g(x) \in \mathbb{R}$.

Case 1: $f(x) \leq g(x)$. Then $f(x) \vee g(x) = g(x)$ and $|f(x) - g(x)| = g(x) - f(x)$.

Case 2: $f(x) \geq g(x)$. Then $f(x) \vee g(x) = f(x)$ and $|f(x) - g(x)| = f(x) - g(x)$.

In either case,

$$f(x) \vee g(x) = \frac{1}{2}\big(f(x) + g(x) + |f(x) - g(x)|\big)$$

which proves the claim. $\qquad\qquad\qquad\qquad\qquad\qquad\qquad\qquad\qquad\qquad$ $\square$

There is also a satisfactory formation of suprema and infima of sequences, as the next theorem shows.

Theorem 2.8 *Let $(f_n)_{n\in\mathbb{N}}$ and $(g_n)_{n\in\mathbb{N}}$ be sequences in $\overline{\mathbb{R}}^X$. Take $\alpha \in \mathbb{R}$ and $f \in \overline{\mathbb{R}}^X$. Then the following hold whenever the relevant operations are defined.*

(a)
$$\bigvee_{n\in\mathbb{N}} (f_n \vee g_n) = \left(\bigvee_{n\in\mathbb{N}} f_n\right) \vee \left(\bigvee_{n\in\mathbb{N}} g_n\right),$$
$$\bigwedge_{n\in\mathbb{N}} (f_n \wedge g_n) = \left(\bigwedge_{n\in\mathbb{N}} f_n\right) \wedge \left(\bigwedge_{n\in\mathbb{N}} g_n\right).$$

(b)
$$\alpha\left(\bigvee_{n\in\mathbb{N}} f_n\right) = \begin{cases} \bigvee_{n\in\mathbb{N}} (\alpha f_n) & \text{if } \alpha \geq 0 \\ \bigwedge_{n\in\mathbb{N}} (\alpha f_n) & \text{if } \alpha \leq 0. \end{cases}$$
$$\alpha\left(\bigwedge_{n\in\mathbb{N}} f_n\right) = \begin{cases} \bigwedge_{n\in\mathbb{N}} (\alpha f_n) & \text{if } \alpha \geq 0 \\ \bigvee_{n\in\mathbb{N}} (\alpha f_n) & \text{if } \alpha \leq 0. \end{cases}$$

(c)
$$\bigvee_{n\in\mathbb{N}} f_n = -\bigwedge_{n\in\mathbb{N}} (-f_n),$$
$$\bigwedge_{n\in\mathbb{N}} f_n = -\bigvee_{n\in\mathbb{N}} (-f_n).$$

(d)
$$f + \bigvee_{n\in\mathbb{N}} f_n = \bigvee_{n\in\mathbb{N}} (f + f_n),$$
$$f + \bigwedge_{n\in\mathbb{N}} f_n = \bigwedge_{n\in\mathbb{N}} (f + f_n).$$

(e) *If $(f_n)_{n\in\mathbb{N}}$ and $(g_n)_{n\in\mathbb{N}}$ are increasing, then*
$$\bigvee_{n\in\mathbb{N}} (f_n + g_n) = \left(\bigvee_{n\in\mathbb{N}} f_n\right) + \left(\bigvee_{n\in\mathbb{N}} g_n\right),$$
$$\bigvee_{n\in\mathbb{N}} (f_n \wedge g_n) = \left(\bigvee_{n\in\mathbb{N}} f_n\right) \wedge \left(\bigvee_{n\in\mathbb{N}} g_n\right).$$

(f) *If $(f_n)_{n\in\mathbb{N}}$ and $(g_n)_{n\in\mathbb{N}}$ are decreasing, then*
$$\bigwedge_{n\in\mathbb{N}} (f_n + g_n) = \left(\bigwedge_{n\in\mathbb{N}} f_n\right) + \left(\bigwedge_{n\in\mathbb{N}} g_n\right),$$
$$\bigwedge_{n\in\mathbb{N}} (f_n \vee g_n) = \left(\bigwedge_{n\in\mathbb{N}} f_n\right) \vee \left(\bigwedge_{n\in\mathbb{N}} g_n\right).$$

Proof. We again prove only a selection of the assertions.
   (a) (The first relation) Given $n \in \mathbb{N}$,
$$f_n \leq \bigvee_{n\in\mathbb{N}} f_n \quad \text{and} \quad g_n \leq \bigvee_{n\in\mathbb{N}} g_n$$

and thus

$$f_n \vee g_n \leq \left(\bigvee_{n \in \mathbb{N}} f_n \right) \vee \left(\bigvee_{n \in \mathbb{N}} g_n \right).$$

Then

$$\bigvee_{n \in \mathbb{N}} (f_n \vee g_n) \leq \left(\bigvee_{n \in \mathbb{N}} f_n \right) \vee \left(\bigvee_{n \in \mathbb{N}} g_n \right).$$

Conversely,

$$f_n \leq f_n \vee g_n \leq \bigvee_{n \in \mathbb{N}} (f_n \vee g_n), \quad \text{and}$$

$$g_n \leq f_n \vee g_n \leq \bigvee_{n \in \mathbb{N}} (f_n \vee g_n)$$

for every $n \in \mathbb{N}$. Therefore

$$\bigvee_{n \in \mathbb{N}} f_n \leq \bigvee_{n \in \mathbb{N}} (f_n \vee g_n) \quad \text{and} \quad \bigvee_{n \in \mathbb{N}} g_n \leq \bigvee_{n \in \mathbb{N}} (f_n \vee g_n),$$

which implies that

$$\left(\bigvee_{n \in \mathbb{N}} f_n \right) \vee \left(\bigvee_{n \in \mathbb{N}} g_n \right) \leq \bigvee_{n \in \mathbb{N}} (f_n \vee g_n).$$

(c) $h \in \overline{\mathbb{R}}^X$ is an upper bound for a sequence $(f_n)_{n \in \mathbb{N}}$ in $\overline{\mathbb{R}}^X$ if and only if $-h$ is a lower bound for $(-f_n)_{n \in \mathbb{N}}$. The claim now follows.

(d) If $f(x) = \infty$, then $f_n(x) \neq -\infty$ for every $n \in \mathbb{N}$ and both sides are equal to ∞.

If $f(x) = -\infty$, then $f_n(x) \neq \infty$ for every $n \in \mathbb{N}$ and $\left(\bigvee_{n \in \mathbb{N}} f_n \right)(x) \neq \infty$. Hence both sides are equal to $-\infty$.

Suppose that $f(x) \in \mathbb{R}$. Then given $n \in \mathbb{N}$,

$$f(x) + \sup_{n \in \mathbb{N}} f_n(x) \geq f(x) + f_n(x)$$

and therefore

$$f(x) + \sup_{n \in \mathbb{N}} f_n(x) \geq \sup \left(f(x) + f_n(x) \right).$$

Conversely, for each $n \in \mathbb{N}$,

$$f_n(x) = \left(f(x) + f_n(x) \right) - f(x) \leq \sup_{n \in \mathbb{N}} \left(f(x) + f_n(x) \right) - f(x),$$

which implies that

$$\sup_{n \in \mathbb{N}} f_n(x) \leq \sup_{n \in \mathbb{N}} \left(f(x) + f_n(x) \right) - f(x)$$

and finally

$$f(x) + \sup_{n \in \mathbb{N}} f_n(x) \leq \sup_{n \in \mathbb{N}} \left(f(x) + f_n(x) \right).$$

Similarly,

$$f + \bigwedge_{n \in \mathbb{N}} f_n = \bigwedge_{n \in \mathbb{N}} (f + f_n).$$

(e) (The first relation) If $f_n(x) = \infty$ or $g_n(x) = \infty$ for some $n \in \mathbb{N}$, then both sides are equal to ∞.

If $f_n(x) < \infty$ and $g_n(x) < \infty$ for every $n \in \mathbb{N}$, then

$$f_n(x) + g_n(x) \leq \sup_{n \in \mathbb{N}} f_n(x) + \sup_{n \in \mathbb{N}} g_n(x)$$

for every $n \in \mathbb{N}$. Thus

$$\sup_{n \in \mathbb{N}} \left(f_n(x) + g_n(x) \right) \leq \sup_{n \in \mathbb{N}} f_n(x) + \sup_{n \in \mathbb{N}} g_n(x).$$

Conversely,

$$f_n(x) + g_m(x) \leq f_{n+m}(x) + g_{n+m}(x) \leq \sup_{n \in \mathbb{N}} \left(f_n(x) + g_n(x) \right)$$

for all $n, m \in \mathbb{N}$, since the sequences are monotone. By (d),

$$f_n(x) + \sup_{m \in \mathbb{N}} g_m(x) \leq \sup_{n \in \mathbb{N}} \left(f_n(x) + g_n(x) \right),$$

and finally, appealing to (d) a second time, we find that

$$\sup_{n \in \mathbb{N}} f_n(x) + \sup_{n \in \mathbb{N}} g_n(x) \leq \sup_{n \in \mathbb{N}} \left(f_n(x) + g_n(x) \right).$$

The other assertions can be proved similarly and are left as an exercise for the reader. □

Finally, we introduce more notation. For $f, g \in \overline{\mathbb{R}}^X$, define

$$\{f \leq g\} := \{x \in X \mid f(x) \leq g(x)\}.$$

$\{f < g\}$, $\{f \geq g\}$, $\{f > g\}$, $\{f = g\}$ and $\{f \neq g\}$ are defined analogously.

Exercises

1. Show that $\mathbb{R}^X$ is a conditionally complete lattice, which is not complete whenever $X \neq \emptyset$.

2. Show for $f, g, h \in \overline{\mathbb{R}}^X$:

   (a) $|f| = f^+ \vee f^-$.

   (b) Whenever all the operations are defined,

   (b1) $\big||f| - |g|\big| \leq |f - g| = f \vee g - f \wedge g$;

   (b2) $\big||f| - |g|\big| \leq |f + g|$;

   (b3) $(f + g)^+ \leq f^+ + g^+$, $(f + g)^- \leq f^- + g^-$.

   (c) If $g, h \geq 0$ and if $g - h$ is defined, then $(g - h)^+ \leq g$ and $(g - h)^- \leq h$.

3. For $n \in \mathbb{N}$, define $f_n : \mathbb{R} \to \mathbb{R}$, $x \mapsto x^n$. Determine $\liminf_{n \to \infty} f_n$ and $\limsup_{n \to \infty} f_n$ in $\overline{\mathbb{R}}^{\mathbb{R}}$.

4. Show that for conclusions (e) and (f) of Theorem 2.8 it is necessary to assume that the sequences are increasing and decreasing respectively.

5. Show:

   (a) The sequence $(f_n)_{n \in \mathbb{N}}$ in $\overline{\mathbb{R}}^X$ is order convergent to $f \in \overline{\mathbb{R}}^X$ if and only if the sequence $(f_n(x))_{n \in \mathbb{N}}$ converges to $f(x)$ for every $x \in X$. In this case

   $$\left(\lim_{n \to \infty} f_n \right)(x) = \lim_{n \to \infty} f_n(x) = f(x)$$

   for every $x \in X$.

   Let $(f_n)_{n \in \mathbb{N}}$ and $(g_n)_{n \in \mathbb{N}}$ be order convergent sequences in $\overline{\mathbb{R}}^X$, and let α be a real number. Prove:

   (b) $(f_n \vee g_n)_{n \in \mathbb{N}}$ is order convergent and

   $$\lim_{n \to \infty} (f_n \vee g_n) = \left(\lim_{n \to \infty} f_n \right) \vee \left(\lim_{n \to \infty} g_n \right).$$

   (c) $(f_n \wedge g_n)_{n \in \mathbb{N}}$ is order convergent and

   $$\lim_{n \to \infty} (f_n \wedge g_n) = \left(\lim_{n \to \infty} f_n \right) \wedge \left(\lim_{n \to \infty} g_n \right).$$

   (d) $(\alpha f_n)_{n \in \mathbb{N}}$ is order convergent and

   $$\lim_{n \to \infty} (\alpha f_n) = \alpha \lim_{n \to \infty} f_n.$$

   (e) If all the operations are defined, then $(f_n + g_n)_{n \in \mathbb{N}}$ is order convergent and

   $$\lim_{n \to \infty} (f_n + g_n) = \left(\lim_{n \to \infty} f_n \right) + \left(\lim_{n \to \infty} g_n \right).$$

6. Let $(f_n)_{n \in \mathbb{N}}$ be a sequence in $\overline{\mathbb{R}}^X$. Under which conditions do

   $$\limsup_{n \to \infty} f_n = \limsup_{k \to \infty} f_{n_k} \quad \text{and} \quad \liminf_{n \to \infty} f_n = \liminf_{k \to \infty} f_{n_k}$$

   hold for all subsequences $(f_{n_k})_{k \in \mathbb{N}}$ of $(f_n)_{n \in \mathbb{N}}$?

7. Let $(A_n)_{n \in \mathbb{N}}$ be a sequence of subsets of X, and take $A \subset X$. Show that $A = \lim_{n \to \infty} A_n$ (in the complete lattice $\mathfrak{P}(X)$) if and only if $e_A = \lim_{n \to \infty} e_{A_n}$ (in $\overline{\mathbb{R}}^X$).

2.3 Vector lattices of functions

In this section, we describe vector lattices, the class of function spaces which is of central importance to our approach to integration theory.

Vector lattices can be dealt with as abstract algebraic structures. The final three chapters of this book examine abstract vector lattices in detail. For the moment, however, we restrict attention to function spaces which allow much simpler arguments. This simplification is frequently due to the fact that the functions (and operations on them) are pointwise defined and so permit the reduction of the general situation to the investigation of the totally ordered set $\overline{\mathbb{R}}$.

A vector lattice is a combination of vector space and lattice structures. Thus in order to obtain true vector lattices, we must restrict ourselves to real-valued functions so as to retain the vector space structure, because the addition of extended real-valued functions cannot be completely defined. We shall see, on the basis of our results, that this restriction is more apparent than real.

Let X be a set. Then $\mathcal{F} \subset \mathbb{R}^X$ is a **vector lattice** if and only if

(i) $\mathcal{F}$ is a vector subspace of $\mathbb{R}^X$;

(ii) given $f, g \in \mathcal{F}$, $f \vee g \in \mathcal{F}$ and $f \wedge g \in \mathcal{F}$.

For example, $\mathbb{R}^X$ itself is a vector lattice, as is the set of all constant functions on X.

The following proposition provides another useful way of characterizing this structure.

Proposition 2.9 *For* $\mathcal{F} \subset \mathbb{R}^X$, *the following are equivalent.*

(a) $\mathcal{F}$ *is a vector lattice.*

(b) $\mathcal{F}$ *is a vector subspace of* $\mathbb{R}^X$ *and* $|f| \in \mathcal{F}$ *for every* $f \in \mathcal{F}$.

Proof. (a)$\Rightarrow$(b). $|f| = f^+ + f^- = f \vee 0 + (-f) \vee 0 \in \mathcal{F}$.
(b)$\Rightarrow$(a). By Theorem 2.7(f),

$$f \vee g = \frac{1}{2}(f + g + |f - g|) \in \mathcal{F} \quad \text{and}$$

$$f \wedge g = \frac{1}{2}(f + g - |f - g|) \in \mathcal{F}.$$

$\square$

Given $\mathcal{F} \subset \overline{\mathbb{R}}^X$, define

$$\mathcal{F}_+ := \{f \in \mathcal{F} \mid f \geq 0\}.$$

Thus $\mathcal{F}_+$ is the set of all positive elements of $\mathcal{F}$. Note that $\mathcal{F}_+$ may be empty. If $\mathcal{F}$ is a vector lattice, however, $\mathcal{F}_+$ contains at least the constant function 0.

Theorems 2.7 and 2.8 show that the algebraic structure of $\mathbb{R}^X$ and the ordering on $\mathbb{R}^X$ are compatible. The compatibility properties extend naturally to vector lattices of functions. We do not list these properties again, but leave it to the reader to review them as necessary. Their significance will become apparent in the following sections.

We now present some important examples of vector lattices of functions.

(a) Continuous functions Let X be a topological space and let $\mathcal{C}(X)$ denote the set of all continuous real-valued functions on X.

Theorem 2.10 $C(X)$ *is a vector lattice.*

Proof. We break the proof down into several steps.

Step 1. Given arbitrary $f, g \in \mathbb{R}^X$,

$$\{f > g\} = \bigcup_{\alpha \in \mathbb{Q}} \bigcup_{n \in \mathbb{N}} \left(f^{-1}(]\alpha, \alpha + n[) \cap g^{-1}(]\alpha - n, \alpha[) \right).$$

Take $x \in \{f > g\}$. Then there is an $\alpha \in \mathbb{Q}$ with $g(x) < \alpha < f(x)$ and an $n \in \mathbb{N}$ with

$$f(x) \in]\alpha, \alpha + n[\quad \text{and} \quad g(x) \in]\alpha - n, \alpha[.$$

It follows that

$$x \in f^{-1}(]\alpha, \alpha + n[) \cap g^{-1}(]\alpha - n, \alpha[)$$

and accordingly that x is an element of the right-hand side of the equation.

Conversely, if x is an element of the set on the right, then there is an $\alpha \in \mathbb{Q}$ and an $n \in \mathbb{N}$ with

$$x \in f^{-1}(]\alpha, \alpha + n[) \cap g^{-1}(]\alpha - n, \alpha[).$$

Thus $f(x) \in]\alpha, \alpha + n[$ and $g(x) \in]\alpha - n, \alpha[$ and so $f(x) > \alpha > g(x)$.

Step 2. Given $f, g \in \mathbb{R}^X$ and $A \subset \mathbb{R}$,

$$(f \vee g)^{-1}(A) = \left(f^{-1}(A) \cap g^{-1}(A) \right) \cup \left(f^{-1}(A) \cap \{f > g\} \right)$$
$$\cup \left(g^{-1}(A) \cap \{g > f\} \right),$$

$$(f \wedge g)^{-1}(A) = \left(f^{-1}(A) \cap g^{-1}(A) \right) \cup \left(f^{-1}(A) \cap \{f < g\} \right)$$
$$\cup \left(g^{-1}(A) \cap \{g < f\} \right).$$

It is sufficient to show that the first equality holds.

Take $x \in (f \vee g)^{-1}(A)$. Then $(f \vee g)(x) \in A$.

Case 1: $f(x) < g(x)$. Then $(f \vee g)(x) = g(x)$ and so $x \in g^{-1}(A) \cap \{g > f\}$.

Case 2: $f(x) > g(x)$. Then $(f \vee g)(x) = f(x)$ and so $x \in f^{-1}(A) \cap \{f > g\}$.

Case 3: $f(x) = g(x)$. Then $x \in f^{-1}(A) \cap g^{-1}(A)$.

Now let x be an element of the right-hand side.

Case 1: $x \in f^{-1}(A) \cap \{f > g\}$. Then $(f \vee g)(x) = f(x) \in A$ and thus $x \in (f \vee g)^{-1}(A)$.

Case 2: $x \in g^{-1}(A) \cap \{g > f\}$. Then $(f \vee g)(x) = g(x) \in A$ and thus $x \in (f \vee g)^{-1}(A)$.

If x is not in either set, then $x \in f^{-1}(A) \cap g^{-1}(A)$. Thus $f(x)$ and $g(x)$ must be equal and $x \in (f \vee g)^{-1}(A)$ once again.

Step 3. Given $f \in \mathbb{R}^X$, $\alpha \in \mathbb{R}$ and $A \subset \mathbb{R}$,

$$(\alpha f)^{-1}(A) = \begin{cases} X & \text{if } \alpha = 0 \text{ and } 0 \in A \\ \emptyset & \text{if } \alpha = 0 \text{ and } 0 \notin A \\ f^{-1}(\{\gamma/\alpha \mid \gamma \in A\}) & \text{if } \alpha \neq 0. \end{cases}$$

This is left to the reader.

Step 4. Given $f \in \mathbb{R}^X$, $\alpha \in \mathbb{R}$ and $A \subset \mathbb{R}$,

$$(f + \alpha)^{-1}(A) = f^{-1}(\{\beta - \alpha \mid \beta \in A\}).$$

If $x \in (f + \alpha)^{-1}(A)$, then $f(x) + \alpha \in A$ and there is a $\beta \in A$ for which $f(x) = \beta - \alpha$. If, on the other hand, $x \in f^{-1}(\{\beta - \alpha \mid \beta \in A\})$, then $f(x) = \beta - \alpha$. Thus $f(x) + \alpha = \beta \in A$ and so $x \in (f + \alpha)^{-1}(A)$.

Step 5. Given $f, g \in \mathbb{R}^X$ and $\alpha, \beta \in \mathbb{R}$, $\alpha < \beta$,

$$(f + g)^{-1}(]\alpha, \beta[) = \{\alpha - f < g\} \cap \{f < -g + \beta\}.$$

Take $x \in X$. Then

$$
\begin{aligned}
x \in (f + g)^{-1}(]\alpha, \beta[) &\iff (f + g)(x) \in]\alpha, \beta[\\
&\iff \alpha < f(x) + g(x) < \beta \\
&\iff \alpha - f(x) < g(x) \text{ and } f(x) < -g(x) + \beta \\
&\iff x \in \{\alpha - f < g\} \cap \{f < -g + \beta\}.
\end{aligned}
$$

To summarize, step 1 implies that $\{f > g\}$ is open for all $f, g \in \mathcal{C}(X)$. That $f \vee g \in \mathcal{C}(X)$ and $f \wedge g \in \mathcal{C}(X)$ for all $f, g \in \mathcal{C}(X)$ follows from step 2, and step 3 shows that $\alpha f \in \mathcal{C}(X)$ for every $f \in \mathcal{C}(X)$ and every $\alpha \in \mathbb{R}$. If $f, g \in \mathcal{C}(X)$, then $\alpha - f$ and $-g + \beta$ are continuous by step 4, and step 5 shows the same is true of $f + g$. Thus, we have proved the theorem. $\square$

(b) Continuous functions with compact support The investigation of such functions is of particular interest in the case of locally compact spaces, such as the spaces $\mathbb{R}^n$ ($n \in \mathbb{N}$). Thus, we take X to be a locally compact space.

Given $f \in \mathbb{R}^X$, the set $\overline{\{f \neq 0\}}$ is the **support** of f. (Recall that $\overline{\{f \neq 0\}}$ is the closure of $\{f \neq 0\}$.) We write $\mathcal{K}(X)$ for the set of all functions $f \in \mathcal{C}(X)$ whose support is compact. For our purposes, the importance of the functions in $\mathcal{K}(X)$ rests upon the fact that they can be integrated with respect to every 'reasonable' measure on X, as we show in Chapter 5.

Theorem 2.11 $\mathcal{K}(X)$ *is a vector lattice.*

Proof. This follows directly from Theorem 2.10 in light of the following inclusions for $f, g \in \mathbb{R}^X$:

$$
\begin{aligned}
\overline{\{f \vee g \neq 0\}} &\subset \overline{\{f \neq 0\}} \cup \overline{\{g \neq 0\}}, \\
\overline{\{f \wedge g \neq 0\}} &\subset \overline{\{f \neq 0\}} \cup \overline{\{g \neq 0\}}, \\
\overline{\{f + g \neq 0\}} &\subset \overline{\{f \neq 0\}} \cup \overline{\{g \neq 0\}}, \\
\overline{\{\alpha f \neq 0\}} &\subset \overline{\{f \neq 0\}}.
\end{aligned}
$$

$\square$

We summarize some important properties of the space $\mathcal{K}(X)$.

Theorem 2.12 *Every function $f \in \mathcal{K}(X)$ is bounded. In other words, for each $f \in \mathcal{K}(X)$ there is an $\alpha \in \mathbb{R}_+$ with $|f| \leq \alpha$. Moreover, there are $x, y \in X$ such that*

$$f(x) = \sup_{z \in X} f(z) \quad and \quad f(y) = \inf_{z \in X} f(z).$$

Proof. This follows immediately from the fact that the continuous image of a compact set is compact. □

Theorem 2.13 *Let $(f_\iota)_{\iota \in I}$ be a non-empty downward directed family in $\mathcal{K}(X)$ such that $\bigwedge_{\iota \in I} f_\iota = 0$. Given $\iota \in I$, put*

$$\alpha_\iota := \sup_{x \in X} f_\iota(x).$$

Then $\inf_{\iota \in I} \alpha_\iota = 0$.

Proof. Take $\alpha \in \mathbb{R}$, $\alpha > 0$. Given $\iota \in I$, put $K_\iota := \{f_\iota \geq \alpha\}$. The sets K_ι are compact and $\bigcap_{\iota \in I} K_\iota = \emptyset$. There is therefore a finite set $J \subset I$ with $\bigcap_{\iota \in J} K_\iota = \emptyset$. Since $(f_\iota)_{\iota \in I}$ is directed down, we can find $\lambda \in I$ such that $f_\lambda \leq f_\iota$ for every $\iota \in J$. It follows that $K_\lambda = \emptyset$ and thus $f_\lambda(x) < \alpha$ for every $x \in X$. Hence $\alpha_\lambda \leq \alpha$ and, since α is arbitrary, $\inf_{\iota \in I} \alpha_\iota = 0$. □

Theorem 2.13 thus asserts that a downward directed family – in particular, a decreasing sequence – in $\mathcal{K}(X)$ which converges pointwise to 0 also converges uniformly to 0 – a remarkable property! In the case of decreasing sequences, Theorem 2.13 is known as Dini's theorem.

(c) Spaces of step functions We now introduce the systems of sets which later serve as the domains of measures.

Let X be a set. Then the subset $\mathfrak{R}$ of $\mathfrak{P}(X)$ is called a **ring of sets** if and only if for all $A, B \in \mathfrak{R}$:

(i) $A \cup B \in \mathfrak{R}$;

(ii) $A \setminus B \in \mathfrak{R}$;

(iii) $\emptyset \in \mathfrak{R}$.

If $\mathfrak{R}$ is a ring of sets and $A, B \in \mathfrak{R}$ then, in addition,

$A \cap B \in \mathfrak{R}$, as $A \cap B = A \setminus (A \setminus B)$;

$A \triangle B := (A \setminus B) \cup (B \setminus A) \in \mathfrak{R}$ (**symmetric difference**).

Simple examples of rings of sets are $\mathfrak{P}(X)$, $\{A \subset X \,|\, A$ is finite$\}$ and $\{A \subset X \,|\, A \subset B\}$ where B is a fixed subset of X.

Step functions are closely connected to rings of sets. In order to investigate spaces of step functions we must first study the dissection property of these rings of sets. This dissection property is of fundamental importance.

Theorem 2.14 (Dissection Theorem) *Let $\mathfrak{R}$ be a ring of sets and $\mathfrak{S}$ be a finite subset of $\mathfrak{R}$. Then there is a finite subset $\mathfrak{S}'$ of $\mathfrak{R}$ consisting of pairwise disjoint sets such that*

$$A = \bigcup_{\substack{B \in \mathfrak{S}' \\ B \subset A}} B$$

*for every $A \in \mathfrak{S}$. $\mathfrak{S}'$ is called a **disjoint dissection** of $\mathfrak{S}$.*

Proof. We proceed by complete induction on the number of elements of $\mathfrak{S}$. If $\mathfrak{S}$ contains precisely one element, then the claim is trivial. Assume that the claim is true for all sets $\mathfrak{S}$ with n elements. Let $\mathfrak{S} \subset \mathfrak{R}$ contain $n+1$ elements. Choose $A \in \mathfrak{S}$. By assumption, there is a disjoint dissection $\mathfrak{S}'_n$ for $\mathfrak{S} \setminus \{A\}$. Then

$$\mathfrak{S}'_{n+1} := \{A \cap B \,|\, B \in \mathfrak{S}'_n\} \cup \{B \setminus A \,|\, B \in \mathfrak{S}'_n\} \cup \{A \setminus \bigcup_{B \in \mathfrak{S}'_n} B\}$$

is such a dissection for $\mathfrak{S}$. (Note that some of the sets in $\mathfrak{S}'_{n+1}$ may be empty.) $\qquad\square$

Now let X be a set and $\mathfrak{R} \subset \mathfrak{P}(X)$ be a ring of sets. An **$\mathfrak{R}$-step function** is a finite linear combination of characteristic functions of sets in $\mathfrak{R}$. $\mathcal{L}(\mathfrak{R})$ denotes the set of all $\mathfrak{R}$-step functions. Then

$$f \in \mathcal{L}(\mathfrak{R}) \iff f = \sum_{k=1}^{n} \alpha_k e_{A_k}$$

where $n \in \mathbb{N}$, $\alpha_k \in \mathbb{R}$ and $A_k \in \mathfrak{R}$ for $1 \le k \le n$.

Theorem 2.15 *$\mathcal{L}(\mathfrak{R})$ is a vector lattice.*

Proof. It follows directly from the definition that $\mathcal{L}(\mathfrak{R})$ is a vector subspace of $\mathbb{R}^X$. Take $f \in \mathcal{L}(\mathfrak{R})$, $f = \sum_{k=1}^{n} \alpha_k e_{A_k}$ and let $\{B_1, \dots, B_l\}$ be a disjoint dissection of $\{A_1, \dots, A_n\}$. For $i \in \{1, \dots, l\}$ put

$$\beta_i := \sum_{\substack{k \le n \\ B_i \subset A_k}} \alpha_k.$$

Then $f = \sum_{i=1}^{l} \beta_i e_{B_i}$, and since the B_i are pairwise disjoint,

$$|f| = \sum_{i=1}^{l} |\beta_i| e_{B_i} \in \mathcal{L}(\mathfrak{R}).$$

The conclusion now follows from Proposition 2.9. $\qquad\square$

We continue our summary of the properties of $\mathcal{L}(\mathfrak{R})$.

Proposition 2.16 *Given $f \in \mathcal{L}(\mathfrak{R})$ and $\alpha \in \mathbb{R}_+$,*

(a) $\{f > \alpha\} \in \mathfrak{R}$;

(b) if $\alpha > 0$, then $\{f \geq \alpha\} \in \mathfrak{R}$;

(c) $f \wedge \alpha \in \mathcal{L}(\mathfrak{R})$.

Proof. Write f as a sum $f = \sum_{i=1}^{l} \beta_i e_{B_i}$, where the sets $B_i \in \mathfrak{R}$ are pairwise disjoint. Then $\{f > \alpha\} = \bigcup_{\beta_i > \alpha} B_i \in \mathfrak{R}$ and $\{f \geq \alpha\} = \bigcup_{\beta_i \geq \alpha} B_i \in \mathfrak{R}$ if $\alpha > 0$. Since $B := \bigcup_{i \leq l} B_i \in \mathfrak{R}$, it follows that $f \wedge \alpha = f \wedge (\alpha e_B) \in \mathcal{L}(\mathfrak{R})$. □

Exercises

1. Let X be a set. Show that the following sets are vector lattices:

   (a) $\ell^\infty(X) := \{f \in \mathbb{R}^X \mid f \text{ is bounded}\}$.
   (b) $\mathcal{F}(X) := \{f \in \mathbb{R}^X \mid \{f \neq 0\} \text{ is finite}\}$.
   (c) $c_0(X) := \{f \in \mathbb{R}^X \mid \{|f| > \varepsilon\} \text{ is finite for each } \varepsilon > 0\}$.
   (d) $c(X) := \{f \in \mathbb{R}^X \mid \exists \alpha \in \mathbb{R}, \{|f - \alpha| > \varepsilon\} \text{ is finite for each } \varepsilon > 0\}$.
   (e) $c_f(X) := \{f \in \mathbb{R}^X \mid \exists \alpha \in \mathbb{R}, \{f \neq \alpha\} \text{ is finite}\}$.
   (f) $\ell^1(X) := \{f \in \mathbb{R}^X \mid \sup_{A \subset X, A \text{ finite}} \sum_{x \in A} |f(x)| < \infty\}$.
   (g) $\ell_f(X) := \{f \in \mathbb{R}^X \mid f(X) \text{ is finite}\}$.
   (h) $\ell_{x_0}(X) := \{f \in \mathbb{R}^X \mid \{f \neq f(x_0)\} \text{ is countable}\}$ for fixed $x_0 \in X$.

   Which of these vector lattices $\mathcal{L}$ satisfy the condition 'If $f \in \mathcal{L}$ and $g \in \mathbb{R}^X$, $|g| \leq |f|$, then $g \in \mathcal{L}$'?

2. Let $X := [a, b]$ be a closed interval in $\mathbb{R}$. Determine which of the following sets are vector lattices:

   (a) $\{f \in \mathbb{R}^X \mid f \text{ is linear}\}$.
   (b) $\{f \in \mathbb{R}^X \mid f \text{ is piecewise linear}\}$.
   (c) $\{f \in \mathbb{R}^X \mid f \text{ is a polynomial}\}$.
   (d) $\{f \in \mathbb{R}^X \mid f \text{ is differentiable}\}$.
   (e) $\{f \in \mathbb{R}^X \mid f \text{ is constant}\}$.

3. Let $\mathfrak{R}$ be a non-empty set of sets with the properties

   (i) $A, B \in \mathfrak{R} \implies A \cup B \in \mathfrak{R}$ (ii) $A \in \mathfrak{R}, B \subset A \implies B \in \mathfrak{R}$.

   Let $\mathcal{L} \subset \mathbb{R}^X$ be a vector lattice and define

   $$\mathcal{L}' := \{f \in \mathbb{R}^X \mid \exists g \in \mathcal{L}, \{f \neq g\} \in \mathfrak{R}\}.$$

   Prove that $\mathcal{L}'$ is a vector lattice with $\mathcal{L} \subset \mathcal{L}'$.

4. Show that the set of all Riemann integrable functions on the interval $[a, b]$ is a vector lattice.
   Hint: Use Darboux sums as defined in the introduction to Chapter 3.

5. Define $f : X \to X$, $x \mapsto x$. Does f belong to $\mathcal{K}(X)$, for (a) $X = [0, 1]$, (b) $X =]0, 1[$?

6. Describe $\mathcal{K}(X)$, where X is taken with its discrete topology.

7. Determine which of the following systems are rings of sets:

   (a) $\{\emptyset\}$; $\{X\}$; $\{\emptyset, X\}$.
   (b) $\{A \subset X \,|\, A$ is finite or $X \setminus A$ is finite$\}$; $\{A \subset X \,|\, A$ is countable$\}$; $\{A \subset X \,|\, A$ is countable or $X \setminus A$ is countable$\}$; $\{A \subset X \,|\, A$ is infinite$\}$; $\{A \subset X \,|\, A$ is infinite or $X \setminus A$ is infinite$\}$ for X infinite.
   (c) $\{A \times B \,|\, A \in \mathfrak{R}, \, B \in \mathfrak{S}\}$ where $\mathfrak{R}, \mathfrak{S}$ are rings of sets.
   (d) $\{A \subset X \,|\, A$ is open$\}$; $\{A \subset X \,|\, A$ is closed$\}$; $\{A \subset X \,|\, A$ is compact$\}$; $\{A \subset X \,|\, A$ is relatively compact$\}$; $\{A \subset X \,|\, A$ is open and closed$\}$; $\{A \subset X \,|\, A$ is dense$\}$; $\{A \subset X \,|\, A$ is nowhere dense$\}$ where X is a topological space.

8. Show that every ring of sets is a lattice with respect to inclusion.

9. Prove that every ring of sets is a ring in the algebraic sense, with addition given by $A + B := A \triangle B$ and multiplication given by $A \cdot B := A \cap B$.

10. Show that if $\varphi : X \to Y$ is a map and $\mathfrak{R} \subset \mathfrak{P}(Y)$ is a ring of sets, then $\varphi^{-1}(\mathfrak{R}) \subset \mathfrak{P}(X)$ is also a ring of sets. Is it true that if $\mathfrak{S} \subset \mathfrak{P}(X)$ is a ring of sets, then so too is $\varphi(\mathfrak{S}) \subset \mathfrak{P}(Y)$?

11. Let X be a set and $\mathfrak{R} \subset \mathfrak{P}(X)$ a ring of sets. Let f be a step function on X. Prove that the following are equivalent.

    (a) f is an $\mathfrak{R}$-step function.
    (b) $\{f > \alpha\} \in \mathfrak{R}$ for any $\alpha > 0$.
    (c) $\{f \geq \alpha\} \in \mathfrak{R}$ for any $\alpha > 0$.

12. Given a set X, define $\mathfrak{R} := \{A \subset X \,|\, A$ is finite or $X \setminus A$ is finite$\}$. Then $\mathfrak{R}$ is a ring of sets on X. Show that

$$\mathcal{L}(\mathfrak{R}) = \{\alpha e_X + g \,|\, \alpha \in \mathbb{R}, \, g \in \mathbb{R}^X, \, \{g \neq 0\} \text{ finite}\}.$$

13. Take $\mathfrak{R}$-step functions f, g. Show that fg and f^n are $\mathfrak{R}$-step functions $(n \in \mathbb{N})$.

2.4 Functionals

Take $\mathcal{F} \subset \overline{\mathbb{R}}^X$. A **functional** on $\mathcal{F}$ is a map $\ell : \mathcal{F} \to \mathbb{R}$. The functional ℓ is

- **linear** if $\ell(\alpha f + \beta g) = \alpha \ell(f) + \beta \ell(g)$ for all $f, g \in \mathcal{F}$ and $\alpha, \beta \in \mathbb{R}$ for which $\alpha f + \beta g \in \mathcal{F}$;
- **positive** if $\ell(f) \geq 0$ for every $f \in \mathcal{F}_+$;
- **increasing** if $\ell(f) \leq \ell(g)$ for all $f, g \in \mathcal{F}$ with $f \leq g$;
- **null-continuous** if $\ell(f_n) \downarrow 0$ whenever $(f_n)_{n \in \mathbb{N}}$ is a sequence in $\mathcal{F}$ and $f_n \downarrow 0$.

Positive linear functionals on vector lattices of functions are of particular interest to us. They are in fact the starting point for the entire theory of integration.

Henceforth $\mathcal{F}$ will always denote a vector lattice.

Proposition 2.17 *Given a linear functional ℓ on $\mathcal{F}$, the following are equivalent.*

(a) ℓ is positive.

(b) ℓ is increasing.

Proof. (a)$\Rightarrow$(b). $f, g \in \mathcal{F}$, $f \leq g$ implies that $g - f \geq 0$ and therefore $\ell(g) - \ell(f) = \ell(g - f) \geq 0$. Thus $\ell(g) \geq \ell(f)$.

(b)$\Rightarrow$(a) follows since $0 \in \mathcal{F}$ and $\ell(0) = 0$. $\square$

The next theorem summarizes some basic properties of positive linear functionals on vector lattices.

Theorem 2.18 *Let ℓ be a positive linear functional on $\mathcal{F}$. Then:*

(a) $\ell(|f|) \geq |\ell(f)|$ for every $f \in \mathcal{F}$.

(b) Given $f, g \in \mathcal{F}$,

$$\ell(f \vee g) \geq \ell(f) \vee \ell(g) \quad and \quad \ell(f \wedge g) \leq \ell(f) \wedge \ell(g).$$

(c) Take $\mathcal{G} \subset \mathcal{F}$ with $g := \bigvee \mathcal{G} \in \mathcal{F}$. Then

$$\ell(g) \geq \sup\{\ell(h) \mid h \in \mathcal{G}\}.$$

Similarly, if $g' := \bigwedge \mathcal{G} \in \mathcal{F}$, then

$$\ell(g') \leq \inf\{\ell(h) \mid h \in \mathcal{G}\}.$$

(d) Given $f, g \in \mathcal{F}$,

$$\big||\ell(f)| - |\ell(g)|\big| \leq \ell(|f - g|).$$

Proof. (a) Take $f \in \mathcal{F}$. Then $-|f| \leq f \leq |f|$ and it follows that $-\ell(|f|) \leq \ell(f) \leq \ell(|f|)$.

(b) Take $f, g \in \mathcal{F}$. Then $\ell(f) \leq \ell(f \vee g)$ and $\ell(g) \leq \ell(f \vee g)$. Thus $\ell(f) \vee \ell(g) \leq \ell(f \vee g)$. The second claim follows similarly.

(c) This can be proven similarly to (b).

(d) Given $f, g \in \mathcal{F}$, $f = (f - g) + g$. Thus $\ell(f) = \ell(f - g) + \ell(g)$ and hence

$$|\ell(f)| \leq |\ell(f - g)| + |\ell(g)| \leq \ell(|f - g|) + |\ell(g)|.$$

Therefore

$$|\ell(f)| - |\ell(g)| \leq \ell(|f - g|).$$

Similarly

$$|\ell(g)| - |\ell(f)| \leq \ell(|g - f|) = \ell(|f - g|).$$

These inequalities together verify the claim. $\square$

While the significance of null-continuity will become apparent in the next chapter, we take this opportunity to point out that null-continuity is synonymous with 'continuity with respect to taking arbitrary monotone limits'.

Theorem 2.19 *Let ℓ be a positive linear functional on $\mathcal{F}$. Then the following are equivalent.*

(a) ℓ is null-continuous.

(b) Given a sequence $(f_n)_{n\in\mathbb{N}}$ in $\mathcal{F}$ for which $f := \uparrow f_n \in \mathcal{F}$,

$$\ell(f) = \uparrow\ell(f_n).$$

(c) Given a sequence $(f_n)_{n\in\mathbb{N}}$ in $\mathcal{F}$ for which $f := \downarrow f_n \in \mathcal{F}$,

$$\ell(f) = \downarrow\ell(f_n).$$

Proof. (a)$\Rightarrow$(b). $f := \uparrow f_n$ implies that $0 = \downarrow (f - f_n)$ and consequently $0 = \downarrow\ell(f - f_n) = \downarrow(\ell(f) - \ell(f_n))$. Thus $\ell(f) = \uparrow\ell(f_n)$.

(b)$\Rightarrow$(c). $f = \downarrow f_n$ implies that $-f = \uparrow(-f_n)$. Then $-\ell(f) = \uparrow(-\ell(f_n))$. Thus $\ell(f) = \downarrow\ell(f_n)$.

(c)$\Rightarrow$(a) is trivial. $\qquad\square$

As an illustration, we consider positive linear functionals on the vector lattices $\mathcal{K}(X)$ and $\mathcal{L}(\mathfrak{R})$ introduced in Section 2.3. These are the kinds of functionals to appear most frequently in integration theory and its applications.

(a) The Riemann integral on $\mathcal{K}([a,b])$ Consider a closed interval $[a,b]$ in $\mathbb{R}$ ($a \le b$). $[a,b]$ is a compact Hausdorff space and so $\mathcal{C}([a,b]) = \mathcal{K}([a,b])$. Given $f \in \mathcal{K}([a,b])$, let

$$\ell(f) := \int_a^b f(x)\,dx$$

be the Riemann integral. It is common knowledge that ℓ is positive and linear. The null-continuity of ℓ is less commonly known. It is, however, a simple result of Theorem 2.13. For let $(f_n)_{n\in\mathbb{N}}$ be a sequence in $\mathcal{K}([a,b])$ with $0 = \downarrow f_n$. Given $n \in \mathbb{N}$, put $\alpha_n := \sup_{x\in[a,b]} f_n(x)$. By Theorem 2.13, $0 = \downarrow\alpha_n$. But $f_n \le \alpha_n e_{[a,b]}$ for every $n \in \mathbb{N}$, so that

$$\ell(f_n) \le \alpha_n\ell(e_{[a,b]}) = \alpha_n(b-a).$$

It follows that $0 = \downarrow\ell(f_n)$.

This simple example invites a substantial generalization, which is the content of (b).

(b) Positive linear functionals on spaces of continuous functions with compact support Let X be a locally compact space. We consider positive linear functionals on $\mathcal{K}(X)$. The next result is a particularly important one.

Theorem 2.20 *Every positive linear functional ℓ on $\mathcal{K}(X)$ is null-continuous.*

Proof. Let $(f_n)_{n\in\mathbb{N}}$ be a sequence in $\mathcal{K}(X)$ for which $0 = \downarrow f_n$. Given $n \in \mathbb{N}$, put $\alpha_n := \sup_{x\in X} f_n(x)$ (cf. Theorem 2.12). For each $n \in \mathbb{N}$, $\sqrt{f_n} \in \mathcal{K}(X)$ and

$$f_n = \sqrt{f_n}\sqrt{f_n} \le \sqrt{\alpha_n}\sqrt{f_1}.$$

Hence

$$\ell(f_n) \le \sqrt{\alpha_n}\,\ell(\sqrt{f_1}).$$

Theorem 2.13 implies that $0 = \downarrow\ell(f_n)$. $\qquad\qquad\square$

The spaces $\mathcal{K}(\mathbb{R}^n)$ ($n \in \mathbb{N}$) are important examples. We shall return to them later. The following example, however, is also an important one.

Take an arbitrary set X with its discrete topology. The compact sets in this topology are precisely the finite subsets of X. Therefore

$$\mathcal{F}(X) := \{f \in \mathbb{R}^X \mid \{f \ne 0\} \text{ is finite}\} = \mathcal{K}(X).$$

Let ℓ be a positive linear functional on $\mathcal{F}(X)$. For $x \in X$, put $g(x) := \ell(e_{\{x\}})$. g is a positive function on X. For each $f \in \mathcal{F}(X)$,

$$f = \sum_{x\in\{f\ne 0\}} f(x)e_{\{x\}}.$$

It follows that

$$\ell(f) = \sum_{x\in\{f\ne 0\}} f(x)\ell(e_{\{x\}}) = \sum_{x\in\{f\ne 0\}} f(x)g(x).$$

Conversely, if we consider a positive function $g \in \mathbb{R}^X$, then the map defined by

$$\ell(f) := \sum_{x\in\{f\ne 0\}} f(x)g(x)$$

is a positive linear functional on $\mathcal{F}(X)$. g and ℓ are uniquely determined by each other. The functional determined by g is denoted by ℓ_g. Thus every positive linear functional on $\mathcal{F}(X)$ is of the form ℓ_g for some positive function $g \in \mathbb{R}^X$. We shall return later to this example as well.

(c) Positive contents, and functionals on spaces of step functions
Functionals derived from contents are of particular significance in various applications of integration theory and especially for probability theory. This section is devoted to the study of such functionals. We commence with several comments on contents.

Let $\mathfrak{R}$ be a ring of subsets of a set X. A **positive content** on $\mathfrak{R}$ is a map $\mu : \mathfrak{R} \to \mathbb{R}$ such that

(i) $\mu(A \cup B) = \mu(A) + \mu(B)$ for $A, B \in \mathfrak{R}$, $A \cap B = \emptyset$ (**additivity**);
(ii) $\mu(A) \ge 0$ for $A \in \mathfrak{R}$ (**positivity**).

Using complete induction, property (i) extends to any finite collection of pairwise disjoint sets.

A positive content μ on $\mathfrak{R}$ is a **positive measure** on $\mathfrak{R}$ if

(iii) $\mu(A_n) \downarrow 0$ for each sequence $(A_n)_{n\in\mathbb{N}}$ in $\mathfrak{R}$ for which $A_n \downarrow \emptyset$.

Property (iii) is called **null-continuity**, just like the corresponding property for functionals. A triple $(X, \mathfrak{R}, \mu)$ is called a **positive measure space** if $\mathfrak{R}$ is a ring of subsets of X and μ is a positive measure on $\mathfrak{R}$.

We summarize several simple properties of positive contents.

Proposition 2.21 *Let μ be a positive content on $\mathfrak{R}$. Then:*

(a) $\mu(\emptyset) = 0$.

(b) $\mu(B) \leq \mu(A)$ *for all* $A, B \in \mathfrak{R}$, $B \subset A$ (**monotonicity**).

(c) $\mu(A \cup B) \leq \mu(A) + \mu(B)$ *for all* $A, B \in \mathfrak{R}$ (**subadditivity**).

(d) $\mu(A \setminus B) = \mu(A) - \mu(A \cap B)$ *for all* $A, B \in \mathfrak{R}$.

(e) $\mu(A\cup B)+\mu(A\cap B) = \mu(A)+\mu(B)$ *for all* $A, B \in \mathfrak{R}$ (**modularity**).

Proof. Since $\emptyset = \emptyset \cup \emptyset$, (a) follows from the additivity of μ.

Given $A, B \in \mathfrak{R}$,
$$A = (A \cap B) \cup (A \setminus B),$$
and $(A \cap B) \cap (A \setminus B) = \emptyset$. Thus
$$\mu(A) = \mu(A \cap B) + \mu(A \setminus B), \tag{1}$$
and (d) follows. If $B \subset A$, then (b) follows immediately since $\mu(A\setminus B) \geq 0$. Observe that
$$A \cup B = (A \setminus B) \cup B$$
and $(A \setminus B) \cap B = \emptyset$. Thus
$$\mu(A \cup B) = \mu(A \setminus B) + \mu(B) \leq \mu(A) + \mu(B). \tag{2}$$

Finally, (e) follows from (1) and from the first relation in (2). $\qquad\square$

Property (c) can be extended to any finite number of elements of $\mathfrak{R}$.

As is the case for functionals, the null-continuity of a positive content implies the continuity of the content with respect to taking monotone limits.

Theorem 2.22 *Take a positive content μ on $\mathfrak{R}$. Then the following are equivalent.*

(a) μ *is a positive measure.*

(b) $\mu(A) = \uparrow \mu(A_n)$ *for every increasing sequence $(A_n)_{n\in\mathbb{N}}$ in $\mathfrak{R}$ for which* $A := \uparrow A_n \in \mathfrak{R}$.

(c) $\mu(A) = \downarrow \mu(A_n)$ *for every decreasing sequence $(A_n)_{n\in\mathbb{N}}$ in $\mathfrak{R}$ for which* $A := \downarrow A_n \in \mathfrak{R}$.

Proof. (a)⇒(b). If $A := \uparrow A_n$, then $\downarrow (A \setminus A_n) = \emptyset$. Thus

$$0 = \downarrow \mu(A \setminus A_n) = \downarrow (\mu(A) - \mu(A_n))$$

and consequently

$$\mu(A) = \uparrow \mu(A_n).$$

(b)⇒(c). If $A := \downarrow A_n$, then $\uparrow (A_1 \setminus A_n) = (A_1 \setminus A)$. Hence

$$\mu(A_1) - \mu(A) = \uparrow (\mu(A_1) - \mu(A_n)),$$

and it follows that

$$\mu(A) = \downarrow \mu(A_n).$$

(c)⇒(a) is trivial. □

We now assign to each positive content μ on $\mathfrak{R}$ a positive linear functional ℓ_μ on $\mathcal{L}(\mathfrak{R})$. We wish to do this in a manner which ensures that $\ell_\mu(e_A) = \mu(A)$ for all $A \in \mathfrak{R}$, having in the back of our minds the idea of the 'area below the graph of e_A'. Since we want ℓ_μ to be linear, the only possible definition is

$$\ell_\mu(f) := \sum_{k=1}^{n} \alpha_k \mu(A_k) \tag{3}$$

for $f = \sum_{k=1}^{n} \alpha_k e_{A_k} \in \mathcal{L}(\mathfrak{R})$. Note that the decomposition of f into this form is not unique. In order for $\ell_\mu(f)$ to be well defined, the sum in its definition must be independent of the decomposition chosen. This is indeed the case, as we next show.

Proposition 2.23 *Let* $f = \sum_{k \in K} \alpha_k e_{A_k} = \sum_{l \in L} \beta_l e_{B_l} \in \mathcal{L}(\mathfrak{R})$. *Then*

$$\sum_{k \in K} \alpha_k \mu(A_k) = \sum_{l \in L} \beta_l \mu(B_l).$$

Proof. Let $\mathfrak{S} := \{A_k \mid k \in K\} \cup \{B_l \mid l \in L\}$. By Theorem 2.14, there is a disjoint dissection $\mathfrak{C}$ for $\mathfrak{S}$. Given any $C \in \mathfrak{C}$,

$$\sum_{\substack{k \in K \\ C \subset A_k}} \alpha_k = \sum_{\substack{l \in L \\ C \subset B_l}} \beta_l$$

and so

$$\sum_{k \in K} \alpha_k \mu(A_k) = \sum_{k \in K} \alpha_k \left(\sum_{\substack{C \in \mathfrak{C} \\ C \subset A_k}} \mu(C) \right) = \sum_{C \in \mathfrak{C}} \left(\sum_{\substack{k \in K \\ C \subset A_k}} \alpha_k \right) \mu(C)$$

$$= \sum_{C \in \mathfrak{C}} \left(\sum_{\substack{l \in L \\ C \subset B_l}} \beta_l \right) \mu(C) = \sum_{l \in L} \beta_l \left(\sum_{\substack{C \in \mathfrak{C} \\ C \subset B_l}} \mu(C) \right) = \sum_{l \in L} \beta_l \mu(B_l).$$

□

Definition (3) is indeed meaningful and so we may proceed to our next theorem.

Theorem 2.24 *Given a positive content μ on $\mathfrak{R}$, ℓ_μ is a positive linear functional on $\mathcal{L}(\mathfrak{R})$. Moreover, ℓ_μ is null-continuous if and only if μ is a positive measure.*

Proof. The linearity of ℓ_μ is trivial. Take $f \in \mathcal{L}(\mathfrak{R})$, $f \geq 0$. Then there are a disjoint family $(A_k)_{k \in K}$ in $\mathfrak{R}$ and real numbers α_k $(k \in K)$ such that

$$f = \sum_{k \in K} \alpha_k e_{A_k} \geq 0.$$

Since the A_k are pairwise disjoint, it follows that $\alpha_k \geq 0$ for all $k \in K$. Thus

$$\ell_\mu(f) = \sum_{k \in K} \alpha_k \mu(A_k) \geq 0.$$

Now let μ be a positive measure and therefore null-continuous. Let $(f_n)_{n \in \mathbb{N}}$ be a sequence in $\mathcal{L}(\mathfrak{R})$ with $0 = \downarrow f_n$. Take $\varepsilon > 0$. For each $n \in \mathbb{N}$, we put $B_n := \{f_n > \varepsilon\}$ and $B := \{f_1 > 0\}$. By Proposition 2.16, B and all the B_n are in $\mathfrak{R}$, and $\emptyset = \downarrow B_n$. Hence $0 = \downarrow \mu(B_n)$. Setting $\alpha := \sup_{x \in X} f_1(x)$, we see that for each $n \in \mathbb{N}$

$$f_n \leq \varepsilon e_B + \alpha e_{B_n}.$$

Choose $m \in \mathbb{N}$ such that $\mu(B_m) < \varepsilon$. Then

$$\ell_\mu(f_n) \leq \varepsilon \mu(B) + \alpha \varepsilon$$

for each $n \geq m$. Since $\varepsilon > 0$ is arbitrary,

$$0 = \downarrow \ell_\mu(f_n).$$

Conversely, let ℓ_μ be null-continuous and take a sequence $(A_n)_{n \in \mathbb{N}}$ in $\mathfrak{R}$ such that $\emptyset = \downarrow A_n$. Then $0 = \downarrow e_{A_n}$ and thus

$$0 = \downarrow \ell_\mu(e_{A_n}) = \downarrow \mu(A_n).$$

Hence μ is also null-continuous.   □

Thus, recapitulating the second statement of the theorem, a positive content μ is null-continuous if and only if the associated functional ℓ_μ is null-continuous. This result, while very natural, is nevertheless remarkable.

Our first example is the ring $\mathfrak{F}(X)$ of all finite subsets of the set X. For an arbitrary positive function $g \in \mathbb{R}^X$, let

$$\mu^g : \mathfrak{F}(X) \longrightarrow \mathbb{R}, \quad A \longmapsto \sum_{x \in A} g(x).$$

It is easy to show that μ^g is a positive measure on $\mathfrak{F}(X)$. Furthermore, it is clear that $\mathcal{L}(\mathfrak{F}(X)) = \mathcal{F}(X)$ (Example (b)). For $f \in \mathcal{F}(X)$, $f =$

$\sum_{x \in \{f \neq 0\}} f(x) e_{\{x\}}$, we have

$$\ell_g(f) = \sum_{x \in \{f \neq 0\}} f(x)g(x) = \ell_{\mu^g}(f).$$

Consequently $\ell_{\mu^g} = \ell_g$ and by Theorem 2.24, ℓ_g is null-continuous.

Put $g = e_X$. Then μ^g is called **counting measure** on X. Counting measure assigns to each finite subset A of X the number of elements of A.

Extending positive measures – a process to be discussed later – gives rise to positive measures on systems of sets with better 'stability' properties than those of rings. This is one of the goals of extension. We list the most important types of systems. A ring $\mathfrak{R}$ of subsets of a set X is a

- **δ-ring** if $\bigcap_{n \in \mathbb{N}} A_n \in \mathfrak{R}$ for every sequence $(A_n)_{n \in \mathbb{N}}$ in $\mathfrak{R}$;

- **σ-ring** if $\bigcup_{n \in \mathbb{N}} A_n \in \mathfrak{R}$ for every sequence $(A_n)_{n \in \mathbb{N}}$ in $\mathfrak{R}$;

- **σ-algebra** on X if $\mathfrak{R}$ is a σ-ring and $X \in \mathfrak{R}$.

Note that every σ-ring $\mathfrak{R}$ is a δ-ring, since

$$\bigcap_{n \in \mathbb{N}} A_n = A_1 \setminus \bigcup_{n \in \mathbb{N}} (A_1 \setminus A_n) \in \mathfrak{R}$$

for every sequence $(A_n)_{n \in \mathbb{N}}$ in $\mathfrak{R}$.

If $\mathfrak{R}$ is a δ-ring and $(A_n)_{n \in \mathbb{N}}$ is a sequence in $\mathfrak{R}$ such that there is an $A \in \mathfrak{R}$ containing all of the A_n, then $\bigcup_{n \in \mathbb{N}} A_n \in \mathfrak{R}$. In fact, we have

$$\bigcup_{n \in \mathbb{N}} A_n = A \setminus \bigcap_{n \in \mathbb{N}} (A \setminus A_n).$$

All of these systems of sets share a common feature. The intersection of systems of a given type is also a system of that type.

Proposition 2.25 *Let Φ be a non-empty set of rings of sets. Then $\bigcap_{\mathfrak{R} \in \Phi} \mathfrak{R}$ is also a ring of sets. The corresponding statements for δ-rings, σ-rings and σ-algebras on a set X are each true as well.*

The proofs are trivial.

We have thus established the existence of smallest systems containing a given set $\mathfrak{S}$ of sets – the system of a given type which is generated by $\mathfrak{S}$ is the intersection of all systems of this type which contain $\mathfrak{S}$. Note that there is always at least one ring containing $\mathfrak{S}$, namely $\mathfrak{P}(\bigcup_{A \in \mathfrak{S}} A)$, and similarly for the other kinds of systems. We call the smallest ring of sets containing $\mathfrak{S}$ the **ring of sets generated by** $\mathfrak{S}$ and denote it by $\mathfrak{S}_r$. Similarly, we have the **δ-ring $\mathfrak{S}_\delta$ generated by** $\mathfrak{S}$, the **σ-ring $\mathfrak{S}_\sigma$ generated by** $\mathfrak{S}$, and the **σ-algebra** on $X \supset \bigcup_{A \in \mathfrak{S}} A$ **generated by** $\mathfrak{S}$.

(d) Stieltjes measures and functionals In practice, contents cannot always be directly defined on a ring of sets. Often only a system with weaker properties is available. The definition must then be extended to the generated ring of sets. An important example of this is provided by the Stieltjes measures on $\mathbb{R}$. They are first defined on a system of sets called a semi-ring of sets. Since these semi-rings of sets are also important in other contexts, we now discuss them briefly.

A set $\mathfrak{S} \neq \emptyset$ of sets is a **semi-ring of sets** if the following assertions hold.

(i) If $A, B \in \mathfrak{S}$, then $A \cap B \in \mathfrak{S}$.

(ii) Given $A, B \in \mathfrak{S}$, there is a finite set $\mathfrak{C} \subset \mathfrak{S}$ of pairwise disjoint sets such that
$$A \setminus B = \bigcup_{C \in \mathfrak{C}} C.$$

Fortunately, the ring of sets generated by a semi-ring has a very simple structure.

Proposition 2.26 *Given a semi-ring of sets $\mathfrak{S}$, $\mathfrak{S}_r$ is the set of all unions of finite subsets $\mathfrak{C}$ of $\mathfrak{S}$ consisting of pairwise disjoint sets.*

Proof. Let $A = \bigcup_{k \in K} A_k$ and $B = \bigcup_{l \in L} B_l$ be such unions. Then
$$A \cap B = \bigcup_{\substack{k \in K \\ l \in L}} (A_k \cap B_l) \tag{4}$$

is also a finite union of disjoint sets from $\mathfrak{S}$. Moreover,
$$A \setminus B = \bigcup_{k \in K} \left(\bigcap_{l \in L} (A_k \setminus B_l) \right).$$

The sets $\bigcap_{l \in L}(A_k \setminus B_l)$ are pairwise disjoint for $k \in K$. By hypothesis (ii) and relation (4), they are finite unions of pairwise disjoint sets from $\mathfrak{S}$. Thus $A \setminus B$ is also such a union. Since
$$A \cup B = (A \cap B) \cup (A \setminus B) \cup (B \setminus A),$$

$A \cup B$ also has this property. The set of all such finite unions of disjoint sets from $\mathfrak{S}$ is thus a ring of sets. Clearly, it is contained in every ring of sets containing $\mathfrak{S}$. It is therefore $\mathfrak{S}_r$. $\square$

Our next theorem confirms that a map from a semi-ring into $\mathbb{R}_+$ can be extended (uniquely) to a content on the ring of sets generated whenever it has the additivity property in the definition of a content. We formulate this result in slightly greater generality.

Theorem 2.27 *Let $\mathfrak{S}$ be a semi-ring of sets. Let $\nu : \mathfrak{S} \to \mathbb{R}$ be a map such that if $\mathfrak{A} \subset \mathfrak{S}$ is a finite set of pairwise disjoint sets for which $\bigcup_{A \in \mathfrak{A}} A \in \mathfrak{S}$, then*
$$\nu\left(\bigcup_{A \in \mathfrak{A}} A \right) = \sum_{A \in \mathfrak{A}} \nu(A).$$

Then there is a unique additive map μ on $\mathfrak{S}_r$ with $\mu|_{\mathfrak{S}} = \nu$. If in addition $\nu(A) \geq 0$ whenever $A \in \mathfrak{S}$, then μ is a positive content.

Proof. Let $\mathfrak{A}$ and $\mathfrak{B}$ be finite subsets of $\mathfrak{S}$ consisting of pairwise disjoint sets such that $\bigcup_{A \in \mathfrak{A}} A = \bigcup_{B \in \mathfrak{B}} B$. Then

$$\nu(A) = \sum_{B \in \mathfrak{B}} \nu(A \cap B) \quad \text{for every } A \in \mathfrak{A},$$

$$\nu(B) = \sum_{A \in \mathfrak{A}} \nu(A \cap B) \quad \text{for every } B \in \mathfrak{B}.$$

Thus

$$\sum_{A \in \mathfrak{A}} \nu(A) = \sum_{A \in \mathfrak{A}} \sum_{B \in \mathfrak{B}} \nu(A \cap B) = \sum_{B \in \mathfrak{B}} \nu(B).$$

Given finite subsets $\mathfrak{A}$ of $\mathfrak{S}$ of pairwise disjoint sets, define

$$\mu\left(\bigcup_{A \in \mathfrak{A}} A\right) := \sum_{A \in \mathfrak{A}} \nu(A).$$

By Proposition 2.26, μ is defined on the entire ring of sets $\mathfrak{S}_r$. The above argument shows that the definition does not depend on the choice of the representation of $\bigcup_{A \in \mathfrak{A}} A$. The conclusions of the theorem now follow easily. □

We now consider the set of right half-open intervals of $\mathbb{R}$ as an example of a semi-ring of sets. Let $\alpha, \beta \in \mathbb{R}$, $\alpha \leq \beta$. Then we call

$$[\alpha, \beta[:= \{x \in \mathbb{R} \mid \alpha \leq x < \beta\}$$

the **right half-open interval** of $\mathbb{R}$ with endpoints α and β. We write $\mathfrak{J}_0$ for the set of all such right half-open intervals of $\mathbb{R}$.

Proposition 2.28 $\mathfrak{J}_0$ *is a semi-ring of sets.*

Proof. Take $\alpha, \beta, \gamma, \delta \in \mathbb{R}$, $\alpha \leq \beta$, $\gamma \leq \delta$. Then

$$[\alpha, \beta[\cap [\gamma, \delta[= \begin{cases} \emptyset & \text{if } \beta \leq \gamma \\ \emptyset & \text{if } \delta \leq \alpha \\ [\alpha \vee \gamma, \beta \wedge \delta[& \text{if } \alpha < \delta \text{ and } \gamma < \beta, \end{cases}$$

and

$$[\alpha, \beta[\setminus [\gamma, \delta[= \begin{cases} [\alpha, \beta[& \text{if } \gamma \leq \delta \leq \alpha \leq \beta \\ [\delta, \beta[& \text{if } \gamma \leq \alpha \leq \delta \leq \beta \\ [\alpha, \gamma[\cup [\delta, \beta[& \text{if } \alpha \leq \gamma \leq \delta \leq \beta \\ [\alpha, \gamma[& \text{if } \alpha \leq \gamma \leq \beta \leq \delta \\ [\alpha, \beta[& \text{if } \alpha \leq \beta \leq \gamma \leq \delta \\ \emptyset & \text{if } \gamma \leq \alpha \leq \beta \leq \delta. \end{cases}$$

The proof follows from these relations. □

By Proposition 2.26, the ring of sets generated by $\mathfrak{J}_0$, which we denote by $\mathfrak{J}$, consists of all finite unions of pairwise disjoint intervals from $\mathfrak{J}_0$. We call such sets **interval forms** on $\mathbb{R}$. Each interval form can thus be written as

$$\bigcup_{k=1}^{n} [\alpha_k, \beta_k[$$

for suitable choices of n and α_k, β_k ($k \in \{1, \ldots, n\}$). Furthermore, the α_k and the β_k can be chosen so that

$$\alpha_1 \leq \beta_1 \leq \alpha_2 \leq \beta_2 \leq \cdots \leq \alpha_n \leq \beta_n.$$

We now consider an increasing function $g : \mathbb{R} \to \mathbb{R}$. Given $\alpha, \beta \in \mathbb{R}$ with $\alpha \leq \beta$, define

$$\mu_g([\alpha, \beta[) := g(\beta) - g(\alpha).$$

It follows immediately that $\mu_g([\alpha, \beta[) \geq 0$. If $\bigcup_{k=1}^{n} [\alpha_k, \beta_k[$ is an interval form on $\mathbb{R}$ such that

$$\bigcup_{k=1}^{n} [\alpha_k, \beta_k[= [\alpha, \beta[\in \mathfrak{J}_0,$$

then the intervals can be arranged in such a way that $\alpha_k = \beta_{k-1}$ for $k \in \{2, \ldots, n\}$. Thus

$$\mu_g([\alpha, \beta[) = g(\beta) - g(\alpha) = \sum_{k=1}^{n} (g(\beta_k) - g(\alpha_k)) = \sum_{k=1}^{n} \mu_g([\alpha_k, \beta_k[).$$

The hypotheses of Theorem 2.27 are thereby satisfied and we have:

Theorem 2.29 *Given an increasing function $g : \mathbb{R} \to \mathbb{R}$ there is a unique positive content μ_g on $\mathfrak{J}$ such that $\mu_g([\alpha, \beta[) = g(\beta) - g(\alpha)$ for every $[\alpha, \beta[\in \mathfrak{J}_0$.*

Conversely, given a positive content μ on $\mathfrak{J}$, there is an increasing function $g : \mathbb{R} \to \mathbb{R}$ with $\mu = \mu_g$. Moreover, given any other increasing function $\tilde{g} : \mathbb{R} \to \mathbb{R}$ with $\mu = \mu_{\tilde{g}}$, there is a $\gamma \in \mathbb{R}$ with $\tilde{g} = g + \gamma$.

Proof. Only the second part of the theorem still requires proof. Given a positive content μ on $\mathfrak{J}$, define

$$g(\alpha) := \begin{cases} \mu([0, \alpha[) & \text{if } \alpha \geq 0 \\ -\mu([\alpha, 0[) & \text{if } \alpha < 0. \end{cases}$$

We leave it to the reader to verify that g is increasing and that $\mu_g([\alpha, \beta[) = \mu([\alpha, \beta[)$ for any $\alpha, \beta \in \mathbb{R}$ with $\alpha \leq \beta$. Thus μ_g and μ coincide on $\mathfrak{J}_0$ and hence also on $\mathfrak{J}$. Given any increasing function $\tilde{g}$ with $\mu = \mu_{\tilde{g}}$, put $\gamma := \tilde{g}(0) - g(0)$. Then, if $\alpha > 0$,

$$g(\alpha) + \gamma = g(\alpha) - g(0) + \tilde{g}(0) - \tilde{g}(\alpha) + \tilde{g}(\alpha)$$
$$= \mu_g([0, \alpha[) - \mu_{\tilde{g}}([0, \alpha[) + \tilde{g}(\alpha) = \tilde{g}(\alpha),$$

and similarly for $\alpha < 0$. □

μ_g is called the **Stieltjes content on $\mathbb{R}$ generated by g**. The question arises: under what conditions is μ_g a positive measure?

Assume for the moment that μ_g is a positive measure. Take $\alpha \in \mathbb{R}$ and let $(\alpha_n)_{n \in \mathbb{N}}$ be a sequence in $\mathbb{R}$ such that $\alpha_n \leq \alpha$ for every $n \in \mathbb{N}$ and $\lim_{n \to \infty} \alpha_n = \alpha$. Then $(\beta_n)_{n \in \mathbb{N}}$ defined by

$$\beta_n := \inf_{m \geq n} \alpha_m \qquad (n \in \mathbb{N})$$

is an increasing sequence in $\mathbb{R}$. Moreover, $\beta_n \leq \alpha_n \leq \alpha$ for each $n \in \mathbb{N}$ and $\uparrow \beta_n = \alpha$. Thus $[\beta_n, \alpha[\downarrow \emptyset$ and hence

$$g(\alpha) - \left(\uparrow g(\beta_n) \right) = \downarrow \left(g(\alpha) - g(\beta_n) \right) = \downarrow \mu_g \left([\beta_n, \alpha[\right) = 0.$$

We conclude that $g(\alpha) = \uparrow g(\beta_n) = \lim_{n \to \infty} g(\alpha_n)$.

A function $g \in \mathbb{R}^X$ is said to be **left continuous** if

$$g(\alpha) = \lim_{n \to \infty} g(\alpha_n)$$

for every $\alpha \in \mathbb{R}$ and every sequence $(\alpha_n)_{n \in \mathbb{N}}$ such that $\alpha_n \leq \alpha$ for each $n \in \mathbb{N}$ and $\alpha = \lim_{n \to \infty} \alpha_n$. (Note that g is not required to be increasing.)

Functions g which generate positive measures μ_g are therefore left continuous. The next theorem shows that the left continuity of g is also sufficient for μ_g to be a positive measure.

Theorem 2.30 *Given an increasing function $g : \mathbb{R} \to \mathbb{R}$, the following are equivalent.*

(a) g is left continuous.

(b) μ_g is a positive measure.

Proof. Only (a)⇒(b) remains to be proven. Let $(A_n)_{n \in \mathbb{N}}$ be a sequence in $\mathfrak{I}$ such that $\emptyset = \downarrow A_n$. We may assume that each A_n is non-empty, for otherwise $\mu_g(A_n) \downarrow 0$ is trivial. Given $n \in \mathbb{N}$, let

$$A_n = \bigcup_{k \in K_n} A_{nk}$$

where $(A_{nk})_{k \in K_n}$ is finite, consisting of pairwise disjoint, non-empty sets of the form $A_{nk} = [\alpha_{nk}, \beta_{nk}[$. Let k_n be the number of elements in K_n.

Take $\varepsilon > 0$. Then for each $n \in \mathbb{N}$ and $k \in K_n$ there is a $\gamma_{nk} \in [\alpha_{nk}, \beta_{nk}[$ such that

$$\mu_g \left([\gamma_{nk}, \beta_{nk}[\right) < \frac{\varepsilon}{2^n k_n}.$$

(We have used the left continuity of g for this.) Given n, set

$$C_n := \bigcup_{k \in K_n} [\alpha_{nk}, \gamma_{nk}[.$$

The sets $\overline{C_n}$ are compact and $\bigcap_{n \in \mathbb{N}} \overline{C_n} = \emptyset$. Thus, there is an $m \in \mathbb{N}$ for which $\bigcap_{n \leq m} \overline{C_n} = \emptyset$ and hence $\bigcap_{n \leq m} C_n = \emptyset$. Thus

$$A_m \subset \bigcup_{n \leq m} (A_n \setminus C_n) \cup \bigcap_{n \leq m} C_n = \bigcup_{n \leq m} (A_n \setminus C_n),$$

and it follows that

$$\mu_g(A_m) \leq \sum_{n \leq m} \mu_g(A_n \setminus C_n) < \varepsilon.$$

Since $\varepsilon > 0$ is arbitrary, $0 = \downarrow \mu_g(A_n)$.   □

Given a left continuous increasing g, the positive Stieltjes content μ_g is called a **positive Stieltjes measure**. The functionals ℓ_{μ_g} on the $\mathfrak{I}$-step functions are called **Stieltjes functionals**. As mentioned earlier, it is one of our main goals in the following chapters to extend these measures (functionals) to a large class of subsets of $\mathbb{R}$ (functions on $\mathbb{R}$) in some 'reasonable' manner.

One special case deserves particular attention, namely Lebesgue measure. It is the best-known and mostly thoroughly investigated measure. Historically all of measure theory and integration theory developed from its study. The **Lebesgue measure** λ is defined to be the Stieltjes measure μ_{id} of the identity function $id : \mathbb{R} \to \mathbb{R}, \ \alpha \mapsto \alpha$. Thus the Lebesgue measure of an interval $[\alpha, \beta[$ is precisely its length, no more, no less: $\lambda([\alpha, \beta[) = \beta - \alpha$. We cannot yet assign Lebesgue measure to intervals of the form $]\alpha, \beta[, \ ]\alpha, \beta]$, or $[\alpha, \beta]$, since these intervals do not belong to the ring of interval forms $\mathfrak{I}$. But once we have completed the extension process, we shall see that – as expected – the Lebesgue measure of these intervals is again their length $\beta - \alpha$. This geometric property lies at the heart of the great significance of the Lebesgue measure in practical problems.

Exercises

1. Let $\mathcal{L}$ be a vector lattice in $\mathbb{R}^X$. Take $x \in X$ and $\alpha \in \mathbb{R}$. Determine whether the functional

$$\mathcal{L} \longrightarrow \mathbb{R}, \quad f \longmapsto \alpha f(x)$$

   is linear, positive or null-continuous.

2. Consider the space $c_0(X)$.

   (a) Are there any positive linear functionals on $c_0(X)$ which are not null-continuous?

(b) Show that $\ell : c_0(X) \to \mathbb{R}$ is a positive linear functional if and only if there is a $g \in \ell^1(X)_+$ such that $\ell(f) = \sum_{x \in X} f(x)g(x)$ for each $f \in c_0(X)$. In this case $g(x) = \ell(e_{\{x\}})$ for every $x \in X$, so that ℓ determines g uniquely.

3. Prove the following statements about the vector lattice $c_f(X)$.

(a) $c_f(X) = \{\alpha e_X + g \mid \alpha \in \mathbb{R}, \ g \in \mathcal{F}(X)\}$.
(b) If X is finite, then $c_f(X) = \mathcal{F}(X)$.
(c) If X is countably infinite, then
$$c_f(X) \longrightarrow \mathbb{R}, \quad \alpha e_X + g \longmapsto \alpha$$
is a positive linear functional which is not null-continuous.
(d) If X is uncountable, then every positive linear functional on $c_f(X)$ is null-continuous.

4. Let $\mathcal{L}$ be a vector lattice in $\mathbb{R}^X$, ℓ a positive linear functional on $\mathcal{L}$ and define
$$d_\ell : \mathcal{L} \times \mathcal{L} \longrightarrow \mathbb{R}, \quad (f, g) \longmapsto \ell(|f - g|).$$
Given $f, g, h \in \mathcal{L}$, $\alpha \in \mathbb{R}$, show that d_ℓ has the following properties:

(a) $d_\ell(f, g) \geq 0$.
(b) $d_\ell(f, f) = 0$.
(c) $d_\ell(f, g) = d_\ell(g, f)$.
(d) $d_\ell(f, h) \leq d_\ell(f, g) + d_\ell(g, h)$.
(e) $d_\ell(f, h) = d_\ell(f, g) + d_\ell(g, h)$ whenever $f \leq g \leq h$.
(f) $d_\ell(f, g) \leq d_\ell(f, h)$ whenever $f \leq g \leq h$.
(g) $d_\ell(f + g, f + h) = d_\ell(g, h)$.
(h) $d_\ell(\alpha f, \alpha g) = |\alpha| d_\ell(f, g)$.
(i) $|\ell(f) - \ell(g)| \leq d_\ell(f, g)$.

By properties (a)–(d), d_ℓ is a pseudometric on $\mathcal{L}$.
The uniform continuity of ℓ with respect to d_ℓ follows from property (i).

5. Take $x \in X$ and define
$$\delta_x : \mathfrak{P}(X) \longrightarrow \mathbb{R}, \quad A \longmapsto \begin{cases} 1 & \text{if } x \in A \\ 0 & \text{if } x \notin A. \end{cases}$$

Show that δ_x is a positive measure (the **Dirac measure** at x), and describe ℓ_{δ_x}.

6. Let X be an infinite set. Put $\mathfrak{R} := \{A \subset X \mid A \text{ is finite or } X \setminus A \text{ is finite}\}$ and
$$\mu : \mathfrak{R} \longrightarrow \mathbb{R}, \quad A \longmapsto \begin{cases} 0 & \text{if } A \text{ is finite} \\ 1 & \text{if } X \setminus A \text{ is finite}. \end{cases}$$

Show that μ is a positive content and that μ is a positive measure if and only if X is uncountable. Describe ℓ_μ.

7. Determine all positive measures on $\mathfrak{P}(\mathbb{N})$.

8. Let μ be a positive content on the ring of sets $\mathfrak{R}$.

   (a) Given $A, B, C \in \mathfrak{R}$, prove that

   $$\mu(A \cup B \cup C) = \mu(A) + \mu(B) + \mu(C)$$
   $$- \mu(A \cap B) - \mu(A \cap C) - \mu(B \cap C) + \mu(A \cap B \cap C).$$

   (b) Formulate and prove the analogous result for n sets $A_1, \ldots, A_n \in \mathfrak{R}$ $(n \in \mathbb{N})$.

9. Prove each of the following statements.

   (a) Let $\mathfrak{R}, \mathfrak{S}$ be rings of sets and μ a positive measure on $\mathfrak{R}$. If $\mathfrak{S} \subset \mathfrak{R}$, then $\mu|_{\mathfrak{S}}$ is a positive measure on $\mathfrak{S}$.

   (b) Let $(X, \mathfrak{R}, \mu)$ be a positive measure space. Take $Y \subset X$ and define

   $$\mathfrak{R}|_Y := \{A \in \mathfrak{R} \mid A \subset Y\} \qquad \text{and} \qquad \mu|_Y := \mu|_{(\mathfrak{R}|_Y)}.$$

   Then $(Y, \mathfrak{R}|_Y, \mu|_Y)$ is a positive measure space. It is called the **restriction** of $(X, \mathfrak{R}, \mu)$ to Y.

   (c) Let $(X, \mathfrak{R}, \mu)$ be a positive measure space. Let Y be a set and $\varphi : X \to Y$ a mapping. Define

   $$\varphi(\mathfrak{R}) := \{A \subset Y \mid \varphi^{-1}(A) \in \mathfrak{R}\},$$
   $$\varphi(\mu) : \varphi(\mathfrak{R}) \longrightarrow \mathbb{R}, \quad A \longmapsto \mu(\varphi^{-1}(A)).$$

   Then $(Y, \varphi(\mathfrak{R}), \varphi(\mu))$ is a positive measure space. It is called the **image** of $(X, \mathfrak{R}, \mu)$ under φ.

10. Let $A_1, \ldots, A_n$ be disjoint non-empty sets. Let $\mathfrak{R}$ denote the set of all unions $A_{i_1} \cup \cdots \cup A_{i_r}$, where $i_1, \ldots, i_r \in \{1, \ldots, n\}$. Prove:

    (a) $\mathfrak{R}$ is the ring of sets generated by $\{A_k \mid 1 \leq k \leq n\}$.

    (b) $\mathfrak{R}$ contains exactly 2^n elements.

    (c) If $\alpha_1, \ldots, \alpha_n$ are positive real numbers, then there is a unique positive content μ on $\mathfrak{R}$ such that $\mu(A_k) = \alpha_k$ for each $k \leq n$. Moreover, this μ is a positive measure on $\mathfrak{R}$.

11. A **lattice of sets** is a set $\mathfrak{R}$ of sets such that $\emptyset \in \mathfrak{R}$ and $A \cup B \in \mathfrak{R}$, $A \cap B \in \mathfrak{R}$ for all $A, B \in \mathfrak{R}$. Let $\mathfrak{R}$ be a lattice of sets. Define

    $$\mathfrak{R}' := \{A \setminus B \mid A, B \in \mathfrak{R}\},$$
    $$\mathfrak{R}'' := \left\{\bigcup_{\iota \in I} A_\iota \mid (A_\iota)_{\iota \in I} \text{ is a finite family in } \mathfrak{R}'\right\}.$$

    Show that $\mathfrak{R}'' = \mathfrak{R}_r$.

12. Determine which kinds of system (lattices of sets, semi-rings, rings of sets, δ-rings, σ-rings or σ-algebras) are formed by the systems in Exercise 7 of Section 2.3.

13. Let $\mathfrak{A}$ be the set of all singletons $\{x\}$ of an uncountable set X. Determine $\mathfrak{A}_r$, $\mathfrak{A}_\delta$, $\mathfrak{A}_\sigma$ and the σ-algebra on X generated by $\mathfrak{A}$.

14. Compare the δ-ring generated by $\{]\alpha,\beta] \,|\, \alpha,\beta \in \mathbb{Q}, \ \alpha < \beta\}$ with that generated by $\mathfrak{J}_0$.

15. Take $\mathfrak{R} \subset \mathfrak{P}(X)$, for some set X. Show that the following are equivalent.

   (a) $\mathfrak{R}$ is a σ-algebra on X.
   (b) $\emptyset \in \mathfrak{R}$ and $\bigcup_{n\in\mathbb{N}} A_n \cup (X \setminus B) \in \mathfrak{R}$ for every sequence $(A_n)_{n\in\mathbb{N}}$ in $\mathfrak{R}$ and every $B \in \mathfrak{R}$.
   (c) $X \in \mathfrak{R}$ and $\bigcap_{n\in\mathbb{N}} A_n \cap (X \setminus B) \in \mathfrak{R}$ for every sequence $(A_n)_{n\in\mathbb{N}}$ in $\mathfrak{R}$ and every $B \in \mathfrak{R}$.
   (d) $\mathfrak{R} \neq \emptyset$, $\bigcup_{n\in\mathbb{N}} A_n \in \mathfrak{R}$ and $X \setminus B \in \mathfrak{R}$ for every sequence $(A_n)_{n\in\mathbb{N}}$ in $\mathfrak{R}$ and every $B \in \mathfrak{R}$.
   (e) $\mathfrak{R} \neq \emptyset$, $\bigcap_{n\in\mathbb{N}} A_n \in \mathfrak{R}$ and $X \setminus B \in \mathfrak{R}$ for every sequence $(A_n)_{n\in\mathbb{N}}$ in $\mathfrak{R}$ and every $B \in \mathfrak{R}$.

16. Let $\mathfrak{R}$ be a ring of sets and take $A \in \mathfrak{R}_\delta$. Prove that there is a $B \in \mathfrak{R}$ with $A \subset B$.
   Hint: Let $\mathfrak{S} := \{C \in \mathfrak{R}_\delta \,|\, \exists\, B \in \mathfrak{R}, \ C \subset B\}$ and show that $\mathfrak{S}$ is a δ-ring.

17. Let $\mathfrak{R}$ be a δ-ring. Show that

$$\mathfrak{R}_\sigma = \left\{ \bigcup_{n\in\mathbb{N}} A_n \,\Big|\, (A_n)_{n\in\mathbb{N}} \text{ is a sequence in } \mathfrak{R} \right\}.$$

18. Let $\mathfrak{S}$ be a set of sets. Prove the following.

   (a) $\mathfrak{S}_r = \{A \,|\, \exists\, \mathfrak{R} \subset \mathfrak{S}, \ \mathfrak{R} \text{ finite}, \ A \in \mathfrak{R}_r\}$.
   (b) $\mathfrak{S}_\delta = \{A \,|\, \exists\, \mathfrak{R} \subset \mathfrak{S}, \ \mathfrak{R} \text{ countable}, \ A \in \mathfrak{R}_\delta\}$.
   (c) $\mathfrak{S}_\sigma = \{A \,|\, \exists\, \mathfrak{R} \subset \mathfrak{S}, \ \mathfrak{R} \text{ countable}, \ A \in \mathfrak{R}_\sigma\}$.

   Hint: First show that the set of sets with the property in (a) ((b),(c)) is a ring of sets (δ-ring, σ-ring) and then show that it must in fact coincide with $\mathfrak{S}_r$ ($\mathfrak{S}_\delta, \mathfrak{S}_\sigma$).

19. Is there an infinite σ-ring with only countably many elements?

20. A positive measure μ on $\mathfrak{R}$ is called **bounded** if $\sup_{A\in\mathfrak{R}} \mu(A) < \infty$. Show that every positive measure on a σ-ring is bounded.

21. Let $\mathfrak{R}$ be a δ-ring, $(A_n)_{n\in\mathbb{N}}$ a sequence in $\mathfrak{R}$ with $\bigcup_{n\in\mathbb{N}} A_n \in \mathfrak{R}$ and μ a positive measure on $\mathfrak{R}$. Prove the following.

   (a) $\bigcup_{m\in\mathbb{N}} \bigcap_{n\geq m} A_n \in \mathfrak{R}$ and $\bigcap_{m\in\mathbb{N}} \bigcup_{n\geq m} A_n \in \mathfrak{R}$.

(b) $\mu\left(\bigcup_{m\in\mathbb{N}}\bigcap_{n\geq m} A_n\right) \leq \liminf_{n\to\infty} \mu(A_n) \leq \limsup_{n\to\infty} \mu(A_n)$
$\leq \mu\left(\bigcap_{m\in\mathbb{N}}\bigcup_{n\geq m} A_n\right)$.

(c) $\mu\left(\bigcup_{n\in\mathbb{N}} A_n\right) \leq \sum_{n\in\mathbb{N}} \mu(A_n)$.

(d) If $\sum_{n\in\mathbb{N}} \mu(A_n) < \infty$, then $\mu\left(\bigcap_{m\in\mathbb{N}}\bigcup_{n\geq m} A_n\right) = 0$.

22. Let $\mathfrak{R}$ be a ring of sets and $(\mu_n)_{n\in\mathbb{N}}$ a sequence of positive measures on $\mathfrak{R}$. Prove the following.

(a) If $(\mu_n(A))_{n\in\mathbb{N}}$ is a convergent increasing sequence for each $A \in \mathfrak{R}$, then
$$\mathfrak{R} \longrightarrow \mathbb{R}, \quad A \longmapsto \lim_{n\to\infty} \mu_n(A)$$
defines a positive measure.

(b) If $\sum_{n\in\mathbb{N}} \mu_n(A)$ converges for each $A \in \mathfrak{R}$, then
$$\mathfrak{R} \longrightarrow \mathbb{R}, \quad A \longmapsto \sum_{n\in\mathbb{N}} \mu_n(A)$$
defines a positive measure.

23. Give an ε–δ description of left continuity.

24. This exercise deals with left continuous, discontinuous functions.

(a) Find simple examples of increasing left continuous, discontinuous functions.

(b) Prove that an increasing function has at most countably many points of discontinuity.

(c) Construct an increasing left continuous function on $\mathbb{R}$ whose set of discontinuities is precisely $\mathbb{Q}$.

25. Let μ be a positive measure on $\mathfrak{J}$ and take $I_1,\ldots,I_n \in \mathfrak{J}_0$. Show that there is a subset $\{J_1,\ldots,J_m\}$ of $\{I_1,\ldots,I_n\}$ such that the J_j are pairwise disjoint and $\mu\left(\bigcup_{i=1}^n I_i\right) \leq 2\sum_{j=1}^m \mu(J_j)$.
Hint: It may be assumed that no I_k is covered by $\bigcup_{i=1,\,i\neq k}^n I_i$. (Why?) Moreover, it may also be assumed that $I_i = [\alpha_i,\beta_i[$ with $\alpha_1 \leq \cdots \leq \alpha_n$. Now show that $\beta_i < \beta_{i+1}$ and $\beta_i < \alpha_{i+2}$ for every i.

2.5 Daniell spaces

This chapter has examined the most important objects required for the construction of integrals, namely vector lattices and positive linear functionals. All that remains for now is to summarize.

A **Daniell space** is a triple $(X, \mathcal{L}, \ell)$ consisting of a set X, a vector lattice $\mathcal{L} \subset \mathbb{R}^X$, and a null-continuous positive linear functional ℓ on $\mathcal{L}$. The extension theory described in the following chapter concerns Daniell spaces. We only construct the extension for null-continuous functionals. There is currently no completely satisfactory integration theory for functionals which are not null-continuous. Our aim is, of course, to obtain a class of functions containing $\mathcal{L}$ – namely, the class $\mathcal{L}^1(\ell)$ of 'ℓ-integrable functions' – and a functional extending ℓ – namely, the 'integral' $\int_\ell$ – displaying much more convenient properties than the original objects $\mathcal{L}$ and ℓ, for example with respect to convergence.

There is a natural partial order relation which enables us to compare Daniell spaces. We define it next.

Let X be a set. Let $\mathcal{F}_1$ and $\mathcal{F}_2$ be non-empty subsets of $\overline{\mathbb{R}}^X$, and ℓ_1 and ℓ_2 functionals on $\mathcal{F}_1$ and $\mathcal{F}_2$ respectively. We define

$$(X, \mathcal{F}_1, \ell_1) \preccurlyeq (X, \mathcal{F}_2, \ell_2) \quad :\Longleftrightarrow \quad \mathcal{F}_1 \subset \mathcal{F}_2 \text{ and } \ell_2|_{\mathcal{F}_1} = \ell_1.$$

In this case we call $(X, \mathcal{F}_2, \ell_2)$ an **extension** of $(X, \mathcal{F}_1, \ell_1)$ and $(X, \mathcal{F}_1, \ell_1)$ a **restriction** of $(X, \mathcal{F}_2, \ell_2)$.

The notion of 'Daniell space' is formulated more generally in [CW]. There the vector lattice $\mathcal{L}$ is replaced by a 'Riesz lattice' which may also contain extended real-valued functions. This leads to a more coherent theory. But there are also technical details which require attention. We therefore restrict ourselves to vector lattices of real-valued functions. These are sufficient for an introductory course such as this.

3

Extension of Daniell spaces

We next develop Daniell's extension process. The idea behind the procedure is best illustrated by the customary introduction of the Riemann integral in terms of the upper and lower Darboux sums, which we recall. Let f be a bounded function defined on the closed interval $[\alpha, \beta]$ of $\mathbb{R}$, $\alpha \le \beta$. Let

$$\alpha = \gamma_1 \le \gamma_2 \le \gamma_3 \le \cdots \le \gamma_n = \beta$$

be a subdivision of $[\alpha, \beta]$. For each such subdivision 3, we define the upper Darboux sum

$$\ell^*(f, 3) := \sum_{k=1}^{n-1} \left(\sup_{x \in [\gamma_k, \gamma_{k+1}]} f(x) \right) (\gamma_{k+1} - \gamma_k)$$

and the lower Darboux sum

$$\ell_*(f, 3) := \sum_{k=1}^{n-1} \left(\inf_{x \in [\gamma_k, \gamma_{k+1}]} f(x) \right) (\gamma_{k+1} - \gamma_k).$$

f is called **Riemann integrable** if, given any $\varepsilon > 0$, there is a subdivision 3 such that

$$\ell^*(f, 3) - \ell_*(f, 3) < \varepsilon,$$

and in this case the uniquely determined real number r satisfying $\ell_*(f, 3) \le r \le \ell^*(f, 3)$, for every subdivision 3, is called the **Riemann integral** of f on $[\alpha, \beta]$ and is denoted by $\int_\alpha^\beta f(x)\, dx$.

This can be interpreted slightly differently. For the subdivision 3, define the step functions

$$f^*(3) := \sum_{k=1}^{n-2} \left(\sup_{x \in [\gamma_k, \gamma_{k+1}]} f(x) \right) e_{[\gamma_k, \gamma_{k+1}[} + \left(\sup_{x \in [\gamma_{n-1}, \gamma_n]} f(x) \right) e_{[\gamma_{n-1}, \gamma_n]},$$

$$f_*(3) := \sum_{k=1}^{n-2} \left(\inf_{x \in [\gamma_k, \gamma_{k+1}]} f(x) \right) e_{[\gamma_k, \gamma_{k+1}[} + \left(\inf_{x \in [\gamma_{n-1}, \gamma_n]} f(x) \right) e_{[\gamma_{n-1}, \gamma_n]}.$$

These step functions are finite sums of Riemann integrable functions and they are therefore Riemann integrable themselves. Now

$$\int_\alpha^\beta \big(f^*(\mathfrak{Z})\big)(x)\,dx = \ell^*(f,\mathfrak{Z}) \quad \text{and} \quad \int_\alpha^\beta \big(f_*(\mathfrak{Z})\big)(x)\,dx = \ell_*(f,\mathfrak{Z}),$$

and $f_*(\mathfrak{Z}) \le f \le f^*(\mathfrak{Z})$. Calling the functions of the form $f^*(\mathfrak{Z})$ **upper (Darboux) functions** and those of the form $f_*(\mathfrak{Z})$ **lower (Darboux) functions** of f, we are led to the following description of the Riemann integrability of the function f.

f is Riemann integrable on $[\alpha,\beta]$ if and only if for every $\varepsilon > 0$ there is an upper function $f^*(\mathfrak{Z})$ and a lower function $f_*(\mathfrak{Z})$ of f such that

$$\int_\alpha^\beta \big(f^*(\mathfrak{Z})\big)(x)\,dx - \int_\alpha^\beta \big(f_*(\mathfrak{Z})\big)(x)\,dx < \varepsilon.$$

Thus Riemann integrability can be characterized in terms of functionals on upper and lower functions.

Daniell's idea was to replace the step functions $f^*(\mathfrak{Z})$ and $f_*(\mathfrak{Z})$ used to approximate f from above and from below by more general upper and lower functions. Indeed, it is reasonable to expect that extending the family of upper and lower functions will extend the family of integrable functions.

In what follows, we first describe the upper and lower functions in Daniell's sense and then define integrable functions as we did above for the case of the Riemann integral.

We have already mentioned that we only carry out the extension for Daniell spaces. Hence the functional ℓ to be extended must be null-continuous. Because of the deficiencies of the Riemann integral, we would like to be able to apply this procedure to the space of Riemann integrable functions on $[\alpha,\beta]$ as well. But is this possible? For an affirmative answer we have to ensure that the Riemann integral is a null-continuous functional. We have already seen that this is true if we restrict ourselves to the space of continuous functions on $[\alpha,\beta]$ (cf. Section 2.2). Fortunately this result can be generalized, in fact, to the space of all Riemann integrable functions. But this is only proved in Theorem 4.20, where there is a detailed discussion of the extension procedure for the Riemann integral.

Until otherwise specified, $(X, \mathcal{L}, \ell)$ is always a Daniell space.

3.1 Upper functions

Daniell's idea was to commence with increasing sequences of elements of a given vector lattice $\mathcal{L}$. We therefore define

$$\mathcal{L}^\uparrow := \{f \in \overline{\mathbb{R}}^X \mid f = \uparrow f_n, \text{ where } f_n \in \mathcal{L} \text{ for all } n \in \mathbb{N}\}.$$

The functions in $\mathcal{L}^\uparrow$ are called **upper functions**. They can take values in $]-\infty,\infty]$.

Proposition 3.1 *Let $(f_n)_{n \in \mathbb{N}}$ and $(g_n)_{n \in \mathbb{N}}$ be sequences in $\mathcal{L}$ with $\uparrow f_n \leq \uparrow g_n$. Then $\uparrow \ell(f_n) \leq \uparrow \ell(g_n)$.*

Proof. By hypothesis, if $m \in \mathbb{N}$ then $f_m = \uparrow (f_m \wedge g_n)$. Theorem 2.19 implies that

$$\ell(f_m) = \uparrow \ell(f_m \wedge g_n) \leq \uparrow \ell(g_n).$$

Since m is arbitrary, the claim now follows. $\square$

Corollary 3.2 *Take $f \in \mathcal{L}^{\uparrow}$. Let $(f_n)_{n \in \mathbb{N}}$ and $(g_n)_{n \in \mathbb{N}}$ be sequences in $\mathcal{L}$ with $f = \uparrow f_n = \uparrow g_n$. Then $\uparrow \ell(f_n) = \uparrow \ell(g_n)$.*

Given $f \in \mathcal{L}^{\uparrow}$ and a sequence $(f_n)_{n \in \mathbb{N}}$ in $\mathcal{L}$ with $f = \uparrow f_n$, define the functional $\ell^{\uparrow}$ on $\mathcal{L}^{\uparrow}$ by

$$\ell^{\uparrow}(f) := \uparrow \ell(f_n).$$

Corollary 3.2 ensures that this functional is well defined. It is independent of the choice of the sequence $(f_n)_{n \in \mathbb{N}}$. It should be emphasized that it is precisely in the proof of the independence that we rely on the null-continuity of ℓ. Observe that $\ell^{\uparrow}(f) = \infty$ is not ruled out.

As an example, consider the Daniell space $(\mathbb{R}, \mathcal{L}(\mathfrak{I}), \ell_{\lambda})$, where λ denotes Lebesgue measure on $\mathfrak{I}$. Take $\alpha, \beta \in \mathbb{R}, \alpha < \beta$. Then, in view of $e_{]\alpha, \beta[} = \uparrow e_{[\alpha + \frac{1}{n}, \beta[}$, we have that $e_{]\alpha, \beta[} \in \mathcal{L}(\mathfrak{I})^{\uparrow}$. Moreover,

$$\ell_{\lambda}^{\uparrow}(e_{]\alpha, \beta[}) = \sup_{n \in \mathbb{N}} \left(\beta - \left(\alpha + \frac{1}{n} \right) \right) = \beta - \alpha.$$

On the other hand, $e_{]\alpha, \beta]} \notin \mathcal{L}(\mathfrak{I})^{\uparrow}$. Suppose the contrary, i.e. that $e_{]\alpha, \beta]} \in \mathcal{L}(\mathfrak{I})^{\uparrow}$. Then there is a sequence $(f_n)_{n \in \mathbb{N}}$ in $\mathcal{L}(\mathfrak{I})_+$ such that $e_{]\alpha, \beta]} = \uparrow f_n$. Thus there is an $m \in \mathbb{N}$ with $f_m(\beta) > 0$. Hence, since f_m is of the form $\sum_{\iota \in I} \alpha_{\iota} e_{[\gamma_{\iota}, \delta_{\iota}[}$, we see that $\beta \in [\gamma, \delta[\subset \{f_m \neq 0\}$ for suitable γ, δ. Fixing $x \in]\beta, \delta[$, we obtain $\sup_{n \in \mathbb{N}} f_n(x) \geq f_m(x) > 0$, contradicting the fact that $e_{]\alpha, \beta]}(x) = 0$.

We next summarize several properties of $\mathcal{L}^{\uparrow}$ and $\ell^{\uparrow}$.

Theorem 3.3

(a) $\mathcal{L} \subset \mathcal{L}^{\uparrow}$ and $\ell^{\uparrow}(f) = \ell(f)$ for every $f \in \mathcal{L}$.

(b) $f + g \in \mathcal{L}^{\uparrow}$ and $\ell^{\uparrow}(f + g) = \ell^{\uparrow}(f) + \ell^{\uparrow}(g)$ for all $f, g \in \mathcal{L}^{\uparrow}$.

(c) $\alpha f \in \mathcal{L}^{\uparrow}$ and $\ell^{\uparrow}(\alpha f) = \alpha \ell^{\uparrow}(f)$ for every $f \in \mathcal{L}^{\uparrow}$ and $\alpha \in \mathbb{R}$, $\alpha \geq 0$.

(d) $f \vee g \in \mathcal{L}^{\uparrow}$ and $f \wedge g \in \mathcal{L}^{\uparrow}$ for all $f, g \in \mathcal{L}^{\uparrow}$.

(e) Given $f, g \in \mathcal{L}^{\uparrow}$, if $f \leq g$ then $\ell^{\uparrow}(f) \leq \ell^{\uparrow}(g)$.

(f) Given $f, g \in \mathcal{L}^{\uparrow}$, if $\ell^{\uparrow}(f) < \infty$ and $\ell^{\uparrow}(g) < \infty$, then $\ell^{\uparrow}(f \vee g) < \infty$.

(g) Given an increasing sequence $(f_n)_{n \in \mathbb{N}}$ in $\mathcal{L}^{\uparrow}$,

$$\uparrow f_n \in \mathcal{L}^{\uparrow} \quad \text{and} \quad \ell^{\uparrow}(\uparrow f_n) = \uparrow \ell^{\uparrow}(f_n).$$

Proof. (a) Take $f \in \mathcal{L}$. Given $n \in \mathbb{N}$, put $f_n := f$. Then $f = \uparrow f_n \in \mathcal{L}^\uparrow$ and $\ell^\uparrow(f) = \uparrow \ell(f_n) = \ell(f)$.

(b),(c),(d) Take $f, g \in \mathcal{L}^\uparrow$ and let $(f_n)_{n \in \mathbb{N}}$ and $(g_n)_{n \in \mathbb{N}}$ be sequences in $\mathcal{L}$ such that $f = \uparrow f_n$ and $g = \uparrow g_n$. Then

$$f + g = \uparrow(f_n + g_n), \qquad \alpha f = \uparrow \alpha f_n \quad (\alpha \geq 0),$$
$$f \vee g = \uparrow(f_n \vee g_n), \qquad f \wedge g = \uparrow(f_n \wedge g_n).$$

It follows immediately that $f + g \in \mathcal{L}^\uparrow$, $\alpha f \in \mathcal{L}^\uparrow$ for $\alpha \geq 0$, $f \vee g \in \mathcal{L}^\uparrow$, and $f \wedge g \in \mathcal{L}^\uparrow$. Furthermore,

$$\ell^\uparrow(f+g) = \uparrow\ell(f_n+g_n) = \uparrow\left(\ell(f_n)+\ell(g_n)\right) = \uparrow\ell(f_n)+ \uparrow\ell(g_n) = \ell^\uparrow(f)+\ell^\uparrow(g)$$

and

$$\ell^\uparrow(\alpha f) = \uparrow\ell(\alpha f_n) = \uparrow\alpha\ell(f_n) = \alpha \uparrow\ell(f_n) = \alpha\ell^\uparrow(f).$$

(e) follows immediately from Proposition 3.1.

(f) Let $(f_n)_{n \in \mathbb{N}}$ and $(g_n)_{n \in \mathbb{N}}$ be as above. Then $f - f_1 \in (\mathcal{L}^\uparrow)_+$, $g - g_1 \in (\mathcal{L}^\uparrow)_+$ and

$$f \vee g \leq (f - f_1 + |f_1|) \vee (g - g_1 + |g_1|) \leq f + g - (f_1 + g_1) + |f_1| + |g_1|.$$

From (a),(b),(d) and (e), we conclude that

$$\ell^\uparrow(f \vee g) \leq \ell^\uparrow(f) + \ell^\uparrow(g) - \ell(f_1 + g_1) + \ell(|f_1| + |g_1|) < \infty.$$

(g) Put $f := \uparrow f_n$. Given $n \in \mathbb{N}$, let $(f_{nk})_{k \in \mathbb{N}}$ be a sequence in $\mathcal{L}$ such that $f_n = \uparrow_k f_{nk}$. For $k \in \mathbb{N}$, define

$$h_k := \bigvee_{n \leq k} f_{nk}.$$

$(h_k)_{k \in \mathbb{N}}$ is increasing and if $n \leq k$, then

$$f_{nk} \leq h_k \leq f_k \leq f.$$

Let $k \to \infty$. Then

$$f_n \leq \uparrow h_k \leq f.$$

Let $n \to \infty$. Then

$$f \leq \uparrow h_k \leq f.$$

Thus $f = \uparrow h_k$. Further, if $n \leq k$, then

$$\ell(f_{nk}) \leq \ell(h_k) \leq \ell^\uparrow(f_k).$$

Let $k \to \infty$. Then

$$\ell^\uparrow(f_n) \leq \uparrow\ell(h_k) \leq \uparrow\ell^\uparrow(f_k).$$

Let $n \to \infty$. Then

$$\uparrow\ell^\uparrow(f_n) \leq \ell^\uparrow(f) \leq \uparrow\ell^\uparrow(f_n).$$

Hence $\ell^\uparrow(f) = \uparrow\ell^\uparrow(f_n)$. $\square$

Exercises

1. Let $(X, \mathcal{L}, \ell)$ be a Daniell space. If $\ell \neq 0$, show that there is an $f \in \mathcal{L}^\uparrow$ with $\ell^\uparrow(f) = \infty$.

2. Let $(X, \mathcal{L}, \ell)$ be a Daniell space. Prove that

$$\ell^\uparrow(f) = \sup_{\substack{g \in \mathcal{L} \\ g \leq f}} \ell(g)$$

   whenever $f \in \mathcal{L}^\uparrow$.

3. Let $(X, \mathcal{L}, \ell)$ be a Daniell space. Show that $\bigvee_{n \in \mathbb{N}} f_n \in \mathcal{L}^\uparrow$ for each sequence $(f_n)_{n \in \mathbb{N}}$ in $\mathcal{L}^\uparrow$, but in general $\sup_{n \in \mathbb{N}} \ell^\uparrow(f_n) < \ell^\uparrow(\bigvee_{n \in \mathbb{N}} f_n)$ if $(f_n)_{n \in \mathbb{N}}$ is not increasing. Find examples of $f, g \in \mathcal{L}^\uparrow$ such that $\ell^\uparrow(f \vee g) > \ell^\uparrow(f) \vee \ell^\uparrow(g)$ and $\ell^\uparrow(f \wedge g) < \ell^\uparrow(f) \wedge \ell^\uparrow(g)$.

4. Which of the following functions belong to $\mathcal{L}(\mathfrak{I})^\uparrow$: $e_{[0,1]}$; e_U with $U \subset \mathbb{R}$ open; $e_{\{x\}}$ with $x \in \mathbb{R}$; $-e_{\mathbb{R}}$; $-e_{[0,1]}$; $-e_{]0,1[}$; $-e_{\{x\}}$ with $x \in \mathbb{R}$; $id_{\mathbb{R}}$; $id_{\mathbb{R}} \vee 0$; $f \in \mathcal{C}(\mathbb{R})_+$; $f \in \mathcal{C}(\mathbb{R})$ with $f^- \in \mathcal{K}(\mathbb{R})$; $f \in \mathcal{C}(\mathbb{R})$?

5. Prove that $\mathcal{K}(\mathbb{R}) \subset \mathcal{K}(\mathbb{R})^\uparrow \subset \mathcal{L}(\mathfrak{I})^\uparrow$ (with both inclusions strict).

6. Determine $\mathcal{L}(\mathfrak{P}(X))^\uparrow$.

7. Determine $\ell^\infty(X)^\uparrow$.

3.2 Lower functions

Our definition of lower functions is similar to that of upper functions. We define

$$\mathcal{L}^\downarrow := \{ f \in \overline{\mathbb{R}}^X \mid f = \downarrow f_n, \text{ where } f_n \in \mathcal{L} \text{ for every } n \in \mathbb{N} \}.$$

The functions in $\mathcal{L}^\downarrow$ are called **lower functions**. They take values in $[-\infty, \infty[$.

Proposition 3.4 *Given* $f \in \overline{\mathbb{R}}^X$, $f \in \mathcal{L}^\downarrow$ *if and only if* $-f \in \mathcal{L}^\uparrow$. *For each sequence* $(f_n)_{n \in \mathbb{N}}$ *in* $\mathcal{L}$ *with* $\downarrow f_n = f \in \mathcal{L}^\downarrow$,

$$\downarrow \ell(f_n) = -\ell^\uparrow(-f).$$

Proof. Let $(f_n)_{n \in \mathbb{N}}$ be a sequence in $\mathcal{L}$ with $f = \downarrow f_n \in \mathcal{L}^\downarrow$. Then

$$-f = - \downarrow f_n = \uparrow(-f_n) \in \mathcal{L}^\uparrow$$

and

$$\ell^\uparrow(-f) = \uparrow \ell(-f_n) = \uparrow(-\ell(f_n)) = - \downarrow \ell(f_n).$$

Similarly, if $-f \in \mathcal{L}^\uparrow$, then $f \in \mathcal{L}^\downarrow$. $\square$

This proposition justifies our next definition. Given $f \in \mathcal{L}^{\downarrow}$, define

$$\ell^{\downarrow}(f) := \downarrow \ell(f_n),$$

where $(f_n)_{n \in \mathbb{N}}$ is an arbitrary sequence in $\mathcal{L}$ with $f = \downarrow f_n$.

Consider $(\mathbb{R}, \mathcal{L}(\mathfrak{I}), \ell_\lambda)$ again. Given $\alpha, \beta \in \mathbb{R}, \alpha < \beta, e_{[\alpha,\beta]} \in \mathcal{L}(\mathfrak{I})^{\downarrow}$ (because $e_{[\alpha,\beta]} = \downarrow e_{[\alpha,\beta+\frac{1}{n}[}$) and $\ell_\lambda^{\downarrow}(e_{[\alpha,\beta]}) = \beta - \alpha$. On the other hand, $e_{]\alpha,\beta]}$ is not in $\mathcal{L}(\mathfrak{I})^{\downarrow}$, for if $(f_n)_{n \in \mathbb{N}}$ is a decreasing sequence in $\mathcal{L}(\mathfrak{I})$ such that $e_{]\alpha,\beta]} \leq f_n$ for every $n \in \mathbb{N}$, then $\inf_{n \in \mathbb{N}} f_n(\alpha) \geq 1$.

Our next theorem is similar to Theorem 3.3.

Theorem 3.5

(a) $\mathcal{L} \subset \mathcal{L}^{\downarrow}$ and $\ell^{\downarrow}(f) = \ell(f)$ for every $f \in \mathcal{L}$.

(b) $f + g \in \mathcal{L}^{\downarrow}$ and $\ell^{\downarrow}(f+g) = \ell^{\downarrow}(f) + \ell^{\downarrow}(g)$ for all $f, g \in \mathcal{L}^{\downarrow}$.

(c) $\alpha f \in \mathcal{L}^{\downarrow}$ and $\ell^{\downarrow}(\alpha f) = \alpha \ell^{\downarrow}(f)$ for every $f \in \mathcal{L}^{\downarrow}$ and $\alpha \in \mathbb{R}, \alpha \geq 0$.

(d) $f \vee g \in \mathcal{L}^{\downarrow}$ and $f \wedge g \in \mathcal{L}^{\downarrow}$ for all $f, g \in \mathcal{L}^{\downarrow}$.

(e) Given $f, g \in \mathcal{L}^{\downarrow}$, if $f \leq g$ then $\ell^{\downarrow}(f) \leq \ell^{\downarrow}(g)$.

(f) Given $f, g \in \mathcal{L}^{\downarrow}$, if $\ell^{\downarrow}(f) > -\infty$ and $\ell^{\downarrow}(g) > -\infty$, then $\ell^{\downarrow}(f \wedge g) > -\infty$.

(g) Given a decreasing sequence $(f_n)_{n \in \mathbb{N}}$ in $\mathcal{L}^{\downarrow}$,

$$\downarrow f_n \in \mathcal{L}^{\downarrow} \quad \text{and} \quad \ell^{\downarrow}(\downarrow f_n) = \downarrow \ell^{\downarrow}(f_n).$$

Proof. These conclusions follow from Theorem 3.3 with the help of Proposition 3.4. We leave the details as an exercise for the reader. □

Theorem 3.6 *Take $f \in \mathcal{L}^{\downarrow}$ and $g \in \mathcal{L}^{\uparrow}$. Then*

(a) $g - f \in \mathcal{L}^{\uparrow}$ and $\ell^{\uparrow}(g - f) = \ell^{\uparrow}(g) - \ell^{\downarrow}(f)$;

(b) if $f \leq g$, then $\ell^{\downarrow}(f) \leq \ell^{\uparrow}(g)$.

Proof. (a) By Proposition 3.4 and Theorem 3.3(b),

$$g - f = g + (-f) \in \mathcal{L}^{\uparrow}$$

and

$$\ell^{\uparrow}(g - f) = \ell^{\uparrow}(g) + \ell^{\uparrow}(-f) = \ell^{\uparrow}(g) - \ell^{\downarrow}(f).$$

(b) $f \leq g$ implies that $g - f \geq 0$. By (a), $g - f \in \mathcal{L}^{\uparrow}$ and

$$\ell^{\uparrow}(g) - \ell^{\downarrow}(f) = \ell^{\uparrow}(g - f) \geq \ell^{\uparrow}(0) = 0.$$

Thus, $\ell^{\downarrow}(f) \leq \ell^{\uparrow}(g)$. □

Exercises

1. Solve Exercises 1–7 in Section 3.1 for $\mathcal{L}^{\downarrow}$ and $\ell^{\downarrow}$ (first giving the proper formulation where necessary).

3.3 The closure of $(X, \mathcal{L}, \ell)$

We now know enough about upper and lower functions to be able to define an extension using the method described in the introduction to this chapter.

Take $\varepsilon > 0$. An ε-**bracket** for $f \in \overline{\mathbb{R}}^X$ is a pair of functions (f', f'') satisfying:

(i) $f' \in \mathcal{L}^{\downarrow}$, $f'' \in \mathcal{L}^{\uparrow}$, $f' \le f \le f''$;

(ii) $\ell^{\uparrow}(f'') \in \mathbb{R}$ and $\ell^{\downarrow}(f') \in \mathbb{R}$;

(iii) $\ell^{\uparrow}(f'') - \ell^{\downarrow}(f') < \varepsilon$.

Let $\overline{\mathcal{L}}(\ell)$ be the set of all functions $f \in \overline{\mathbb{R}}^X$ which admit an ε-bracket for every $\varepsilon > 0$.

We extend ℓ to $\overline{\mathcal{L}}(\ell)$. In doing so we obtain an extension $(X, \overline{\mathcal{L}}(\ell), \overline{\ell})$ which, as we shall see, has substantially more convenient properties than $(X, \mathcal{L}, \ell)$. Nevertheless, $\overline{\mathcal{L}}(\ell)$ is still not the space of ℓ-integrable functions. We need a final minor extension. But the essential step is the construction of $(X, \overline{\mathcal{L}}(\ell), \overline{\ell})$. The second extension merely serves to incorporate the concept of a null set. We shall discuss this again later.

Proposition 3.7 *Given* $f \in \overline{\mathcal{L}}(\ell)$, *define*

$$\alpha_f^{\downarrow} := \sup\{\ell^{\downarrow}(f') \,|\, f' \in \mathcal{L}^{\downarrow},\ f' \le f\}$$

and

$$\alpha_f^{\uparrow} := \inf\{\ell^{\uparrow}(f'') \,|\, f'' \in \mathcal{L}^{\uparrow},\ f'' \ge f\}.$$

Then $\alpha_f^{\downarrow} = \alpha_f^{\uparrow}$ *and this is finite.*

Proof. '$\le$' follows immediately from Theorem 3.6(b), as does the fact that $\alpha_f^{\downarrow}$ and $\alpha_f^{\uparrow}$ are real. Conversely, by the definition of $\overline{\mathcal{L}}(\ell)$, there are ε-brackets (f', f'') of f, for any $\varepsilon > 0$. It follows that $0 \le \alpha_f^{\uparrow} - \alpha_f^{\downarrow} < \varepsilon$ for every $\varepsilon > 0$. Hence $\alpha_f^{\uparrow} = \alpha_f^{\downarrow}$. $\qquad\square$

Given $f \in \overline{\mathcal{L}}(\ell)$, define

$$\overline{\ell}(f) := \alpha_f^{\uparrow} = \alpha_f^{\downarrow},$$

where $\alpha_f^{\uparrow}$ and $\alpha_f^{\downarrow}$ are as in the last proposition. Ours is the procedure used by Riemann for his integral, except for one decisive difference. For the approximation of f from above and below Riemann allowed only step functions, whereas we allow a more general class of functions, namely the upper and lower functions.

We have seen in the preceding two sections that if $\alpha < \beta$, then $e_{]\alpha,\beta]}$ belongs to neither $\mathcal{L}(\mathfrak{J})^{\uparrow}$ nor $\mathcal{L}(\mathfrak{J})^{\downarrow}$. But $e_{]\alpha,\beta]} \in \overline{\mathcal{L}}(\ell_\lambda)$, since if $\varepsilon > 0$, then $\left(e_{[\alpha+\frac{\varepsilon}{3},\beta]},\ e_{]\alpha,\beta+\frac{\varepsilon}{3}[} \right)$ is an ε-bracket for $e_{]\alpha,\beta]}$. Moreover, as expected $\overline{\ell}_\lambda(e_{]\alpha,\beta]}) = \beta - \alpha$. On the other hand, it is not easy to find a characteristic function of a bounded set *not* belonging to $\overline{\mathcal{L}}(\ell_\lambda)$. We shall not meet our first such example until Section 4.2. Note, however, that no constant non-zero function on $\mathbb{R}$ belongs to $\overline{\mathcal{L}}(\ell_\lambda)$.

We now turn to the basic properties of $\overline{\mathcal{L}}(\ell)$ and $\overline{\ell}$.

Theorem 3.8

   (a) Take $f, g \in \overline{\mathcal{L}}(\ell)$. If $f + g$ is defined, then $f + g \in \overline{\mathcal{L}}(\ell)$ and $\overline{\ell}(f + g) = \overline{\ell}(f) + \overline{\ell}(g)$.

   (b) Given $f \in \overline{\mathcal{L}}(\ell)$ and $\alpha \in \mathbb{R}$, $\alpha f \in \overline{\mathcal{L}}(\ell)$ and $\overline{\ell}(\alpha f) = \alpha \overline{\ell}(f)$.

   (c) Take $f, g \in \overline{\mathcal{L}}(\ell)$. If $f \leq g$, then $\overline{\ell}(f) \leq \overline{\ell}(g)$.

   (d) Given $f, g \in \overline{\mathcal{L}}(\ell)$, $f \vee g \in \overline{\mathcal{L}}(\ell)$ and $f \wedge g \in \overline{\mathcal{L}}(\ell)$.

Proof. (a),(b),(d) Take $f, g \in \overline{\mathcal{L}}(\ell)$. For $\varepsilon > 0$, let (f', f'') and (g', g'') be $\varepsilon/2$-brackets of f and g respectively. Then

$$f' + g' \leq f + g \leq f'' + g''$$

whenever $f + g$ is defined. Moreover,

$$\ell^\downarrow(f' + g') = \ell^\downarrow(f') + \ell^\downarrow(g') \in \mathbb{R},$$
$$\ell^\uparrow(f'' + g'') = \ell^\uparrow(f'') + \ell^\uparrow(g'') \in \mathbb{R}$$

and

$$\ell^\uparrow(f'' + g'') - \ell^\downarrow(f' + g') = \left(\ell^\uparrow(f'') - \ell^\downarrow(f')\right) + \left(\ell^\uparrow(g'') - \ell^\downarrow(g')\right) < \varepsilon.$$

Thus $f + g \in \overline{\mathcal{L}}(\ell)$ and since

$$\ell^\downarrow(f' + g') \leq \overline{\ell}(f) + \overline{\ell}(g) \leq \ell^\uparrow(f'' + g''),$$

it follows that

$$\overline{\ell}(f + g) = \overline{\ell}(f) + \overline{\ell}(g).$$

Similarly, given $\alpha \in \mathbb{R}$, $\alpha f \in \overline{\mathcal{L}}(\ell)$ and $\overline{\ell}(\alpha f) = \alpha \overline{\ell}(f)$. Furthermore,

$$f' \vee g' \leq f \vee g \leq f'' \vee g''$$

and

$$f' \wedge g' \leq f \wedge g \leq f'' \wedge g''.$$

Thus

$$\left(\ell^\uparrow(f'' \vee g'') - \ell^\downarrow(f' \vee g')\right) + \left(\ell^\uparrow(f'' \wedge g'') - \ell^\downarrow(f' \wedge g')\right)$$
$$= \left(\ell^\uparrow(f'' \vee g'') + \ell^\uparrow(f'' \wedge g'')\right) - \left(\ell^\downarrow(f' \vee g') + \ell^\downarrow(f' \wedge g')\right)$$
$$= \ell^\uparrow(f'' \vee g'' + f'' \wedge g'') - \ell^\downarrow(f' \vee g' + f' \wedge g')$$
$$= \ell^\uparrow(f'' + g'') - \ell^\downarrow(f' + g')$$
$$= \left(\ell^\uparrow(f'') - \ell^\downarrow(f')\right) + \left(\ell^\uparrow(g'') - \ell^\downarrow(g')\right) < \varepsilon.$$

Hence

$$\ell^\uparrow(f'' \vee g'') \in \mathbb{R}, \ \ell^\uparrow(f'' \wedge g'') \in \mathbb{R}, \ \ell^\downarrow(f' \vee g') \in \mathbb{R} \text{ and } \ell^\downarrow(f' \wedge g') \in \mathbb{R}.$$

Since both of the first bracketed expressions above must be positive,

$$\ell^\uparrow(f'' \vee g'') - \ell^\downarrow(f' \vee g') < \varepsilon$$

and

$$\ell^\uparrow(f'' \wedge g'') - \ell^\downarrow(f' \wedge g') < \varepsilon.$$

Since ε is arbitrary, it follows that $f \vee g \in \overline{\mathcal{L}}(\ell)$ and $f \wedge g \in \overline{\mathcal{L}}(\ell)$.

(c) is an immediate result of Proposition 3.7 and the definition of $\overline{\ell}$. $\quad\square$

Our next result shows that many (but not all!) of the functions in $\mathcal{L}^\uparrow$ belong to $\overline{\mathcal{L}}(\ell)$ and that the values of $\overline{\ell}$ and $\ell^\uparrow$ for these functions coincide (with the corresponding result holding for $\mathcal{L}^\downarrow$).

Proposition 3.9

(a) *If* $f \in \mathcal{L}^\uparrow$, $\ell^\uparrow(f) < \infty$, *then* $f \in \overline{\mathcal{L}}(\ell)$ *and* $\overline{\ell}(f) = \ell^\uparrow(f)$.

(b) *If* $f \in \mathcal{L}^\downarrow$, $\ell^\downarrow(f) > -\infty$, *then* $f \in \overline{\mathcal{L}}(\ell)$ *and* $\overline{\ell}(f) = \ell^\downarrow(f)$.

Proof. (a) Take a sequence $(f_n)_{n\in\mathbb{N}}$ in $\mathcal{L}$ with $f = \uparrow f_n$. If $\ell^\uparrow(f) < \infty$, then for each $\varepsilon > 0$ there is an $n \in \mathbb{N}$ with $\ell^\uparrow(f) - \ell(f_n) < \varepsilon$. Then (f_n, f) is an ε-bracket of f and the claim follows.

(b) is proved analogously. $\quad\square$

Essentially all that is left to do at this point is to verify the null-continuity of $\overline{\ell}$. This is done in the proof of the following theorem.

Theorem 3.10

(a) $\overline{\ell}$ *is a null-continuous positive linear functional on* $\overline{\mathcal{L}}(\ell)$.

(b) $(X, \overline{\mathcal{L}}(\ell) \cap \mathbb{R}^X, \overline{\ell})$ *is a Daniell space and*

$$(X, \mathcal{L}, \ell) \preccurlyeq (X, \overline{\mathcal{L}}(\ell) \cap \mathbb{R}^X, \overline{\ell}) \preccurlyeq (X, \overline{\mathcal{L}}(\ell), \overline{\ell}).$$

(For simplicity's sake we use $\overline{\ell}$ *to denote the restriction of* $\overline{\ell}$ *to* $\overline{\mathcal{L}}(\ell) \cap \mathbb{R}^X$. *The formal ambiguity will not cause problems.)*

Proof. (a) By Proposition 3.9, $0 \in \overline{\mathcal{L}}(\ell)$ and $\overline{\ell}(0) = 0$. Hence, by Theorem 3.8(c), $\overline{\ell}$ is positive. Linearity also follows from Theorem 3.8. To prove the null-continuity of $\overline{\ell}$, let $(f_n)_{n\in\mathbb{N}}$ be a sequence in $\overline{\mathcal{L}}(\ell)$ with $\downarrow f_n = 0$, and take $\varepsilon > 0$. Given $n \in \mathbb{N}$, there is a $g_n \in \mathcal{L}^\downarrow$ with $0 \le g_n \le f_n$, such that $\overline{\ell}(f_n) \le \ell^\downarrow(g_n) + \varepsilon/2^n$. Given $n \in \mathbb{N}$, define $h_n := \bigwedge_{k\le n} g_k$. By Theorem 3.5(d) and Proposition 3.9(b), $h_n \in \overline{\mathcal{L}}(\ell)$ for $n \in \mathbb{N}$ and $\ell^\downarrow(h_n) = \overline{\ell}(h_n)$. Furthermore, $\downarrow h_n = 0$ and thus, by Theorem 3.5(g), $\downarrow \ell^\downarrow(h_n) = 0$. Noting that $h_n \in \mathbb{R}^X$ for $n \in \mathbb{N}$ and

$$\bar{\ell}(f_n) - \ell^{\downarrow}(h_n) = \bar{\ell}(f_n - h_n) = \bar{\ell}\Big(\bigwedge_{k \leq n} f_k - \bigwedge_{k \leq n} g_k \Big)$$

$$\leq \bar{\ell}\Big(\sum_{k \leq n}(f_k - g_k) \Big) = \sum_{k \leq n} \bar{\ell}(f_k - g_k) < \varepsilon,$$

we conclude that $\downarrow\bar{\ell}(f_n) \leq \varepsilon$. But ε is arbitrary, so that $\downarrow\bar{\ell}(f_n) = 0$.

(b) It is a consequence of (a) and Theorem 3.8 that $(X, \overline{\mathcal{L}}(\ell) \cap \mathbb{R}^X, \bar{\ell})$ is a Daniell space. In the light of Theorem 3.3(a), the extension property follows from Proposition 3.9(a). ☐

We call $(X, \overline{\mathcal{L}}(\ell), \bar{\ell})$ the **closure** of $(X, \mathcal{L}, \ell)$. It can be shown that $(X, \overline{\mathcal{L}}(\ell), \bar{\ell})$ is a Daniell space in the more general sense mentioned in Section 2.5 (see [CW] for details). The 'classical' Daniell construction is complete with the introduction of the closure $(X, \overline{\mathcal{L}}(\ell), \bar{\ell})$. However, as mentioned earlier, we undertake another extension below, the reason for this becoming clear in the examples of Section 3.7.

We are now at a convenient spot for the introduction of the Cantor set, whose fame rests in part on its being a source of examples and counter-examples in analysis. We suggest that the reader follow the construction by drawing a diagram!

Start with the interval $[0, 1]$. Remove an open interval, I_{11}, of length $1/3$ centred on $1/2$, leaving two disjoint closed intervals, J_{11} and J_{12}. Now remove the open interval I_{2i} of length $1/3^2$ centred on the midpoint of J_{1i} $(i = 1, 2)$ from J_{1i}. This yields four pairwise disjoint closed intervals, $J_{21}, J_{22}, J_{23}, J_{24}$. Next remove each of the open intervals I_{3i} of length $1/3^3$ centred on the midpoints of J_{2i} respectively. This leaves 8 $(= 2^3)$ pairwise disjoint closed intervals J_{3i} $(i \in \{1, \ldots, 8\})$. Continue in this manner to obtain for each $n \in \mathbb{N}$ a family of 2^n pairwise disjoint closed intervals J_{ni} $(i \in \{1, \ldots, 2^n\})$. The set

$$C := \bigcap_{n \in \mathbb{N}} \bigcup_{i=1}^{2^n} J_{ni}$$

is called the **Cantor set**, or the **Cantor discontinuum**. The set C is compact, being the intersection of a family of compact sets.

Let C_l (C_r) denote the set of left (right) endpoints of the intervals J_{ni} $(n \in \mathbb{N}, i \in \{1, \ldots, 2^n\})$. It is easy to see that C_l and C_r are countably infinite disjoint subsets of C. It is tempting to guess that $C = C_l \cup C_r$. But this is not true, as we shall soon show. C contains many more elements than just the endpoints of the intervals J_{ni}!

Note that the length of J_{ni} is $1/3^n$. (This follows by induction on n.) We next establish the following additional properties.

(i) Given $x \in C \setminus C_l$ and $\varepsilon > 0$, $]x - \varepsilon, x[\cap C_l \neq \emptyset$ and $]x - \varepsilon, x[\cap C_r \neq \emptyset$.

(ii) Given $x \in C \backslash C_r$ and $\varepsilon > 0$, $]x, x + \varepsilon[\cap C_l \neq \emptyset$ and $]x, x + \varepsilon[\cap C_r \neq \emptyset$.

To prove (i), take $x \in C \backslash C_l$ and $\varepsilon > 0$. Choose $n \in \mathbb{N}$ with $1/3^n < \varepsilon$. Then $x \in J_{ni}$ for some i. The left endpoint a of J_{ni} belongs to $]x - \varepsilon, x[\cap C_l$. Now choose an $m > n$ with $1/3^m < x - a$. Then a is also the left endpoint of some J_{mk}. We conclude that the right endpoint of J_{mk} belongs to $]x - \varepsilon, x[\cap C_r$. The proof of (ii) is similar.

It follows from (i) and (ii) that each point of C is an accumulation point of both C_l and C_r.

The Cantor set has no interior points. To verify this, take $x \in C$ and suppose that C contains the interval $]x - \varepsilon, x + \varepsilon[$. Then $x \notin C_l \cup C_r$. By (i), there is a $y \in]x - \varepsilon, x[\cap C_r$. But then $]y, x[$ contains points not belonging to C, which is a contradiction.

We next show that the cardinality of C is $2^{\aleph_0}$, i.e. C has the same cardinality as the set $\{1, 2\}^{\mathbb{N}}$ of all sequences $(\alpha_n)_{n \in \mathbb{N}}$ for which α_n takes only the values 1 and 2. First note that for each $x \in C$ and for each $n \in \mathbb{N}$ there is a unique $k_n(x) \in \{1, \ldots, 2^n\}$ such that $x \in J_{n, k_n(x)}$. Put $x_n := 1$ if $k_n(x)$ is odd, and $x_n := 2$ if $k_n(x)$ is even. This defines a mapping

$$\varphi : C \longrightarrow \{1, 2\}^{\mathbb{N}}, \quad x \longmapsto (x_n)_{n \in \mathbb{N}}.$$

Then φ is injective, for if $x, y \in C, x < y$, then there is a smallest $n \in \mathbb{N}$ with $k_n(x) \neq k_n(y)$. It follows that $k_n(x) + 1 = k_n(y)$, i.e. $x_n \neq y_n$. To verify that φ is surjective, take $(\alpha_n)_{n \in \mathbb{N}} \in \{1, 2\}^{\mathbb{N}}$. We use recursion to construct a sequence $(k_n)_{n \in \mathbb{N}}$ of natural numbers. Define

$$k_1 := \begin{cases} 1 & \text{if } \alpha_1 = 1 \\ 2 & \text{if } \alpha_1 = 2, \end{cases}$$

$$k_n := \begin{cases} 2k_{n-1} - 1 & \text{if } \alpha_n = 1 \\ 2k_{n-1} & \text{if } \alpha_n = 2 \end{cases} \quad \text{for } n > 1.$$

The set $\bigcap_{n \in \mathbb{N}} J_{n, k_n}$ contains precisely one point x. (Why?) Since $k_n(x) = k_n$ for every $n \in \mathbb{N}$, we see that $\varphi(x) = (\alpha_n)_{n \in \mathbb{N}}$. Thus φ is bijective, proving our statement about the cardinality of C.

Now $\{1, 2\}^{\mathbb{N}}$ is uncountable, and hence the same is true of C. In particular C contains many more elements than just those of $C_l \cup C_r$!

Let λ again denote Lebesgue measure on $\mathcal{J}$. We saw in Section 3.2 that $e_{J_{nk}} \in \mathcal{L}(\mathcal{J})^{\downarrow}$ and $\ell_\lambda^{\uparrow}(e_{J_{nk}}) = 1/3^n$ for every $n \in \mathbb{N}$ and every $k \in \{1, \ldots, 2^n\}$. Thus $f_n := \sum_{k=1}^{2^n} e_{J_{nk}} \in \mathcal{L}(\mathcal{J})^{\downarrow}$ and $\ell_\lambda^{\uparrow}(f_n) = 2^n/3^n$, by Theorem 3.5(b). Using Theorem 3.5(g) and Proposition 3.9(b), we conclude that

$$e_C = \downarrow f_n \in \overline{\mathcal{L}}(\ell_\lambda) \quad \text{and} \quad \overline{\ell_\lambda}(e_C) = \inf_{n \in \mathbb{N}} (2^n/3^n) = 0.$$

This result may be reformulated in anticipation of later terminology, by saying that the Cantor set is a Lebesgue null set.

The next section shows what we have actually accomplished by introducing $(X, \overline{\mathcal{L}}(\ell), \bar{\ell})$.

Exercises

1. Let $(X, \mathcal{L}, \ell)$ be a Daniell space. Show that for each $f \in \overline{\mathcal{L}}(\ell)$ there is an increasing sequence $(f_n)_{n \in \mathbb{N}}$ in $\mathcal{L}_+$ with $\{f \neq 0\} \subset \bigcup_{n \in \mathbb{N}} \{f_n > 0\}$.

2. Let X be a set and ℓ a null-continuous positive linear functional on $\ell^\infty(X)$. Prove the following.

   (a) If $f \in \ell^\infty(X)^\uparrow$ and $\alpha \in \mathbb{R}$, then $f \wedge \alpha \in \ell^\infty(X)$ and $\ell^\uparrow(f) = \sup_{\alpha \in \mathbb{R}} \ell(f \wedge \alpha)$.

   (b) If $f \in \ell^\infty(X)^\downarrow$ and $\alpha \in \mathbb{R}$, then $f \vee \alpha \in \ell^\infty(X)$ and $\ell^\downarrow(f) = \inf_{\alpha \in \mathbb{R}} \ell(f \vee \alpha)$.

   (c) $\overline{\mathcal{L}}(\ell) = \{f \in \overline{\mathbb{R}}^X \mid \sup_{\alpha \in \mathbb{R}} \ell(|f| \wedge \alpha) < \infty\}$.

   (d) If $f \in \overline{\mathcal{L}}(\ell)$, then
   $$\bar{\ell}(f) = \sup_{\alpha \in \mathbb{R}} \ell(f^+ \wedge \alpha) - \sup_{\alpha \in \mathbb{R}} \ell(f^- \wedge \alpha).$$

   (e) If $f \in \overline{\mathcal{L}}(\ell)$, $g \in \overline{\mathbb{R}}^X$ and $|g| \leq |f|$, then $g \in \overline{\mathcal{L}}(\ell)$.

3. Take $x, a, b \in \mathbb{R}$, $a < b$. Put $\mathcal{L} := \mathcal{L}(\mathfrak{J})$. Let g be an increasing left continuous function on $\mathbb{R}$ and put $\mu := \mu_g$. Prove the following.

   (a) $e_{\{x\}}, e_{[a,b]}, e_{]a,b]}, e_{]a,b[} \in \overline{\mathcal{L}}(\ell_\mu)$.

   (b) $\overline{\ell_\mu}(e_{\{x\}}) = 0$ if and only if g is continuous at x.

   (c) If g is continuous at a and b, then
   $$\overline{\ell_\mu}(e_{[a,b]}) = \overline{\ell_\mu}(e_{]a,b]}) = \overline{\ell_\mu}(e_{]a,b[}) = \overline{\ell_\mu}(e_{[a,b[}) = g(b) - g(a).$$

   (d) If $(]a_n, b_n[)_{n \in \mathbb{N}}$ is a sequence of pairwise disjoint open intervals such that $U := \bigcup_{n \in \mathbb{N}}]a_n, b_n[$ is bounded and if g is continuous at each a_k and each b_j, then $e_U \in \mathcal{L}^\uparrow$ and
   $$\ell_\mu{}^\uparrow(e_U) = \sum_{n \in \mathbb{N}} \big(g(b_n) - g(a_n)\big) < \infty.$$

4. This exercise continues our investigation of the Cantor set C. Prove the following.

   (a) There is a unique increasing real function f on $\mathbb{R}$ with the following properties: $f(x) = 0$ if $x \in]-\infty, 0[$, $f(x) = 1$ if $x \in]1, \infty[$, and for each $n \in \mathbb{N}$ and $i \in \{1, \ldots, 2^{n-1}\}$, $f(x) = \frac{2i-1}{2^n}$ if $x \in I_{ni}$.

   (b) f is continuous, $f(0) = 0$ and $f(1) = 1$.

   (c) $\overline{\ell_{\mu_f}}(e_C) = 1$.

   (d) For each $\alpha \in [0, 1]$ there is a continuous increasing function g on $\mathbb{R}$ with $\overline{\ell_{\mu_g}}(e_C) = \alpha$, $g(0) = 0$ and $g(1) = 1$.

Hint for (a): First define f on

$$U := \,]-\infty, 0[\, \cup \bigcup_{n \in \mathbb{N}} \bigcup_{i=1}^{2^{n-1}} I_{ni} \, \cup \,]1, \infty[,$$

and show that f is increasing on U. Next define

$$f(x) := \sup\{f(u) \mid u \in U, \, u \le x\}$$

for $x \in \mathbb{R}$.

5. Let $(X, \mathcal{L}, \ell)$ be a Daniell space. Take $f \in \overline{\mathbb{R}}^X$ and $\alpha \in \mathbb{R}$. Show that the following are equivalent.

(a) $f \in \overline{\mathcal{L}}(\ell)$ and $\overline{\ell}(f) = \alpha$.

(b) For every $\varepsilon > 0$, there is an ε-bracket (f', f'') of f such that
$$\ell^{\downarrow}(f') \le \alpha \le \ell^{\uparrow}(f'').$$
If $f \ge 0$, then there is such a bracket in $\mathcal{L}^{\downarrow}_+ \times \mathcal{L}^{\uparrow}_+$.

(c) There is an increasing sequence $(f'_n)_{n \in \mathbb{N}}$ in $\mathcal{L}^{\downarrow}$ and a decreasing sequence $(f''_n)_{n \in \mathbb{N}}$ in $\mathcal{L}^{\uparrow}$ such that the sequences $\left(\ell^{\downarrow}(f'_n)\right)_{n \in \mathbb{N}}$ and $\left(\ell^{\uparrow}(f''_n)\right)_{n \in \mathbb{N}}$ both lie in $\mathbb{R}$,
$$\bigvee_{n \in \mathbb{N}} f'_n \le f \le \bigwedge_{n \in \mathbb{N}} f''_n$$
and
$$\sup_{n \in \mathbb{N}} \ell^{\downarrow}(f'_n) = \alpha = \inf_{n \in \mathbb{N}} \ell^{\uparrow}(f''_n).$$
If $f \ge 0$, then such sequences can be found in $\mathcal{L}^{\downarrow}_+$ and $\mathcal{L}^{\uparrow}_+$, respectively.

(d) $\sup\{\ell^{\downarrow}(g) \mid g \in \mathcal{L}^{\downarrow}, \, g \le f\} = \alpha = \inf\{\ell^{\uparrow}(g) \mid g \in \mathcal{L}^{\uparrow}, \, g \ge f\}$.

(e) Given $\varepsilon > 0$, $\mathcal{L}$ contains an increasing sequence $(f'_n)_{n \in \mathbb{N}}$ and a decreasing sequence $(f''_n)_{n \in \mathbb{N}}$ such that
$$\bigvee_{n \in \mathbb{N}} (f'_n \wedge f) = f = \bigwedge_{n \in \mathbb{N}} (f''_n \vee f),$$
$$-\infty < \inf_{n \in \mathbb{N}} \ell(f''_n) \le \alpha \le \sup_{n \in \mathbb{N}} \ell(f'_n) < \infty,$$
$$\sup_{n \in \mathbb{N}} \ell(f'_n) - \inf_{n \in \mathbb{N}} \ell(f''_n) \le \varepsilon.$$
If $f \ge 0$, then $\mathcal{L}_+$ contains such sequences.

6. Let $(X, \mathcal{L}, \ell)$ be a Daniell space such that $f \wedge \alpha \in \mathcal{L}$ for all $f \in \mathcal{L}$ and $\alpha > 0$. Take $f \in \overline{\mathcal{L}}(\ell)$, and let α, β be real numbers such that $\alpha \le f \le \beta$ and $\alpha \le 0 \le \beta$. Prove that for any $\varepsilon > 0$ there are functions $f' \in \mathcal{L}^{\downarrow} \cap \overline{\mathcal{L}}(\ell)$ and $f'' \in \mathcal{L}^{\uparrow} \cap \overline{\mathcal{L}}(\ell)$ fulfilling the following conditions:

(i) $\alpha \le f' \le f \le f'' \le \beta$;

(ii) $\bar{\ell}(f') \ge \bar{\ell}(f) - \varepsilon$;

(iii) $\bar{\ell}(f'') \le \bar{\ell}(f) + \varepsilon$.

7. Let $\mathfrak{R}$ be a ring of sets on X, $\mathcal{L} := \mathcal{L}(\mathfrak{R})$ and ℓ a null-continuous positive linear functional on $\mathcal{L}$. Show that for each $A \subset X$ with $e_A \in \overline{\mathcal{L}}$ and for each $\varepsilon > 0$, there are $B, C \subset X$ such that $B \subset A \subset C$, $e_B \in \mathcal{L}^\downarrow$, $e_C \in \mathcal{L}^\uparrow$ and $\bar{\ell}(e_C) - \bar{\ell}(e_B) < \varepsilon$.

8. Let $\ell := \ell_{\delta_x}$ be the functional introduced in Exercise 5 of Section 2.4 and put $\mathcal{L} := \mathcal{L}(\mathfrak{P}(X))$. Describe $(\overline{\mathcal{L}}(\ell), \bar{\ell})$.

3.4 Convergence theorems for $(X, \overline{\mathcal{L}}(\ell), \bar{\ell})$

The results in this section form one of the central pieces of the theory of integration. They all deal – with minor variations – with the problem of finding conditions to ensure that: (i) the pointwise limit f of a sequence $(f_n)_{n\in\mathbb{N}}$ of integrable functions is also integrable; and (ii) that the integral of f is the limit of the integrals of the functions f_n, or in other words, that taking limits commutes with integration. Finding sufficient conditions is of pre-eminent importance, since in practice one is repeatedly faced precisely with the problem of interchanging limits and integration. We shall see shortly that rather mild conditions are sufficient. (But the pointwise convergence of the sequence $(f_n)_{n\in\mathbb{N}}$ will not do *by itself* – the exercises contain simple counterexamples!) The theorems are formulated here for $\overline{\mathcal{L}}(\ell)$ and not yet for the space $\mathcal{L}^1(\ell)$ of ℓ-integrable functions. But as we have already remarked, it is but a small step to extend the results to $\mathcal{L}^1(\ell)$, which, by the way, coincides in many important cases with $\overline{\mathcal{L}}(\ell)$.

We stress that convergence theorems of the type formulated for the Daniell integral are *not* valid for the Riemann integral. In order for the limit of a sequence of Riemann integrable functions to be Riemann integrable, rather strong hypotheses are required in general, such as the sequence's being uniformly convergent. It is this that makes the Riemann integral so inflexible in practice.

For a simple example showing the inadequacy of the Riemann integral, take an enumeration $(q_k)_{k\in\mathbb{N}}$ of $\mathbb{Q}\cap[0,1]$. Given $n \in \mathbb{N}$, put $f_n := e_{\{q_1,\dots,q_n\}}$. Then $(f_n)_{n\in\mathbb{N}}$ is an increasing sequence of Riemann integrable functions on $[0,1]$, but the pointwise limit function $e_{\mathbb{Q}\cap[0,1]}$ is not Riemann integrable, since every upper Darboux function is greater than or equal to $e_{[0,1]}$, while every lower Darboux function is less than or equal to 0. This example shows that the following results would not be true for the Riemann integral. All of these deficiencies of the Riemann integral are, however, lacking for the Lebesgue integral – see Theorem 4.20 below.

We now proceed with the results promised. The first of our theorems deals with monotone sequences.

Theorem 3.11 (Monotone Convergence Theorem) *Let* $(f_n)_{n \in \mathbb{N}}$ *be a sequence in* $\overline{\mathcal{L}}(\ell)$ *and take* $f \in \overline{\mathbb{R}}^X$.

  (a) If $f = \uparrow f_n$ *and* $\uparrow \overline{\ell}(f_n) < \infty$, *then* $f \in \overline{\mathcal{L}}(\ell)$ *and* $\overline{\ell}(f) = \uparrow \overline{\ell}(f_n)$.

  (b) If $f = \downarrow f_n$ *and* $\downarrow \overline{\ell}(f_n) > -\infty$, *then* $f \in \overline{\mathcal{L}}(\ell)$ *and* $\overline{\ell}(f) = \downarrow \overline{\ell}(f_n)$.

Proof. We need only prove (a). The proof of (b) proceeds analogously. Let $f = \uparrow f_n$ and $\uparrow \overline{\ell}(f_n) < \infty$. Take $\varepsilon > 0$. Given $n \in \mathbb{N}$, there is an $\varepsilon/2^{n+1}$-bracket (f_n', f_n'') of f_n. Thus

$$f_n' \le f_n \le f_n'' \quad \text{and} \quad \ell^\uparrow(f_n'') - \ell^\downarrow(f_n') < \varepsilon/2^{n+1}.$$

Then, given $n \in \mathbb{N}$, $\left(\bigvee_{k \le n} f_k', \bigvee_{k \le n} f_k'' \right)$ is a $\left(\sum_{k \le n} \varepsilon/2^{k+1} \right)$-bracket for $\bigvee_{k \le n} f_k$, that is for f_n. This follows by an inductive argument, similar to the one in the proof of Theorem 3.8(d). In particular,

$$\overline{\ell}(f_n) \le \ell^\uparrow \left(\bigvee_{k \le n} f_k'' \right) \le \overline{\ell}(f_n) + \sum_{k=1}^n \frac{\varepsilon}{2^{k+1}}.$$

Put $f'' := \bigvee_{n \in \mathbb{N}} f_n''$. Then, using Theorem 3.3(g),

$$\uparrow \overline{\ell}(f_n) \le \ell^\uparrow(f'') \le \frac{\varepsilon}{2} + \uparrow \overline{\ell}(f_n).$$

It follows that $\ell^\uparrow(f'')$ is real. The function f'' will be one of the two in our ε-bracket for f. To obtain the other function, take $m \in \mathbb{N}$ with

$$\uparrow \overline{\ell}(f_n) - \frac{\varepsilon}{4} < \overline{\ell}(f_m).$$

Now take $f' \in \mathcal{L}^\downarrow$ such that $f' \le f_m$, $\ell^\downarrow(f') \in \mathbb{R}$, and

$$\overline{\ell}(f_m) - \frac{\varepsilon}{4} < \ell^\downarrow(f') \le \overline{\ell}(f_m).$$

Then

$$\uparrow \overline{\ell}(f_n) - \frac{\varepsilon}{2} < \ell^\downarrow(f') \le \uparrow \overline{\ell}(f_n).$$

Combining inequalities, we obtain that

$$\ell^\downarrow(f') \le \uparrow \overline{\ell}(f_n) \le \ell^\uparrow(f'')$$

and

$$\ell^\uparrow(f'') - \ell^\downarrow(f') < \varepsilon.$$

Thus (f', f'') is an ε-bracket of f. Since ε is arbitrary, it follows that $f \in \overline{\mathcal{L}}(\ell)$ and $\overline{\ell}(f) = \uparrow \overline{\ell}(f_n)$.   □

The last result is also known as **Beppo Levi's theorem**.

In the proof of Lebesgue's convergence theorem we shall use the following result, which is of interest in its own right.

Theorem 3.12 (Fatou's Lemma) *Let* $(f_n)_{n\in\mathbb{N}}$ *be a sequence in* $\overline{\mathcal{L}}(\ell)$.

(a) If there is a $g \in \overline{\mathcal{L}}(\ell)$ *with* $f_n \leq g$ *for every* $n \in \mathbb{N}$ *and if* $\limsup_{n\to\infty} \overline{\ell}(f_n)$
$> -\infty$, *then*

$$\limsup_{n\to\infty} f_n \in \overline{\mathcal{L}}(\ell) \quad and \quad \overline{\ell}\big(\limsup_{n\to\infty} f_n\big) \geq \limsup_{n\to\infty} \overline{\ell}(f_n).$$

(b) If there is a $g \in \overline{\mathcal{L}}(\ell)$ *with* $f_n \geq g$ *for every* $n \in \mathbb{N}$ *and if* $\liminf_{n\to\infty} \overline{\ell}(f_n)$
$< \infty$, *then*

$$\liminf_{n\to\infty} f_n \in \overline{\mathcal{L}}(\ell) \quad and \quad \overline{\ell}\big(\liminf_{n\to\infty} f_n\big) \leq \liminf_{n\to\infty} \overline{\ell}(f_n).$$

Proof. We again only prove (a). By definition,

$$\limsup_{n\to\infty} f_n = \bigwedge_{n\in\mathbb{N}} \bigvee_{m\geq n} f_m.$$

Given $n \in \mathbb{N}$ and $m \in \mathbb{N}$, $m \geq n$, put

$$h_n := \bigvee_{k\geq n} f_k \quad and \quad h_{nm} := \bigvee_{n\leq k\leq m} f_k.$$

Given $n \in \mathbb{N}$, $(h_{nm})_{m\geq n}$ is increasing and $h_{nm} \leq h_n \leq g$. Thus $\overline{\ell}(h_{nm}) \leq$
$\overline{\ell}(g) < \infty$. By Theorem 3.11, $h_n \in \overline{\mathcal{L}}(\ell)$ for every $n \in \mathbb{N}$ and

$$\overline{\ell}(h_n) = \uparrow\overline{\ell}(h_{nm}) \geq \sup_{m\geq n} \overline{\ell}(f_m).$$

$(h_n)_{n\in\mathbb{N}}$ is decreasing, $\downarrow h_n = \limsup_{n\to\infty} f_n$ and

$$\overline{\ell}(h_n) \geq \sup_{m\geq n} \overline{\ell}(f_m) \geq \limsup_{k\to\infty} \overline{\ell}(f_k)$$

for each $n \in \mathbb{N}$. By Theorem 3.11, $\limsup_{n\to\infty} f_n \in \overline{\mathcal{L}}(\ell)$ and

$$\overline{\ell}\big(\limsup_{n\to\infty} f_n\big) = \downarrow\overline{\ell}(h_n) \geq \limsup_{n\to\infty} \overline{\ell}(f_n).$$

$\square$

The following theorem is the most illustrious of the powerful results in this section. It states that the pointwise limit of a sequence of functions in $\overline{\mathcal{L}}(\ell)$ is again in $\overline{\mathcal{L}}(\ell)$ and that taking the limit may be interchanged with integration, provided that there is another function in $\overline{\mathcal{L}}(\ell)$ dominating the absolute values of all the functions of the given sequence. Because of this hypothesis, the theorem is also known as the **dominated convergence theorem**.

Theorem 3.13 (Lebesgue Convergence Theorem) *Let* $(f_n)_{n\in\mathbb{N}}$ *be a sequence in* $\overline{\mathcal{L}}(\ell)$ *which converges pointwise to* $f \in \overline{\mathbb{R}}^X$. *If there is a* $g \in \overline{\mathcal{L}}(\ell)$ *with* $|f_n| \leq g$ *for each* $n \in \mathbb{N}$, *then*

$$f \in \overline{\mathcal{L}}(\ell) \quad and \quad \overline{\ell}(f) = \lim_{n\to\infty} \overline{\ell}(f_n).$$

Proof. $f = \lim_{n \to \infty} f_n$ implies that

$$f = \limsup_{n \to \infty} f_n = \liminf_{n \to \infty} f_n.$$

Thus, by Theorem 3.12, $f \in \overline{\mathcal{L}}(\ell)$. Since

$$\overline{\ell}(f) = \overline{\ell}\left(\limsup_{n \to \infty} f_n \right) \ge \limsup_{n \to \infty} \overline{\ell}(f_n) \ge \liminf_{n \to \infty} \overline{\ell}(f_n) \ge \overline{\ell}\left(\liminf_{n \to \infty} f_n \right) = \overline{\ell}(f),$$

it follows that

$$\overline{\ell}(f) = \limsup_{n \to \infty} \overline{\ell}(f_n) = \liminf_{n \to \infty} \overline{\ell}(f_n).$$

$\square$

While the existence of the common bound g for the terms of the sequence $(f_n)_{n \in \mathbb{N}}$ in Theorem 3.13 is a sufficient condition for its conclusion, it is not necessary. However, it is simpler to abide by this sufficient condition for the moment. But in Chapter 6 we shall return to the problem of finding necessary and sufficient conditions for the case of functionals ℓ which are derived from positive measures, i.e. for functionals of the form ℓ_μ.

Theorem 3.14 *Let* $(f_n)_{n \in \mathbb{N}}$ *be a sequence in* $\overline{\mathcal{L}}(\ell)$. *Given* $x \in X$, *suppose that* $\sum_{n=1}^{\infty} f_n(x)$ *exists in* $\overline{\mathbb{R}}$, *and let* $\sum_{n \in \mathbb{N}} \overline{\ell}(|f_n|) < \infty$. *Define*

$$\sum_{n=1}^{\infty} f_n : X \longrightarrow \overline{\mathbb{R}}, \quad x \longmapsto \sum_{n=1}^{\infty} f_n(x).$$

Then $\sum_{n=1}^{\infty} f_n \in \overline{\mathcal{L}}(\ell)$,

$$\overline{\ell}\left(\sum_{n=1}^{\infty} f_n \right) = \sum_{n=1}^{\infty} \overline{\ell}(f_n),$$

and the series $\sum_{n=1}^{\infty} \overline{\ell}(f_n)$ *converges absolutely.*

Proof. The sequence $\left(\sum_{m \le n} |f_m| \right)_{n \in \mathbb{N}}$ is increasing and

$$\uparrow \overline{\ell}\left(\sum_{m \le n} |f_m| \right) = \uparrow \left(\sum_{m \le n} \overline{\ell}(|f_m|) \right) \le \sum_{n \in \mathbb{N}} \overline{\ell}(|f_n|) < \infty.$$

Thus $\sum_{n \in \mathbb{N}} |f_n| = \uparrow \left(\sum_{m \le n} |f_m| \right) \in \overline{\mathcal{L}}(\ell)$ (Theorem 3.11(a)). But

$$\sum_{n=1}^{\infty} f_n = \lim_{n \to \infty} \left(\sum_{m \le n} f_m \right)$$

and

$$\left| \sum_{m \le n} f_m \right| \le \sum_{n \in \mathbb{N}} |f_n|$$

for each $n \in \mathbb{N}$. Hence, by Theorem 3.13, $\sum_{n=1}^{\infty} f_n \in \overline{\mathcal{L}}(\ell)$ and

$$\overline{\ell}\left(\sum_{n=1}^{\infty} f_n \right) = \lim_{n \to \infty} \overline{\ell}\left(\sum_{m \le n} f_m \right) = \lim_{n \to \infty} \left(\sum_{m \le n} \overline{\ell}(f_m) \right) = \sum_{n=1}^{\infty} \overline{\ell}(f_n).$$

Since $|\bar{\ell}(f_n)| \le \bar{\ell}(|f_n|)$ for each $n \in \mathbb{N}$ and since $\sum_{n \in \mathbb{N}} \bar{\ell}(|f_n|) < \infty$, the series $\sum_{n=1}^{\infty} \bar{\ell}(f_n)$ is absolutely convergent.   □

The convergence theorems formulated provide the foundations for applications of the theory of integration. They can all be formulated in greater generality with the help of the notion of 'almost everywhere'. We shall concern ourselves with this concept and formulate these more general theorems later.

Note that Theorems 3.12–3.14 are all based on the monotone convergence principle as formulated in Theorem 3.11. A Daniell space (in the general sense discussed in Section 2.5) is said to be closed if this convergence principle applies to it. It can be shown (cf. [CW]) that $(X, \overline{\mathcal{L}}(\ell), \bar{\ell})$ is the smallest closed Daniell space extending $(X, \mathcal{L}, \ell)$. This is why $(X, \overline{\mathcal{L}}(\ell), \bar{\ell})$ is called the 'closure of $(X, \mathcal{L}, \ell)$'.

Exercises

1. A bounded function $f \in \mathbb{R}^{\mathbb{R}}$ is said to be **improperly Riemann integrable** if $f|_{[0,n]}$ and $f|_{[-n,0]}$ are both Riemann integrable for each $n \in \mathbb{N}$ and if $\lim_{n \to \infty} \int_0^n f(x)dx$ and $\lim_{n \to \infty} \int_{-n}^0 f(x)dx$ both exist in $\mathbb{R}$. In such a case we define

$$\int f(x)dx := \int_{-\infty}^{\infty} f(x)dx := \lim_{n \to \infty} \int_{-n}^0 f(x)dx + \lim_{n \to \infty} \int_0^n f(x)dx.$$

   Now let λ be Lebesgue measure on $\mathfrak{I}$, and let $f \in \mathbb{R}^{\mathbb{R}}$ be bounded. Prove the following.

   (a) Take $\alpha, \beta \in \mathbb{R}$ with $\alpha \le \beta$. If $f|_{[\alpha,\beta]}$ is Riemann integrable, then $fe_{[\alpha,\beta]} \in \overline{\mathcal{L}}(\ell_\lambda)$ and $\int_\alpha^\beta f(x)dx = \overline{\ell_\lambda}(fe_{[\alpha,\beta]})$.

   (b) If f and $|f|$ are improperly Riemann integrable, then $f \in \overline{\mathcal{L}}(\ell_\lambda)$ and $\int f(x)dx = \overline{\ell_\lambda}(f)$.

   (c) There is a function which is improperly Riemann integrable, but which does not belong to $\overline{\mathcal{L}}(\ell_\lambda)$.

   (d) Define

$$f(x) := \begin{cases} 1 & \text{if } x \in \mathbb{Q} \\ 0 & \text{if } x \in \mathbb{R} \setminus \mathbb{Q}. \end{cases}$$

   Then $f \in \overline{\mathcal{L}}(\ell_\lambda)$, but f is not improperly Riemann integrable even though there is an increasing sequence $(f_n)_{n \in \mathbb{N}}$ of improperly Riemann integrable functions with the properties that $f = \lim_{n \to \infty} f_n$, for each $n \in \mathbb{N}$, f_n vanishes outside $[-n, n]$ and $\int f_n(x)dx = 0$.

2. Let $(X, \mathcal{L}, \ell)$ be a Daniell space, $(f_n)_{n \in \mathbb{N}}$ a uniformly convergent sequence of bounded functions in $\overline{\mathcal{L}}(\ell) \cap \mathbb{R}^X$ and define $f := \lim_{n \to \infty} f_n$. Assume

that $e_X \in \overline{\mathcal{L}}(\ell)$. Show that

$$f \in \overline{\mathcal{L}}(\ell) \qquad \text{and} \qquad \bar{\ell}(f) = \lim_{n \to \infty} \bar{\ell}(f_n).$$

Can the assumption $e_X \in \overline{\mathcal{L}}(\ell)$ be omitted?

3. Let λ be Lebesgue measure on $\mathfrak{J}$. Prove the following.

   (a) Given $n \in \mathbb{N}$, define $f_n := \frac{1}{n} e_{[0,n[}$. Then $(f_n)_{n \in \mathbb{N}}$ converges uniformly to $0 \in \overline{\mathcal{L}}(\ell_\lambda)$, but for each $n \in \mathbb{N}$ $\bar{\ell_\lambda}(f_n) = 1$.

   (b) Define $g_n := \sum_{k=1}^{n} \frac{1}{k} e_{[k-1,k[}$. Then the sequence $(g_n)_{n \in \mathbb{N}}$ converges uniformly but $\lim_{n \to \infty} g_n \notin \overline{\mathcal{L}}(\ell_\lambda)$.

   (c) For each $n \in \mathbb{N}$ there are uniquely determined numbers $k(n), j(n) \in \mathbb{N} \cup \{0\}$ with $j(n) < 2^{k(n)}$ such that $n = 2^{k(n)} + j(n)$. Define

   $$h_n := e_{\left[\frac{j(n)}{2^{k(n)}}, \frac{j(n)+1}{2^{k(n)}} \right[} \qquad \text{for each } n \in \mathbb{N}.$$

   Then for no $x \in [0,1[$ does $(h_n(x))_{n \in \mathbb{N}}$ converge, but $\lim_{n \to \infty} \bar{\ell_\lambda}(h_n) = 0$.

4. Let $(X, \mathfrak{R}, \mu)$ be a positive measure space and define

$$\overline{\mathfrak{R}}(\mu) := \{A \subset X \mid e_A \in \overline{\mathcal{L}}(\ell_\mu)\},$$
$$\bar{\mu} : \overline{\mathfrak{R}}(\mu) \longrightarrow \mathbb{R}, \qquad A \longmapsto \bar{\ell_\mu}(e_A).$$

Prove the following.

   (a) $\overline{\mathfrak{R}}(\mu)$ is a δ-ring and $\bar{\mu}$ is a positive measure on $\overline{\mathfrak{R}}(\mu)$.

   (b) Let $(A_n)_{n \in \mathbb{N}}$ be an increasing sequence in $\overline{\mathfrak{R}}(\mu)$ with $\sup_{n \in \mathbb{N}} \bar{\mu}(A_n) < \infty$. Then $\bigcup_{n \in \mathbb{N}} A_n \in \overline{\mathfrak{R}}(\mu)$ and

   $$\bar{\mu}\left(\bigcup_{n \in \mathbb{N}} A_n \right) = \sup_{n \in \mathbb{N}} \bar{\mu}(A_n).$$

   (c) If $(A_n)_{n \in \mathbb{N}}$ is a disjoint sequence in $\overline{\mathfrak{R}}(\mu)$ with $\sum_{n \in \mathbb{N}} \bar{\mu}(A_n) < \infty$, then $\bigcup_{n \in \mathbb{N}} A_n \in \overline{\mathfrak{R}}(\mu)$ and

   $$\bar{\mu}\left(\bigcup_{n \in \mathbb{N}} A_n \right) = \sum_{n \in \mathbb{N}} \bar{\mu}(A_n).$$

5. Let $(X, \mathcal{L}, \ell)$ be a Daniell space and $(f_n)_{n \in \mathbb{N}}$ a sequence in $\overline{\mathcal{L}}(\ell)_+$ converging to $f \in \mathbb{R}^X$. Assume that $\lim_{n \to \infty} \bar{\ell}(f_n)$ exists in $\mathbb{R}$. Prove that $f \in \overline{\mathcal{L}}(\ell)$ and that

$$\lim_{n \to \infty} \bar{\ell}(|f_n - f|) = \lim_{n \to \infty} \bar{\ell}(f_n) - \bar{\ell}(f).$$

Hint: $|f_n - f| + f - f_n \le 2f$.

6. Let λ be Lebesgue measure on $\mathfrak{J}$ and take $\alpha \in \mathbb{R}$, $\alpha > 1$. Define

$$f_\alpha : \mathbb{R} \longrightarrow \mathbb{R}, \quad t \longmapsto \begin{cases} t^{\alpha-1}e^{-t} & \text{if } t \geq 0 \\ 0 & \text{if } t < 0. \end{cases}$$

(a) Show that $f_\alpha \in \overline{\mathcal{L}}(\ell_\lambda)$.

Define

$$\Gamma(\alpha) := \overline{\ell_\lambda}(f_\alpha).$$

The function $\alpha \mapsto \Gamma(\alpha)$ is called the **gamma function**. Define further

$$g_\alpha : \mathbb{R} \longrightarrow \mathbb{R}, \quad t \longmapsto \begin{cases} \frac{e^{-t}}{1-e^{-t}}t^{\alpha-1} & \text{if } t \geq 0 \\ 0 & \text{if } t < 0. \end{cases}$$

(b) Show that $g_\alpha \in \overline{\mathcal{L}}(\ell_\lambda)$ and that

$$\overline{\ell_\lambda}(g_\alpha) = \Gamma(\alpha) \sum_{n \in \mathbb{N}} n^{-\alpha}.$$

Hint: Use the series representation of $\frac{1}{1-e^{-t}}$.

7. Let $(X, \mathcal{L}, \ell)$ be a Daniell space and T a metric space. Take $a \in T$ and $f \in \overline{\mathbb{R}}^{X \times T}$. Suppose that:

(i) for each $t \in T$, $f(\cdot, t) \in \overline{\mathcal{L}}(\ell)$;

(ii) there is a $g \in \overline{\mathcal{L}}(\ell)$ such that $|f(x, t)| \leq g(x)$ for each $(x, t) \in X \times T$;

(iii) for each $x \in X$, $f(x, \cdot)$ is continuous at a .

Show that the mapping $T \to \mathbb{R}$, $t \mapsto \overline{\ell}(f(\cdot, t))$ is continuous at a. Is this true in arbitrary topological spaces?

8. Let $(X, \mathcal{L}, \ell)$ be a Daniell space and $I =]a, b[$ an open interval of $\mathbb{R}$. Suppose that $f \in \mathbb{R}^{X \times I}$ satisfies the following conditions:

(i) $f(x, \cdot)$ is differentiable (or continuously differentiable) for each $x \in X$;

(ii) $f(\cdot, t) \in \overline{\mathcal{L}}(\ell)$ for each $t \in I$;

(iii) there is a $g \in \overline{\mathcal{L}}(\ell)$ such that $\left|\frac{\partial f}{\partial t}(x, t)\right| \leq g(x)$ for each $(x, t) \in X \times I$.

Define $h : I \to \mathbb{R}$, $t \mapsto \overline{\ell}(f(\cdot, t))$. Prove the following.

(a) Given $t \in I$, $\frac{\partial f}{\partial t}(\cdot, t) \in \overline{\mathcal{L}}(\ell)$.

(b) h is differentiable (or continuously differentiable) and for each $t \in I$

$$\frac{dh}{dt}(t) = \overline{\ell}\left(\frac{\partial f}{\partial t}(\cdot, t)\right).$$

3.5 Examples

We now turn to considering a number of important examples.

(a) Extending $(X, \mathcal{F}(X), \ell_g)$ We retain the notation introduced earlier. Thus X is a set, $\mathcal{F}(X)$ the set of all functions $f \in \mathbb{R}^X$ such that $\{f \neq 0\}$ is finite, $g \in \mathbb{R}_+^X$ and

$$\ell_g(f) := \sum_{x \in \{f \neq 0\}} f(x)g(x).$$

We carry out the extension step by step.

(a1) $f \in \overline{\mathbb{R}}^X$ is contained in $\mathcal{F}(X)^\uparrow$ if and only if $f(x) > -\infty$ for each $x \in X$, $\{f < 0\}$ is finite and $\{f \neq 0\}$ is countable. In particular, $f \in \overline{\mathbb{R}}^X$, $f \geq 0$, is in $\mathcal{F}(X)^\uparrow$ if and only if $\{f \neq 0\}$ is countable.

Take $f \in \mathcal{F}(X)^\uparrow$. Then $f = \uparrow f_n$ for some $(f_n)_{n \in \mathbb{N}}$, where, for each $n \in \mathbb{N}$, $f_n \in \mathcal{F}(X)$. We see that $f \geq f_1$, so that $\{f < 0\} \subset \{f_1 < 0\}$ and $\{f \neq 0\} \subset \bigcup_{n \in \mathbb{N}} \{f_n \neq 0\}$. Thus we have established in turn each of the properties formulated above.

Conversely, assume that they obtain. Put $A := \{f > 0\}$. Then A is countable and there is an increasing sequence $(A_n)_{n \in \mathbb{N}}$ of finite sets with $A = \bigcup_{n \in \mathbb{N}} A_n$. But then

$$f = \uparrow \left((f \wedge ne_{A_n}) + fe_{\{f<0\}} \right) \in \mathcal{F}(X)^\uparrow. \tag{1}$$

(a2) Given $f \in \mathcal{F}(X)^\uparrow$,

$$\ell_g{}^\uparrow(f) = \sum_{x \in \{f<0\}} f(x)g(x) + \sup_{A \in \mathfrak{F}(X)} \sum_{x \in A} f(x)g(x). \tag{2}$$

In particular, if $f \geq 0$ then

$$\ell_g{}^\uparrow(f) = \sup_{A \in \mathfrak{F}(X)} \sum_{x \in A} f(x)g(x),$$

where $\mathfrak{F}(X)$ denotes, as earlier, the set of all finite subsets of X.
Given $A \in \mathfrak{F}(X)$,

$$f \geq \sum_{x \in \{f<0\}} f(x)e_{\{x\}} + \sum_{x \in A} f(x)e_{\{x\}}.$$

Note that

$$\sum_{x \in A} f(x)e_{\{x\}} = \uparrow \left(\sum_{x \in A} (f(x) \wedge n)e_{\{x\}} \right).$$

Then

$$\ell_g{}^\uparrow \left(\sum_{x \in A} f(x)e_{\{x\}} \right) = \uparrow \left(\sum_{x \in A} (f(x) \wedge n)g(x) \right) = \sum_{x \in A} f(x)g(x).$$

Thus

$$\ell_g{}^\uparrow(f) \geq \sum_{x \in \{f < 0\}} f(x)g(x) + \sum_{x \in A} f(x)g(x),$$

and $\geq$ follows for (2). That $\leq$ also holds follows easily from (1).

(a3) The following are dual to (a1) and (a2).

$f \in \overline{\mathbb{R}}^X$ is in $\mathcal{F}(X)^\downarrow$ if and only if $f(x) < \infty$ for each $x \in X$, $\{f > 0\}$ is finite and $\{f \neq 0\}$ is countable. $f \in \overline{\mathbb{R}}^X$, $f \leq 0$, is in $\mathcal{F}(X)^\downarrow$ if and only if $\{f \neq 0\}$ is countable.

Given $f \in \mathcal{F}(X)^\downarrow$,

$$\ell_g{}^\downarrow(f) = \sum_{x \in \{f > 0\}} f(x)g(x) + \inf_{A \in \mathfrak{F}(X)} \sum_{x \in A} f(x)g(x),$$

and if $f \leq 0$ then

$$\ell_g{}^\downarrow(f) = \inf_{A \in \mathfrak{F}(X)} \sum_{x \in A} f(x)g(x).$$

(a4) A simple characterization of $\overline{\mathcal{L}}(\ell_g)$ arises by noting that $f \in \overline{\mathcal{L}}(\ell_g)$ if and only if $|f| \in \overline{\mathcal{L}}(\ell_g)$.

In fact, if $f \in \overline{\mathcal{L}}(\ell_g)$, then $|f| \in \overline{\mathcal{L}}(\ell_g)$. But if, conversely, $|f| \in \overline{\mathcal{L}}(\ell_g)$, then there is an $f'' \in \mathcal{F}(X)^\uparrow$, $f'' \geq |f|$, such that $\ell_g{}^\uparrow(f'') < \infty$. $\{f'' \neq 0\}$ is countable, as we have already seen. Since

$$0 \leq f^+ \leq f'' \quad \text{and} \quad 0 \leq f^- \leq f'',$$

$\{f^+ \neq 0\}$ and $\{f^- \neq 0\}$ are also countable. Thus $f^+, f^- \in \mathcal{F}(X)^\uparrow$. But

$$\ell_g{}^\uparrow(f^+) \leq \ell_g{}^\uparrow(f'') < \infty \quad \text{and} \quad \ell_g{}^\uparrow(f^-) \leq \ell_g{}^\uparrow(f'') < \infty.$$

Then by Proposition 3.9(a), $f^+, f^- \in \overline{\mathcal{L}}(\ell_g)$. Therefore f is also in $\overline{\mathcal{L}}(\ell_g)$.

Thus, we we can characterize $\overline{\mathcal{L}}(\ell_g)$ as follows.

Theorem 3.15 *Given* $f \in \overline{\mathbb{R}}^X$, $f \in \overline{\mathcal{L}}(\ell_g)$ *if and only if* $\{f \neq 0\}$ *is countable and*

$$\sup_{A \in \mathfrak{F}(X)} \sum_{x \in A} |f(x)|g(x) < \infty.$$

Given $f \in \overline{\mathcal{L}}(\ell_g)$,

$$\overline{\ell_g}(f) = \sup_{A \in \mathfrak{F}(X)} \sum_{x \in A} f^+(x)g(x) - \sup_{A \in \mathfrak{F}(X)} \sum_{x \in A} f^-(x)g(x).$$

We consider the special case $g = e_X$. If $f \in \overline{\mathcal{L}}(\ell_{e_X})$, then we say that f is **summable**. A family $(\alpha_\iota)_{\iota \in I}$ of real numbers is **summable** if

$$f : I \longrightarrow \mathbb{R}, \quad \iota \longmapsto \alpha_\iota$$

is summable. In this case, we define

$$\sum_{\iota \in I} \alpha_\iota := \overline{\ell_{e_X}}(f).$$

Thus, the theory of summable families of real numbers is just a special case of the theory of integration.

(b) We now consider the space $\mathcal{K}(X)$ of those continuous functions on a locally compact space X whose support is compact. Note that if X is taken with the discrete topology, then the present case reduces to (a).

Given $x \in X$, let $\mathfrak{U}(x)$ denote the set of all neighbourhoods of x. Given arbitrary $f \in \overline{\mathbb{R}}^X$ and $x \in X$, define

$$\limsup_{y \to x} f(y) := \inf_{U \in \mathfrak{U}(x)} \sup_{y \in U} f(y),$$

$$\liminf_{y \to x} f(y) := \sup_{U \in \mathfrak{U}(x)} \inf_{y \in U} f(y).$$

Then

$$\liminf_{y \to x} f(y) \le f(x) \le \limsup_{y \to x} f(y).$$

A function $f \in \overline{\mathbb{R}}^X$ is

- **lower semicontinuous at $x \in X$** if $f(x) = \liminf_{y \to x} f(y)$;
- **upper semicontinuous at $x \in X$** if $f(x) = \limsup_{y \to x} f(y)$;
- **lower semicontinuous** if it is lower semicontinuous at every $x \in X$;
- **upper semicontinuous** if it is upper semicontinuous at every $x \in X$.

For example, given $a, b \in \mathbb{R}$ with $a < b$, the function $e_{]a,b[}$ is lower semicontinuous, while $e_{[a,b]}$ is upper semicontinuous.

Proposition 3.16 *A function $f \in \overline{\mathbb{R}}^X$ is continuous at $x \in X$ if and only if*

$$f(x) = \liminf_{y \to x} f(y) = \limsup_{y \to x} f(y).$$

Proof. Let f be continuous at $x \in X$ and take $\varepsilon > 0$. Then

$$U := f^{-1}\big(]f(x) - \varepsilon, \, f(x) + \varepsilon[\big) \in \mathfrak{U}(x)$$

and so

$$f(x) - \varepsilon \le \inf_{y \in U} f(y) \le \sup_{y \in U} f(y) \le f(x) + \varepsilon. \tag{3}$$

We conclude that

$$f(x) - \varepsilon \le \liminf_{y \to x} f(y) \le \limsup_{y \to x} f(y) \le f(x) + \varepsilon.$$

Since ε is arbitrary, it follows that

$$f(x) = \liminf_{y \to x} f(y) = \limsup_{y \to x} f(y).$$

For the converse, take $\varepsilon > 0$. Then there is a $U \in \mathfrak{U}(x)$ such that (3) holds. Thus, given $y \in U$, $|f(y) - f(x)| \le \varepsilon$. Since ε is arbitrary, it follows that f is continuous at x. $\square$

In other words, the function $f \in \mathbb{R}^X$ is continuous at a point if and only if it is both lower and upper semicontinuous at this point.

The following important property of the spaces $\mathcal{K}(X)^\uparrow$ and $\mathcal{K}(X)^\downarrow$ now follows easily.

Theorem 3.17 *Let $(f_\iota)_{\iota \in I}$ be a non-empty family in $\mathcal{K}(X)$. Then $\bigvee_{\iota \in I} f_\iota$ is lower semicontinuous and $\bigwedge_{\iota \in I} f_\iota$ is upper semicontinuous.*

Proof. Put $f := \bigvee_{\iota \in I} f_\iota$. Clearly, if $x \in X$, then

$$f(x) \geq \liminf_{y \to x} f(y).$$

On the other hand, by Proposition 3.16,

$$f_\iota(x) = \liminf_{y \to x} f_\iota(y) \leq \liminf_{y \to x} f(y)$$

holds for $\iota \in I$. It follows that

$$f(x) = \sup_{\iota \in I} f_\iota(x) \leq \liminf_{y \to x} f(y).$$

$\square$

Corollary 3.18

  (a) *The functions in $\mathcal{K}(X)^\uparrow$ are lower semicontinuous.*

  (b) *The functions in $\mathcal{K}(X)^\downarrow$ are upper semicontinuous.*

Unfortunately, $\mathcal{K}(X)^\uparrow$ is not always precisely the set of all lower semicontinuous functions, nor is $\mathcal{K}(X)^\downarrow$ always the set of all upper semicontinuous functions. However, we have the following theorem which, in particular, characterizes $\mathcal{K}(\mathbb{R}^n)^\uparrow$ and $\mathcal{K}(\mathbb{R}^n)^\downarrow$ for $n \in \mathbb{N}$.

Theorem 3.19 *Take a locally compact space X.*

  (a) *Let $f \in \overline{\mathbb{R}}^X$ be lower semicontinuous. If there is a $g \in \mathcal{K}(X)$ with $f \geq g$, then*

$$f = \bigvee \{ h \in \mathcal{K}(X) \,|\, h \leq f \}.$$

  (b) *Let $f \in \overline{\mathbb{R}}^X$ be upper semicontinuous. If there is a $g \in \mathcal{K}(X)$ with $f \leq g$, then*

$$f = \bigwedge \{ h \in \mathcal{K}(X) \,|\, h \geq f \}.$$

Suppose that, in addition, X has a countable base for its topology.

  (c) *Let $f \in \overline{\mathbb{R}}^X$ be lower semicontinuous. If $f \geq g$ for some $g \in \mathcal{K}(X)$, then there is a sequence $(f_n)_{n \in \mathbb{N}}$ in $\mathcal{K}(X)$ with $f = \uparrow f_n$.*

  (d) *Let $f \in \overline{\mathbb{R}}^X$ be upper semicontinuous. If $f \leq g$ for some $g \in \mathcal{K}(X)$, then there is a sequence $(f_n)_{n \in \mathbb{N}}$ in $\mathcal{K}(X)$ with $f = \downarrow f_n$.*

Proof. (a) Define $\tilde{f} := f - g$. Then $\tilde{f} \geq 0$, and it is easy to see that $\tilde{f}$ is lower semicontinuous. Take $x \in X$ with $\tilde{f}(x) > 0$. For each $\alpha \in {]0, \tilde{f}(x)[}$,

there is a neighbourhood U of x such that $\tilde{f}(y) \geq \alpha$ for each $y \in U$. By
Urysohn's theorem, there is an $h \in \mathcal{K}(X)$ such that $h(X) \subset [0,\alpha]$, $h(x) = \alpha$
and $h(y) = 0$ for each $y \in X \setminus U$. Then $h \leq \tilde{f}$ and since α is arbitrary, it
follows that

$$\tilde{f}(x) = \sup\{h(x) \,|\, h \in \mathcal{K}(X),\ h \leq \tilde{f}\}.$$

This relation holds trivially if $\tilde{f}(x) = 0$. Therefore,

$$\tilde{f} = \bigvee\{h \in \mathcal{K}(X) \,|\, h \leq \tilde{f}\}$$

and

$$f = g + \tilde{f} = g + \bigvee\{h \in \mathcal{K}(X) \,|\, h \leq \tilde{f}\} = \bigvee\{h \in \mathcal{K}(X) \,|\, h \leq f\}.$$

(b) can be proved similarly.

(c) We first show that $\mathcal{K}(X)$ is a separable normed space with respect to
the supremum norm

$$\|h\| := \sup_{x \in X} |h(x)|.$$

Let $\mathfrak{B}$ be a countable base for the topology on X. Since X is locally
compact, $\mathfrak{U} := \{B \in \mathfrak{B} \,|\, B$ is relatively compact$\}$ is also a countable base
for the topology on X. Every compact set of X is contained in a finite
union of elements of $\mathfrak{U}$. We can therefore construct an increasing sequence
$(U_n)_{n\in\mathbb{N}}$ of open, relatively compact sets such that every compact subset
of X is contained in some U_n. Fix $n \in \mathbb{N}$. The compact space $\overline{U_n}$ has a
countable base and is therefore metrizable. Thus, using the corollary to
the Stone–Weierstrass theorem listed in the preliminaries (p. 7), $C(\overline{U_n})$ is
separable. Then the same is true of

$$\{f \in C(\overline{U_n}) \,|\, \overline{\{f \neq 0\}} \subset U_n\},$$

which therefore has a countable dense subset $\mathcal{F}_n$. Extend each $f \in \mathcal{F}_n$
to a function $\tilde{f} \in C(X)$ by defining $\tilde{f}(x) := 0$ for $x \in X \setminus \overline{U_n}$. Then
$\bigcup_{n\in\mathbb{N}}\{\tilde{f} \,|\, f \in \mathcal{F}_n\}$ is a countable dense subset of $\mathcal{K}(X)$.

Now suppose that $f \in \overline{\mathbb{R}}^X$ is lower semicontinuous and that $f \geq g$ for
some $g \in \mathcal{K}(X)$. We may assume that $f \geq 0$. Then

$$\mathcal{H} := \{h \in \mathcal{K}(X) \,|\, h \leq f\}$$

is also separable with respect to the metric induced by $\|\cdot\|$. Let $\mathcal{G}$ be a
countable dense subset of $\mathcal{H}$. We show that $f = \bigvee\mathcal{G}$. Actually, $\geq$ is trivial.
Take $x \in X$ with $f(x) > 0$. Let α be an arbitrary element of $]0, f(x)[$. Then
there is an $h \in \mathcal{H}$ with $h(x) > \alpha$ and we can find a $g \in \mathcal{G}$ such that

$$|h(x) - g(x)| < h(x) - \alpha.$$

It follows that $g(x) > \alpha$. Since α was arbitrary, we conclude that

$$f(x) = \bigvee\{g(x) \,|\, g \in \mathcal{G}\}.$$

Thus $f = \bigvee \mathcal{G}$. We enumerate the elements of $\mathcal{G}$ by putting $\mathcal{G} = \{g_1, g_2, \dots\}$. Given $n \in \mathbb{N}$, we define

$$f_n := \bigvee_{k \leq n} g_k.$$

Then $(f_n)_{n \in \mathbb{N}}$ is an increasing sequence in $\mathcal{K}(X)$ and $f = \uparrow f_n$.
 (d) can be proved similarly.   □

Corollary 3.20 *Let X be a locally compact space with a countable base. Then:*

 (a) $\mathcal{K}(X)^\uparrow$ is the set of all lower semicontinuous functions f for which there is a $g \in \mathcal{K}(X)$ such that $f \geq g$.

 (b) $\mathcal{K}(X)^\downarrow$ is the set of all upper semicontinuous functions f for which there is a $g \in \mathcal{K}(X)$ such that $f \leq g$.

As we mentioned above, this corollary applies in particular to the separable metric spaces $\mathbb{R}^n$.

If $\mathcal{L} := \mathcal{K}(X)$ and ℓ is a positive linear functional on $\mathcal{L}$, then we do not find as simple a criterion for functions to be in $\overline{\mathcal{L}}(\ell)$ as we did in example (a).

Exercises

1. Let X be a set and ℓ a null-continuous positive linear functional on $c_0(X)$. Verify the following.

 (a) $(X, \overline{\mathcal{L}}(\ell), \overline{\ell}) = (X, \overline{\mathcal{L}}(\ell|_{\mathcal{F}(X)}), \overline{\ell|_{\mathcal{F}(X)}})$.

 (b) $c_0(X) \subset \bigcap \{\overline{\mathcal{L}}(\ell_g) \mid g \in \ell^1(X)_+\} \subset \ell^\infty(X)$.

2. Let X be an uncountable set. Take a fixed $\gamma \in \mathbb{R}_+ \setminus \{0\}$, and define the functional ℓ on $c_f(X)$ by

$$\ell : c_f(X) \longrightarrow \mathbb{R}, \quad \alpha e_X + g \longmapsto \alpha \gamma.$$

 Verify the following.

 (a) $\overline{\mathcal{L}}(\ell) = \{\alpha e_X + g \mid \alpha \in \mathbb{R}, \ \{g \neq 0\} \text{ countable}\}$.
 (b) If $f = \alpha e_X + g \in \overline{\mathcal{L}}(\ell)$, then $\overline{\ell}(f) = \alpha \gamma$.

3. Does Proposition 3.16 also hold for functions in $\overline{\mathbb{R}}^X$?

4. Let X be a topological space and take $f \in \overline{\mathbb{R}}^X$. Show that the following are equivalent.

 (a) f is lower semicontinuous.
 (b) For each $\alpha \in \mathbb{R}$, $\{f \leq \alpha\}$ is closed.
 (c) For each $\alpha \in \mathbb{R}$, $\{f > \alpha\}$ is open.

 Which sets have lower semicontinuous characteristic functions?

5. Prove that if X is compact and $f \in \overline{\mathbb{R}}^X$ is lower semicontinuous, then there is an $x \in X$ with $f(x) = \inf_{y \in X} f(y)$, i.e. f has a minimum.

6. Let (X, d) be a metric space. Take $x \in X$ and $f \in \overline{\mathbb{R}}^X$.

   (a) If $|f(x)| \neq \infty$, show that the following are equivalent.

      (a1) f is lower semicontinuous at x.
      (a2) $\forall \varepsilon > 0 \; \exists \delta > 0 \; \forall y \in B(x, \delta) \implies f(y) > f(x) - \varepsilon$.
        ($B(x, \delta)$ denotes the open ball with centre x and radius δ.)

   (b) Formulate the corresponding criteria for the cases $f(x) = \infty$ and $f(x) = -\infty$.

7. Find a locally compact space X such that $e_X \notin \mathcal{K}(X)^\uparrow$.

8. Let X be a topological space and take $f, g \in \overline{\mathbb{R}}^X$ with f lower semicontinuous and g upper semicontinuous. Show that $\{g < f\}$ is open.

9. Let X be a topological space and let $f, g \in \overline{\mathbb{R}}^X$ be lower semicontinuous. Prove the following.

   (a) If $f + g$ is defined, then $f + g$ is lower semicontinuous.

   (b) $f \wedge g$ is lower semicontinuous.

   (c) If $(f_\iota)_{\iota \in I}$ is a family of lower semicontinuous functions in $\overline{\mathbb{R}}^X$, then $\bigvee_{\iota \in I} f_\iota$ is lower semicontinuous.

3.6 Null functions, null sets and integrability

Let $(X, \mathcal{L}, \ell)$ be a Daniell space. We define

$$\overline{\mathcal{N}}(\ell) := \{f \in \overline{\mathcal{L}}(\ell) \,|\, \overline{\ell}(|f|) = 0\}$$

and

$$\overline{\mathfrak{N}}(\ell) := \{A \subset X \,|\, e_A \in \overline{\mathcal{N}}(\ell)\}.$$

The functions in $\overline{\mathcal{N}}(\ell)$ are elements of $\overline{\mathcal{L}}(\ell)$ which may be 'disregarded'. They belong to the set of ℓ-null functions, which we introduce later in this section.

Before introducing general ℓ-null functions and ℓ-null sets, we list the most important properties of $\overline{\mathcal{N}}(\ell)$ and $\overline{\mathfrak{N}}(\ell)$.

Proposition 3.21

  (a) $\overline{\mathfrak{N}}(\ell)$ is a σ-ring. If $A \in \overline{\mathfrak{N}}(\ell)$ and $B \subset A$, then $B \in \overline{\mathfrak{N}}(\ell)$.

  (b) Given $f \in \overline{\mathbb{R}}^X$, $f \in \overline{\mathcal{N}}(\ell)$ if and only if $\{f \neq 0\} \in \overline{\mathfrak{N}}(\ell)$.

  (c) Given $f \in \overline{\mathbb{R}}^X$, $f \in \overline{\mathcal{N}}(\ell)$ if and only if $\infty|f| \in \overline{\mathcal{N}}(\ell)$.

  (d) Take $f \in \overline{\mathcal{N}}(\ell)$. Given $g \in \overline{\mathbb{R}}^X$, if $|g| \leq |f|$, then $g \in \overline{\mathcal{N}}(\ell)$.

  (e) Given $f, g \in \overline{\mathcal{N}}(\ell)$ and $\alpha \in \overline{\mathbb{R}}$, $|f| + |g| \in \overline{\mathcal{N}}(\ell)$, $f \vee g \in \overline{\mathcal{N}}(\ell)$, $f \wedge g \in \overline{\mathcal{N}}(\ell)$, and $\alpha f \in \overline{\mathcal{N}}(\ell)$.

Proof. (a) Let $(A_n)_{n \in \mathbb{N}}$ be a sequence in $\overline{\mathfrak{N}}(\ell)$. Then for $n \in \mathbb{N}$,

$$0 \leq \bar{\ell}\big(e_{\bigcup_{k \leq n} A_k}\big) = \bar{\ell}\Big(\bigvee_{k \leq n} e_{A_k} \Big) \leq \sum_{k \leq n} \bar{\ell}(e_{A_k}) = 0,$$

and thus $e_{\bigcup_{k \leq n} A_k} \in \overline{\mathcal{N}}(\ell)$. But

$$e_{\bigcup_{n \in \mathbb{N}} A_n} = \uparrow e_{\bigcup_{k \leq n} A_k} \quad \text{and} \quad \sup_{n \in \mathbb{N}} \bar{\ell}\big(e_{\bigcup_{k \leq n} A_k}\big) = 0.$$

Therefore, by the monotone convergence theorem,

$$e_{\bigcup_{n \in \mathbb{N}} A_n} \in \overline{\mathcal{L}}(\ell) \quad \text{and} \quad \bar{\ell}\big(e_{\bigcup_{n \in \mathbb{N}} A_n}\big) = 0.$$

Hence, $\bigcup_{n \in \mathbb{N}} A_n \in \overline{\mathfrak{N}}(\ell)$.

Take $A \in \overline{\mathfrak{N}}(\ell)$ and $B \subset A$. Then $e_A \in \overline{\mathcal{N}}(\ell)$ and for each $\varepsilon > 0$ there is an $f \in \mathcal{L}^\uparrow$, $f \geq e_A$, such that $\ell^\uparrow(f) < \varepsilon$. Thus $f \geq e_B$, and hence $(0, f)$ is an ε-bracket of e_B. Since $\varepsilon > 0$ is arbitrary, it follows that $e_B \in \overline{\mathcal{L}}(\ell)$ and $\bar{\ell}(e_B) = 0$. Thus, $B \in \overline{\mathfrak{N}}(\ell)$.

The remaining assertions of (a) follow easily.

(d) Take $f \in \overline{\mathcal{N}}(\ell)$. Then for each $\varepsilon > 0$ there is an $h \in \mathcal{L}^\uparrow$ such that $|f| \leq h$ and $\ell^\uparrow(h) < \varepsilon$. If $g \in \overline{\mathbb{R}}^X$, $|g| \leq |f|$, then $(0, h)$ is an ε-bracket of $|g|$. Since $\varepsilon > 0$ is arbitrary, it follows that $|g| \in \overline{\mathcal{L}}(\ell)$ and $\bar{\ell}(|g|) = 0$. Hence $g \in \overline{\mathcal{N}}(\ell)$.

(b) Let $f \in \overline{\mathcal{N}}(\ell)$. Then by (d) and the monotone convergence theorem,

$$e_{\{f \neq 0\}} = \uparrow (|nf| \wedge 1) \in \overline{\mathcal{L}}(\ell)$$

and $\bar{\ell}(e_{\{f \neq 0\}}) = 0$. Therefore $\{f \neq 0\} \in \overline{\mathfrak{N}}(\ell)$. Conversely, if $\{f \neq 0\} \in \overline{\mathfrak{N}}(\ell)$, then, by the same argument,

$$|f| = \uparrow \left(|f| \wedge n e_{\{f \neq 0\}} \right) \in \overline{\mathcal{L}}(\ell)$$

and $\bar{\ell}(|f|) = 0$. Hence $f \in \overline{\mathcal{N}}(\ell)$.

(c) follows from (b) and the fact that $\{f \neq 0\} = \{\infty | f| \neq 0\}$.

(e) is clearly a corollary to the above arguments.   □

Taking Lebesgue measure λ on $\mathfrak{J}$, we see that every finite subset of $\mathbb{R}$ belongs to $\overline{\mathfrak{N}}(\ell_\lambda)$ and so, by Proposition 3.21(a), the same is true of every countable subset of $\mathbb{R}$. In particular, $\mathbb{Q}$ is a Lebesgue null set, hence 'small' in the sense of measure theory. Note, however, that $\mathbb{Q}$ is dense in $\mathbb{R}$, hence 'large' in the sense of topology. One could easily be tempted to conjecture that *only* countable subsets of $\mathbb{R}$ can be Lebesgue null sets. But this temptation should be resisted! A prominent counterexample is given by the Cantor set (cf. Section 3.3).

The next theorem indicates in which sense the functions of $\overline{\mathcal{N}}(\ell)$ can be disregarded. Note assertion (b) in particular.

Theorem 3.22

(a) *Given $f \in \overline{\mathcal{L}}(\ell)$, the sets $\{f = \infty\}$, $\{f = -\infty\}$ and $\{|f| = \infty\}$ belong to $\mathfrak{N}(\ell)$.*

(b) *Take $f \in \overline{\mathcal{L}}(\ell)$ and $g \in \overline{\mathbb{R}}^X$ such that $\{f \neq g\} \in \mathfrak{N}(\ell)$. Then $g \in \overline{\mathcal{L}}(\ell)$ and $\overline{\ell}(g) = \overline{\ell}(f)$.*

(c) *Take $f, g \in \overline{\mathcal{L}}(\ell)$ with $f \leq g$, $\overline{\ell}(f) = \overline{\ell}(g)$. Then $\{f \neq g\} \in \mathfrak{N}(\ell)$, and if $h \in \overline{\mathbb{R}}^X$, $f \leq h \leq g$, then $h \in \overline{\mathcal{L}}(\ell)$ and $\overline{\ell}(h) = \overline{\ell}(f) = \overline{\ell}(g)$.*

Proof. (a) We may assume that $f \geq 0$. Then

$$\infty e_{\{f=\infty\}} = \downarrow (1/n) f.$$

That $\infty e_{\{f=\infty\}} \in \overline{\mathcal{N}}(\ell)$ is a result of the monotone convergence theorem. Therefore, by Proposition 3.21, $\{f = \infty\} \in \mathfrak{N}(\ell)$. Because $f \geq 0$ the proof is complete.

(b) Put $A := \{f \neq g\}$. Then $fe_A, ge_A \in \overline{\mathcal{N}}(\ell)$. We show that $fe_{X\setminus A} \in \overline{\mathcal{L}}(\ell)$. Since $f = f^+ - f^-$, it is sufficient to consider the case $f \geq 0$.

Take $\varepsilon > 0$ and choose $\varepsilon/2$-brackets (f', f'') and $(0, h)$ for f and fe_A respectively. Then

$$f' - h \leq fe_{X\setminus A} \leq f''$$

and

$$\ell^\uparrow(f'') - \ell^\downarrow(f' - h) = \ell^\uparrow(f'') - \ell^\downarrow(f') + \ell^\uparrow(h) < \varepsilon.$$

Hence $(f' - h, f'')$ is an ε-bracket of $fe_{X\setminus A}$. But $\varepsilon > 0$ is arbitrary. Thus $fe_{X\setminus A} \in \overline{\mathcal{L}}(\ell)$. It follows that

$$g = ge_A + fe_{X\setminus A} \in \overline{\mathcal{L}}(\ell)$$

and that

$$\overline{\ell}(g) = \overline{\ell}(fe_{X\setminus A}) = \overline{\ell}(fe_{X\setminus A}) + \overline{\ell}(fe_A) = \overline{\ell}(f).$$

(c) By (a),

$$A := \{|f| = \infty\} \cup \{|g| = \infty\} \in \mathfrak{N}(\ell).$$

It follows from (b) that $fe_{X\setminus A}, ge_{X\setminus A} \in \overline{\mathcal{L}}(\ell)$,

$$\overline{\ell}(fe_{X\setminus A}) = \overline{\ell}(f) \quad \text{and} \quad \overline{\ell}(ge_{X\setminus A}) = \overline{\ell}(g).$$

Hence $h := ge_{X\setminus A} - fe_{X\setminus A} \in \overline{\mathcal{L}}(\ell)$ and

$$\overline{\ell}(h) = \overline{\ell}(g) - \overline{\ell}(f) = 0.$$

Thus $\{h \neq 0\} \in \mathfrak{N}(\ell)$ (Proposition 3.21(b)). But

$$\{f \neq g\} \subset A \cup \{h \neq 0\}.$$

Hence $\{f \neq g\} \in \mathfrak{N}(\ell)$. The remaining assertions now follow from (b). □

We call assertion (c) the **sandwich principle**: every function 'sandwiched' between two functions of $\overline{\mathcal{L}}(\ell)$ with the same integral automatically belongs to $\overline{\mathcal{L}}(\ell)$ and has the same integral as the 'upper layer' and the 'lower layer' of the 'sandwich'.

We have already mentioned that our concepts of a null function and null set embrace more than just $\overline{\mathcal{N}}(\ell)$ and $\overline{\mathfrak{M}}(\ell)$. To facilitate our definitions, we put

$$\mathfrak{R}(\mathcal{F}) := \{\{f \neq 0\} \mid f \in \mathcal{F}\}$$

for $\mathcal{F} \subset \overline{\mathbb{R}}^X$ and define

$$\mathfrak{R}(\ell) := \mathfrak{R}(\overline{\mathcal{L}}(\ell)).$$

Proposition 3.23

(a) $\mathfrak{R}(\ell)$ *is a σ-ring and* $\overline{\mathfrak{M}}(\ell) \subset \mathfrak{R}(\ell)$. *Given* $A \in \mathfrak{R}(\ell)$, *there is a* $g \in \overline{\mathcal{L}}(\ell)_+ \cap \mathbb{R}^X$ *for which* $A = \{g > 0\}$.

(b) *Given* $h \in \overline{\mathcal{L}}(\ell)$ *and* $A \in \mathfrak{R}(\ell)$, $he_A \in \overline{\mathcal{L}}(\ell)$.

(c) $f \in \overline{\mathbb{R}}^X$ *is contained in* $\overline{\mathcal{N}}(\ell)$ *if and only if* $f \in \overline{\mathcal{L}}(\ell)$ *and* $fe_A \in \overline{\mathcal{N}}(\ell)$ *for every* $A \in \mathfrak{R}(\ell)$.

(d) $A \subset X$ *is contained in* $\overline{\mathfrak{M}}(\ell)$ *if and only if* $e_A \in \overline{\mathcal{L}}(\ell)$ *and* $A \cap B \in \overline{\mathfrak{M}}(\ell)$ *for every* $B \in \mathfrak{R}(\ell)$.

Proof. (a),(b) For each set $A \in \mathfrak{R}(\ell)$ there is an $f \in \overline{\mathcal{L}}(\ell)$ with $f \geq 0$ such that $A = \{f \neq 0\} = \{f > 0\}$. Then $B := \{f = \infty\} \in \overline{\mathfrak{M}}(\ell)$ and hence $fe_{X \setminus B} \in \overline{\mathcal{L}}(\ell)$. Putting $g := fe_{X \setminus B} + e_B$, we have $g \in \overline{\mathcal{L}}(\ell) \cap \mathbb{R}^X$ and $A = \{g > 0\}$.

Now take $A, B \in \mathfrak{R}(\ell)$ and $f, g \in \overline{\mathcal{L}}(\ell)_+ \cap \mathbb{R}^X$ such that $A = \{f > 0\}$ and $B = \{g > 0\}$. Then

$$A \cup B = \{f \vee g > 0\} \in \mathfrak{R}(\ell),$$
$$A \cap B = \{f \wedge g > 0\} \in \mathfrak{R}(\ell),$$

and for $h \in \overline{\mathcal{L}}(\ell)_+$,

$$he_A = \uparrow(h \wedge nf) \in \overline{\mathcal{L}}(\ell)_+$$

by the monotone convergence theorem. Hence

$$A \setminus B = \{f - fe_{A \cap B} > 0\} \in \mathfrak{R}(\ell),$$

and $he_A = h^+ e_A - h^- e_A \in \overline{\mathcal{L}}(\ell)$, where h is an arbitrary element of $\overline{\mathcal{L}}(\ell)$.

Let $(A_n)_{n \in \mathbb{N}}$ be a sequence in $\mathfrak{R}(\ell)$. Given $n \in \mathbb{N}$, there is an $f_n \in \overline{\mathcal{L}}(\ell)_+$ such that $A_n = \{f_n > 0\}$. We may assume – multiplying by suitable numbers if necessary – that $\overline{\ell}(f_n) < 1/n^2$. By Theorem 3.14, $f := \sum_{n \in \mathbb{N}} f_n \in \overline{\mathcal{L}}(\ell)$ and hence

$$\bigcup_{n \in \mathbb{N}} A_n = \{f \neq 0\} \in \mathfrak{R}(\ell).$$

If $A \in \overline{\mathfrak{M}}(\ell)$, then $A = \{e_A \neq 0\} \in \mathfrak{R}(\ell)$.

(c) Take $f \in \overline{\mathbb{R}}^X$. If $f \in \overline{\mathcal{N}}(\ell)$, then clearly $f \in \overline{\mathcal{L}}(\ell)$. It follows from

$|fe_A| \leq |f|$ that $fe_A \in \overline{\mathcal{N}}(\ell)$ for every $A \in \mathfrak{R}(\ell)$. Conversely, if $f \in \overline{\mathcal{L}}(\ell)$ and $fe_A \in \overline{\mathcal{N}}(\ell)$ for every $A \in \mathfrak{R}(\ell)$, then $f = fe_{\{f \neq 0\}} \in \overline{\mathcal{N}}(\ell)$.

    (d) is proven similarly to (c).   □

Properties (c) and (d) of the proposition suggest the following definitions.

- $f \in \overline{\mathbb{R}}^X$ is an **ℓ-null function** if $fe_A \in \overline{\mathcal{N}}(\ell)$ for every $A \in \mathfrak{R}(\ell)$.
- $A \subset X$ is an **ℓ-null set** if $A \cap B \in \overline{\mathfrak{N}}(\ell)$ for evey $B \in \mathfrak{R}(\ell)$.

Let $\mathcal{N}(\ell)$ denote the set of all ℓ-null functions and $\mathfrak{N}(\ell)$ the set of all ℓ-null sets.

In the literature often only the elements of $\overline{\mathcal{N}}(\ell)$ are called ℓ-null functions. We extend the concept to our later advantage. Similarly, $\overline{\mathfrak{N}}(\ell)$ is only a subset of the set of ℓ-null sets.

The next theorem is an immediate result of Propositions 3.21 and 3.23.

Theorem 3.24

    (a) $\mathfrak{N}(\ell)$ is a σ-ring. $A \in \mathfrak{N}(\ell)$ implies that $B \in \mathfrak{N}(\ell)$ for every $B \subset A$.

    (b) Given $f \in \overline{\mathbb{R}}^X$, $f \in \mathcal{N}(\ell)$ if and only if $\{f \neq 0\} \in \mathfrak{N}(\ell)$.

    (c) Given $f \in \overline{\mathbb{R}}^X$, $f \in \mathcal{N}(\ell)$ if and only if $\infty|f| \in \mathcal{N}(\ell)$.

    (d) Given $A \subset X$, $A \in \mathfrak{N}(\ell)$ if and only if $e_A \in \mathcal{N}(\ell)$.

    (e) Take $f \in \mathcal{N}(\ell)$. Given $g \in \overline{\mathbb{R}}^X$, if $|g| \leq |f|$ then $g \in \mathcal{N}(\ell)$.

    (f) Given $f, g \in \mathcal{N}(\ell)$, $|f| + |g| \in \mathcal{N}(\ell)$, $f \vee g \in \mathcal{N}(\ell)$, $f \wedge g \in \mathcal{N}(\ell)$ and $\alpha f \in \mathcal{N}(\ell)$ for every $\alpha \in \mathbb{R}$.

    (g) $\overline{\mathcal{N}}(\ell) \subset \mathcal{N}(\ell)$ and $\overline{\mathfrak{N}}(\ell) \subset \mathfrak{N}(\ell)$.

Note in particular that the union of countably many null sets is again a null set and that every subset of a null set is necessarily a null set. Assertion (b) establishes the natural relationship between null functions and null sets.

Closely related to the notions of ℓ-null sets and ℓ-null functions is the very useful concept of 'ℓ-almost everywhere'. Let P be a statement about the points of X. We say that P is **ℓ-almost everywhere (ℓ-a.e.)** true if the set of all $x \in X$ for which $P(x)$ is not true or is not defined is an ℓ-null set. For example, we say that f and g are ℓ-almost everywhere equal if and only if $\{f \neq g\} \in \mathfrak{N}(\ell)$. Thus, $e_{\mathbb{Q}} = 0$ ℓ_λ-a.e. Similarly, we say that $f \leq g$ ℓ-a.e. if and only if $\{f > g\} \in \mathfrak{N}(\ell)$, and we define $f < g$ ℓ-a.e., $f \geq g$ ℓ-a.e., $f > g$ ℓ-a.e., etc. correspondingly. Hence we might formulate Theorem 3.24(b) as '$f \in \mathcal{N}(\ell)$ if and only if $f = 0$ ℓ-a.e.'

Proposition 3.25 *The relation $f = g$ ℓ-a.e. is an equivalence relation on $\overline{\mathbb{R}}^X$.*

The proof follows immediately from Theorem 3.24(a).

Proposition 3.26 *The relation $f \leq g$ ℓ-a.e. on $\overline{\mathbb{R}}^X$ has the following properties:*

    (a) If $f \leq g$, then $f \leq g$ ℓ-a.e.

(b) $f \leq f$ ℓ-a.e.

(c) If $f \leq g$ ℓ-a.e. and $g \leq f$ ℓ-a.e., then $f = g$ ℓ-a.e.

(d) If $f \leq g$ ℓ-a.e. and $g \leq h$ ℓ-a.e., then $f \leq h$ ℓ-a.e.

This follows also easily from Theorem 3.24(a).

We now define the notion of an ℓ-integrable function.

• $f \in \overline{\mathbb{R}}^X$ is ℓ-**integrable** if there is a $g \in \overline{\mathcal{L}}(\ell)$ with $f = g$ ℓ-a.e.

We write $\mathcal{L}^1(\ell)$ for the set of all ℓ-integrable functions in $\overline{\mathbb{R}}^X$.

This suggests that the integral of f should be defined as the value $\bar{\ell}(g)$. It must first be shown, however, that $\bar{\ell}(g)$ is independent of the choice of $g \in \overline{\mathcal{L}}(\ell)$ with $f = g$ ℓ-a.e.

Proposition 3.27 Take $g, h \in \overline{\mathcal{L}}(\ell)$ with $g = h$ ℓ-a.e. Then $\bar{\ell}(g) = \bar{\ell}(h)$.

Proof. By assumption, $\{g \neq h\} \in \mathfrak{N}(\ell)$. But

$$\{g \neq h\} \subset \{g \neq 0\} \cup \{h \neq 0\} \in \mathfrak{R}(\ell),$$

and hence, by the definition of ℓ-null sets, $\{g \neq h\} \in \overline{\mathfrak{N}}(\ell)$. Thus, by Theorem 3.22(b), $\bar{\ell}(g) = \bar{\ell}(h)$.   □

Given $f \in \mathcal{L}^1(\ell)$, define the ℓ-**integral of** f by

$$\int_\ell f := \bar{\ell}(g), \quad \text{where } g \in \overline{\mathcal{L}}(\ell), \ f = g \ \ell\text{-a.e.}$$

The triple $(X, \mathcal{L}^1(\ell), \int_\ell)$ is called the **integral** for the Daniell space $(X, \mathcal{L}, \ell)$. This definition completes Daniell's extension procedure.

We now summarize the properties of $\mathcal{L}^1(\ell)$ and $\int_\ell$. Each of the following assertions – all of which are of major importance – is a direct consequence of proven properties of $\overline{\mathcal{L}}(\ell)$ and $\bar{\ell}$, combined with the idea of a property holding ℓ-a.e. (Of course, the difficulties in the proofs were disposed of in the proofs of the corresponding results in $\overline{\mathcal{L}}(\ell)$.) We therefore only prove some of these and leave the rest to the reader. These theorems may be viewed as a summary of the most important results leading to the notion of the integral. Studying the subsequent chapters requires a thorough understanding of them.

Theorem 3.28

(a) $f \in \overline{\mathbb{R}}^X$ is in $\mathcal{L}^1(\ell)$ if and only if there is a $g \in \overline{\mathcal{L}}(\ell) \cap \mathbb{R}^X$ such that $f = g$ ℓ-a.e.

(b) Take $f, g \in \mathcal{L}^1(\ell)$. Then $f(x) + g(x)$ is ℓ-a.e. defined. If $h \in \overline{\mathbb{R}}^X$, $h(x) = f(x) + g(x)$ ℓ-a.e., then $h \in \mathcal{L}^1(\ell)$ and $\int_\ell h = \int_\ell f + \int_\ell g$.

(c) Given $f \in \mathcal{L}^1(\ell)$ and $\alpha \in \mathbb{R}$, $\alpha f \in \mathcal{L}^1(\ell)$ and $\int_\ell (\alpha f) = \alpha \int_\ell f$.

(d) If $f, g \in \mathcal{L}^1(\ell)$, then $f \vee g \in \mathcal{L}^1(\ell)$, $f \wedge g \in \mathcal{L}^1(\ell)$, $f^+ \in \mathcal{L}^1(\ell)$, $f^- \in \mathcal{L}^1(\ell)$, and $|f| \in \mathcal{L}^1(\ell)$.

Proof. We only prove (b). By (a), there are $f_1, g_1 \in \overline{\mathcal{L}}(\ell) \cap \mathbb{R}^X$ such that $f = f_1$ ℓ-a.e. and $g = g_1$ ℓ-a.e. Then

$$\{|f| = \infty\} \cup \{|g| = \infty\} \subset \{f \neq f_1\} \cup \{g \neq g_1\} \in \mathfrak{N}(\ell),$$

and thus $f(x)+g(x)$ is ℓ-a.e. defined. Let A be the set of all $x \in X$ for which $f(x) + g(x)$ is defined and $h(x) \neq f(x) + g(x)$. Then $A \in \mathfrak{N}(\ell)$. Putting $h_1 := f_1 + g_1 \in \overline{\mathcal{L}}(\ell)$, we see that $\overline{\ell}(h_1) = \overline{\ell}(f_1) + \overline{\ell}(g_1) = \int_\ell f + \int_\ell g$ and

$$\{h \neq h_1\} \subset \{f \neq f_1\} \cup \{g \neq g_1\} \cup A.$$

Thus $h = h_1$ ℓ-a.e., and it follows that $h \in \mathcal{L}^1(\ell)$ and

$$\int_\ell h = \overline{\ell}(h_1) = \int_\ell f + \int_\ell g.$$

$\square$

Theorem 3.29

(a) $\int_\ell$ is a null-continuous positive linear functional on $\mathcal{L}^1(\ell)$.

(b) $(X, \mathcal{L}^1(\ell) \cap \mathbb{R}^X, \int_\ell)$ is a Daniell space. (Note that we abuse notation and write $\int_\ell$ for the restriction of $\int_\ell$ to $\mathcal{L}^1(\ell) \cap \mathbb{R}^X$.)

(c) $(X, \mathcal{L}, \ell) \preccurlyeq (X, \overline{\mathcal{L}}(\ell), \overline{\ell}) \preccurlyeq (X, \mathcal{L}^1(\ell), \int_\ell)$.

Proof. We only prove the null-continuity of $\int_\ell$. Let $(f_n)_{n \in \mathbb{N}}$ be a sequence in $\mathcal{L}^1(\ell)$ such that $0 = \downarrow f_n$. For each $n \in \mathbb{N}$ there is a $g_n \in \overline{\mathcal{L}}(\ell)$ with $f_n = g_n$ ℓ-a.e. By Theorem 3.24(a) and Proposition 3.23(a),

$$A := \left(\bigcup_{n \in \mathbb{N}} \{f_n \neq g_n\} \right) \cap \left(\bigcup_{n \in \mathbb{N}} \{g_n \neq 0\} \right) \in \mathfrak{N}(\ell).$$

Putting $B := X \setminus A$, we have $g_n e_B \in \overline{\mathcal{L}}(\ell)$ and $\overline{\ell}(g_n e_B) = \overline{\ell}(g_n) = \int_\ell f_n$ for every $n \in \mathbb{N}$. Moreover, $0 = \downarrow g_n e_B$. Then, by Theorem 3.10(a), $0 = \downarrow \overline{\ell}(g_n e_B) = \downarrow \int_\ell f_n$. $\square$

Theorem 3.30

(a) Given $f \in \mathcal{L}^1(\ell)$, the sets $\{f = \infty\}$, $\{f = -\infty\}$ and $\{|f| = \infty\}$ belong to $\mathfrak{N}(\ell)$.

(b) If $f \in \mathcal{L}^1(\ell)$ and $g \in \overline{\mathbb{R}}^X$, $f = g$ ℓ-a.e., then $g \in \mathcal{L}^1(\ell)$ and $\int_\ell g = \int_\ell f$.

(c) If $f \in \mathcal{N}(\ell)$, then $f \in \mathcal{L}^1(\ell)$ and $\int_\ell f = 0$. If f is positive, then the converse is also true.

(d) Given $f, g \in \mathcal{L}^1(\ell)$, if $f \leq g$ ℓ-a.e., then $\int_\ell f \leq \int_\ell g$.

(e) Take $f, g \in \mathcal{L}^1(\ell)$ with $f \leq g$ ℓ-a.e. and $\int_\ell f = \int_\ell g$. Then the set $\{f \neq g\}$ belongs to $\mathfrak{N}(\ell)$, and if $h \in \overline{\mathbb{R}}^X$, $f \leq h \leq g$ ℓ-a.e., then $h \in \mathcal{L}^1(\ell)$ and $\int_\ell h = \int_\ell f = \int_\ell g$.

(f) Given $f \in \overline{\mathcal{L}}(\ell)$ and $g \in \mathcal{L}^1(\ell)$, if $\{g \neq 0\} \subset \{f \neq 0\}$, then $g \in \overline{\mathcal{L}}(\ell)$.

Proof. (e) There are $f_1, g_1 \in \overline{\mathcal{L}}(\ell)$ such that $f = f_1$ ℓ-a.e. and $g = g_1$ ℓ-a.e. Then $f_1 \leq g_1$ ℓ-a.e. and since $\{f_1 > g_1\} \subset \{f_1 \neq 0\} \cup \{g_1 \neq 0\} \in \mathfrak{R}(\ell)$, we see that $\{f_1 > g_1\} \in \overline{\mathfrak{N}}(\ell)$. Putting $A := X \setminus \{f_1 > g_1\}$, we use Theorem 3.22(b) to conclude that $f_1 e_A$ as well as $g_1 e_A$ belong to $\overline{\mathcal{L}}(\ell)$ and that

$$\overline{\ell}(f_1 e_A) = \overline{\ell}(f_1) = \int_\ell f, \qquad \overline{\ell}(g_1 e_A) = \overline{\ell}(g_1) = \int_\ell g.$$

Then $\{f_1 e_A \neq g_1 e_A\} \in \overline{\mathfrak{N}}(\ell)$ by Theorem 3.22(c), and hence $\{f \neq g\} \in \mathfrak{N}(\ell)$. Since $f_1 \leq h \leq g_1$ ℓ-a.e., the set $B := \{h < f_1\} \cup \{g_1 < h\}$ belongs to $\mathfrak{N}(\ell)$ and $f_1 e_A \leq h e_{A \setminus B} \leq g_1 e_A$. Theorem 3.22(c) now allows us to conclude that

$$h e_{A \setminus B} \in \overline{\mathcal{L}}(\ell) \quad \text{and} \quad \overline{\ell}(h e_{A \setminus B}) = \overline{\ell}(f_1 e_A) = \int_\ell f.$$

Since $h = h e_{A \setminus B}$ ℓ-a.e., the assertion follows from (b).

(f) There is an $h \in \overline{\mathcal{L}}(\ell)$ such that $g = h$ ℓ-a.e. But

$$\{g \neq h\} \subset \{g \neq 0\} \cup \{h \neq 0\} \subset \{f \neq 0\} \cup \{h \neq 0\}.$$

Then, by the definition of $\mathfrak{N}(\ell)$, $\{g \neq h\} \in \overline{\mathfrak{N}}(\ell)$. The conclusion now follows by Theorem 3.22(b). □

We call (e) the **sandwich principle**. Note that (b) can be interpreted as saying that changing the values of an integrable function on a null set does not affect the value of the integral at all!

We now formulate convergence theorems for $\mathcal{L}^1(\ell)$. As we mentioned earlier, these belong to the 'highlights' of integration theory.

Theorem 3.31 (Monotone Convergence Theorem)

(a) Let $(f_n)_{n \in \mathbb{N}}$ be an increasing sequence in $\mathcal{L}^1(\ell)$ satisfying

$$\sup_{n \in \mathbb{N}} \int_\ell f_n < \infty.$$

Then

$$\bigvee_{n \in \mathbb{N}} f_n \in \mathcal{L}^1(\ell) \quad \text{and} \quad \int_\ell \left(\bigvee_{n \in \mathbb{N}} f_n \right) = \sup_{n \in \mathbb{N}} \int_\ell f_n.$$

(b) Let $(f_n)_{n \in \mathbb{N}}$ be a decreasing sequence in $\mathcal{L}^1(\ell)$ satisfying

$$\inf_{n \in \mathbb{N}} \int_\ell f_n > -\infty.$$

Then

$$\bigwedge_{n \in \mathbb{N}} f_n \in \mathcal{L}^1(\ell) \quad \text{and} \quad \int_\ell \left(\bigwedge_{n \in \mathbb{N}} f_n \right) = \inf_{n \in \mathbb{N}} \int_\ell f_n.$$

Theorem 3.32 (Fatou's Lemma) *Let $(f_n)_{n\in\mathbb{N}}$ be a sequence in $\mathcal{L}^1(\ell)$.*

(a) If there is a $g \in \mathcal{L}^1(\ell)$ with $f_n \leq g$ ℓ-a.e. for every $n \in \mathbb{N}$, and if $\limsup_{n\to\infty} \int_\ell f_n > -\infty$, then

$$\limsup_{n\to\infty} f_n \in \mathcal{L}^1(\ell) \quad and \quad \int_\ell \left(\limsup_{n\to\infty} f_n\right) \geq \limsup_{n\to\infty} \int_\ell f_n.$$

(b) If there is a $g \in \mathcal{L}^1(\ell)$ with $f_n \geq g$ ℓ-a.e. for every $n \in \mathbb{N}$, and if $\liminf_{n\to\infty} \int_\ell f_n < \infty$, then

$$\liminf_{n\to\infty} f_n \in \mathcal{L}^1(\ell) \quad and \quad \int_\ell \left(\liminf_{n\to\infty} f_n\right) \leq \liminf_{n\to\infty} \int_\ell f_n.$$

Theorem 3.33 (Lebesgue Convergence Theorem) *Let $(f_n)_{n\in\mathbb{N}}$ be a sequence in $\mathcal{L}^1(\ell)$ converging ℓ-a.e. to $f \in \overline{\mathbb{R}}^X$. If there is a $g \in \mathcal{L}^1(\ell)$ with $|f_n| \leq g$ ℓ-a.e. for every $n \in \mathbb{N}$, then*

$$f \in \mathcal{L}^1(\ell) \quad and \quad \int_\ell f = \lim_{n\to\infty} \int_\ell f_n.$$

Proof. There is an $h \in \overline{\mathcal{L}}(\ell)_+$ such that $g = h$ ℓ-a.e. For each $n \in \mathbb{N}$ there is a $g_n \in \overline{\mathcal{L}}(\ell)$ with $f_n = g_n$ ℓ-a.e. Then $A_1 := \bigcup_{n\in\mathbb{N}}\{|g_n| > h\} \in \mathfrak{N}(\ell)$ and since $\{|g_n| > h\} \subset \{g_n \neq 0\}$ for every $n \in \mathbb{N}$, we see that $A_1 \in \mathfrak{N}(\ell)$. Since $(g_n)_{n\in\mathbb{N}}$ also converges to f ℓ-a.e., the set A_2 of all $x \in X$ for which $(g_n(x))_{n\in\mathbb{N}}$ has no limit in $\overline{\mathbb{R}}$, is an ℓ-null set. But $A_2 \subset \bigcup_{n\in\mathbb{N}}\{g_n \neq 0\}$, and we conclude that $A_2 \in \mathfrak{N}(\ell)$. Putting $B := X \setminus (A_1 \cup A_2)$, we have that $g_n e_B \in \overline{\mathcal{L}}(\ell)$, $\overline{\ell}(g_n e_B) = \overline{\ell}(g_n)$ and $|g_n e_B| \leq h$ for every $n \in \mathbb{N}$. The Lebesgue convergence theorem for $\overline{\mathcal{L}}(\ell)$ implies that

$$\lim_{n\to\infty} g_n e_B \in \overline{\mathcal{L}}(\ell) \quad and \quad \overline{\ell}\left(\lim_{n\to\infty} g_n e_B\right) = \lim_{n\to\infty} \overline{\ell}(g_n).$$

Since $f = \lim_{n\to\infty} g_n e_B$ ℓ-a.e., the conclusion follows from Theorem 3.30(b). $\qquad\square$

This theorem is also called the **dominated convergence theorem**, while Theorem 31 is also known as **Beppo Levi's theorem**.

Theorem 3.34 *Let $(f_n)_{n\in\mathbb{N}}$ be a sequence in $\mathcal{L}^1(\ell)$ such that $\sum_{n\in\mathbb{N}} \int_\ell |f_n| < \infty$. Then $\sum_{n=1}^{\infty} f_n(x)$ is ℓ-a.e. defined. If, moreover, $f \in \overline{\mathbb{R}}^X$, $f(x) = \sum_{n=1}^{\infty} f_n(x)$ ℓ-a.e., then*

$$f \in \mathcal{L}^1(\ell) \quad and \quad \int_\ell f = \sum_{n=1}^{\infty} \int_\ell f_n,$$

and the series $\sum_{n=1}^{\infty} \int_\ell f_n$ converges absolutely.

Proof. Given $n \in \mathbb{N}$, there is a $g_n \in \overline{\mathcal{L}}(\ell)$ with $f_n = g_n$ ℓ-a.e. Put $g := \sum_{n=1}^{\infty} |g_n|$. By the monotone convergence theorem, $g \in \overline{\mathcal{L}}(\ell)$. Hence $\{g = \infty\} \in \mathfrak{N}(\ell)$.

Take $x \in A := \{g < \infty\}$ and $l, m \in \mathbb{N}$ with $l > m$. Then

$$\left| \sum_{n=1}^{l} g_n(x) - \sum_{n=1}^{m} g_n(x) \right| \leq \sum_{n=m+1}^{l} |g_n(x)|.$$

Thus $\left(\sum_{n=1}^{m} g_n(x) \right)_{m \in \mathbb{N}}$ is a Cauchy sequence in $\mathbb{R}$ and so converges. It follows that $\sum_{n=1}^{\infty} f_n(x)$ is well defined ℓ-a.e.

Given $n \in \mathbb{N}$, define

$$h_n : X \longrightarrow \mathbb{R}, \qquad x \longmapsto \begin{cases} g_n(x) & \text{if } x \in A \\ 0 & \text{otherwise.} \end{cases}$$

Then $h_n \in \overline{\mathcal{L}}(\ell)$ for each $n \in \mathbb{N}$. Moreover, $\sum_{n=1}^{\infty} h_n(x)$ is defined for each $x \in X$. Theorem 3.14 thus implies that

$$\sum_{n=1}^{\infty} h_n \in \overline{\mathcal{L}}(\ell), \qquad \overline{\ell}\left(\sum_{n=1}^{\infty} h_n \right) = \sum_{n=1}^{\infty} \overline{\ell}(h_n)$$

and that $\sum_{n=1}^{\infty} \overline{\ell}(h_n)$ converges absolutely. The remaining claims in the statement of the theorem now follow, since if $f \in \mathbb{R}^X$ and if $f(x) = \sum_{n=1}^{\infty} f_n(x)$ ℓ-a.e., then $f = \sum_{n=1}^{\infty} h_n$ ℓ-a.e. $\qquad\square$

The reader may understandably wonder why we choose $(X, \mathcal{L}^1(\ell), \int_\ell)$ for our integral instead of $(X, \overline{\mathcal{L}}(\ell), \overline{\ell})$. We shall answer this question after considering several examples in the next section.

Exercises

1. Let $(X, \mathcal{L}, \ell)$ be a Daniell space. Prove the following.

   (a) If $f, g \in \overline{\mathcal{L}}(\ell)$, then $\{f < g\}, \{f > g\}, \{f \neq g\} \in \mathfrak{R}(\ell)$.
   (b) $\mathfrak{R}(\ell) = \mathfrak{R}(\overline{\mathcal{L}}(\ell) \cap \mathbb{R}^X)$.
   (c) If $f \in \overline{\mathcal{L}}(\ell)$ and $A \in \mathfrak{R}(\ell)$, then $f e_{X \setminus A} \in \overline{\mathcal{L}}(\ell)$.

2. Let $X := \{1, 2\}$, $\mathcal{L} := \{\alpha e_X \mid \alpha \in \mathbb{R}\}$, and $\ell : \mathcal{L} \to \mathbb{R}$, $\alpha e_X \mapsto \alpha$. Show that $(X, \mathcal{L}, \ell)$ is a Daniell space, and determine $\mathfrak{N}(\ell)$, $\mathcal{N}(\ell)$, $\mathcal{L}^1(\ell)$, $\int_\ell$.

3. Let $X := \{1, 2\}$, $\mathcal{L} := \{\alpha e_{\{1\}} \mid \alpha \in \mathbb{R}\}$. Define $\ell : \mathcal{L} \to \mathbb{R}$, $\alpha e_{\{1\}} \mapsto \alpha$. Prove the following.

   (a) $(X, \mathcal{L}, \ell)$ is a Daniell space.
   (b) $(X, \overline{\mathcal{L}}(\ell), \overline{\ell}) = (X, \mathcal{L}, \ell)$. In particular, $\overline{\mathcal{N}}(\ell) = \{0\}$ and $\overline{\mathfrak{N}}(\ell) = \{\emptyset\}$.
   (c) $\mathcal{N}(\ell) = \{\alpha e_{\{2\}} \mid \alpha \in \mathbb{R}\}$ and $\mathfrak{N}(\ell) = \{\emptyset, \{2\}\}$.
   (d) $\mathcal{L}^1(\ell) = \{f \in \overline{\mathbb{R}}^{\{1,2\}} \mid f(1) \in \mathbb{R}\}$ and $\int_\ell f = f(1)$ for any $f \in \mathcal{L}^1(\ell)$.

4. Let $\ell := \ell_{\delta_x}$ be the functional introduced in Exercise 5 of Section 2.4. Determine $\mathfrak{N}(\ell)$, $\mathcal{N}(\ell)$, $\mathcal{L}^1(\ell)$, $\int_\ell$.

5. Let ℓ be a null-continuous positive linear functional on $\ell^\infty(X)$. Prove the following.

   (a) $\mathcal{L}^1(\ell)_+ = \{f \in \overline{\mathbb{R}}_+^X \mid \sup_{\alpha \in \mathbb{R}_+} \ell(f \wedge \alpha) < \infty\}$.

   (b) $\mathcal{N}(\ell) = \{f \in \overline{\mathbb{R}}^X \mid \ell(e_{\{f \neq 0\}}) = 0\}$.

   (c) $\mathfrak{N}(\ell) = \{A \subset X \mid \ell(e_A) = 0\}$.

6. Let $(X, \mathfrak{R}, \mu)$ be a positive measure space. Show that $X \setminus \bigcup_{A \in \mathfrak{R}} A \in \mathfrak{N}(\ell_\mu)$, but in general $X \setminus \bigcup_{A \in \mathfrak{R}} A \notin \overline{\mathfrak{N}}(\ell_\mu)$.

7. Let $(X, \mathcal{L}, \ell)$ be a Daniell space. Take $A \subset X$, $f \in \overline{\mathbb{R}}^X$. Prove the propositions below.

   (a) The following are equivalent:

   (a1) $A \in \overline{\mathfrak{N}}(\ell)$;
   (a2) $\forall \varepsilon > 0 \; \exists (A_n)_{n \in \mathbb{N}}$ in $\{C \subset X \mid e_C \in \overline{\mathcal{L}}(\ell)\}$, $A \subset \bigcup_{n \in \mathbb{N}} A_n$, $\sum_{n \in \mathbb{N}} \overline{\ell}(e_{A_n}) < \varepsilon$.

   (b) The following are equivalent:

   (b1) $A \in \mathfrak{N}(\ell)$;
   (b2) $\forall B \in \mathfrak{R}(\mathcal{L}) \implies A \cap B \in \mathfrak{N}(\ell)$;
   (b3) $\forall B \in \mathfrak{R}(\ell) \; \forall \varepsilon > 0 \; \exists (A_n)_{n \in \mathbb{N}}$ in $\{C \subset X \mid e_C \in \overline{\mathcal{L}}(\ell)\}$, $A \cap B \subset \bigcup_{n \in \mathbb{N}} A_n$, $\sum_{n \in \mathbb{N}} \overline{\ell}(e_{A_n}) < \varepsilon$.

   (c) The following are equivalent:

   (c1) $f \in \mathcal{N}(\ell)$;
   (c2) $\forall B \in \mathfrak{R}(\mathcal{L}) \implies f e_B \in \mathcal{N}(\ell)$.

8. Generalize Theorem 3.31 to ℓ-a.e. increasing (decreasing) sequences.

9. Which of the conclusions in Section 3.4 (including the exercises!) remain true when $\overline{\mathcal{L}}(\ell)$ is replaced by $\mathcal{L}^1(\ell)$?

10. Let $(X, \mathcal{L}, \ell)$ be a Daniell space and take $f \in \mathcal{L}^1(\ell)$. Prove the following propositions.

    (a) Given any $B \in \mathfrak{R}(\ell)$, $f e_B \in \overline{\mathcal{L}}(\ell)$.

    (b) There are disjoint sets $B \in \mathfrak{R}(\ell)$, $C \in \mathfrak{N}(\ell)$ such that $f = f e_B + f e_C$.

    (c) If $f \geq 0$, then $\int_\ell f = \sup_{B \in \mathfrak{R}(\ell)} \int_\ell f e_B$.

    (d) $\int_\ell f^+ = \sup_{B \in \mathfrak{R}(\ell)} \int_\ell f e_B$.

    (e) $\int_\ell f = \sup_{B \in \mathfrak{R}(\ell)} \int_\ell f e_B + \inf_{B \in \mathfrak{R}(\ell)} \int_\ell f e_B$.

3.7 Examples

(a) Extending $(X, \mathcal{F}(X), \ell_g)$ We retain all of the assumptions and notation from Example (a) of Section 3.5. We now characterize $\mathfrak{N}(\ell_g)$ and $\mathcal{N}(\ell_g)$.

Proposition 3.35

(a) $\overline{\mathfrak{N}}(\ell_g) = \{A \subset \{g = 0\} \mid A \text{ is countable}\}$.

(b) $\mathfrak{N}(\ell_g) = \mathfrak{P}(\{g = 0\})$.

(c) $\overline{\mathcal{N}}(\ell_g) = \{f \in \overline{\mathbb{R}}^X \mid \{f \neq 0\} \text{ countable}, \{f \neq 0\} \subset \{g = 0\}\}$.

(d) $\mathcal{N}(\ell_g) = \{f \in \overline{\mathbb{R}}^X \mid \{f \neq 0\} \subset \{g = 0\}\}$.

Proof. (a) and (c) are immediate from Theorem 3.15. (b) and (d) follow from (a) and (c) respectively. ☐

We can therefore describe $\left(X, \mathcal{L}^1(\ell_g), \int_{\ell_g}\right)$ in the following terms.

Theorem 3.36 $f \in \overline{\mathbb{R}}^X$ *is in* $\mathcal{L}^1(\ell_g)$ *if and only if*

$$\sup_{A \in \mathfrak{F}(X)} \sum_{x \in A} |f(x)| g(x) < \infty.$$

Given $f \in \mathcal{L}^1(\ell_g)$,

$$\int_{\ell_g} f = \sup_{A \in \mathfrak{F}(X)} \sum_{x \in A} f^+(x) g(x) - \sup_{A \in \mathfrak{F}(X)} \sum_{x \in A} f^-(x) g(x).$$

Proof. If $f \in \mathcal{L}^1(\ell_g)$, then clearly

$$\sup_{A \in \mathfrak{F}(X)} \sum_{x \in A} |f(x)| g(x) < \infty.$$

Conversely, assume that this inequality holds. Put $h := f e_{\{g > 0\}}$. We have

$$\{h \neq 0\} = \bigcup_{n \in \mathbb{N}} \{|h|g \geq 1/n\}.$$

The sets $\{|h|g \geq 1/n\}$ are finite and therefore $\{h \neq 0\}$ is countable. Furthermore,

$$\sup_{A \in \mathfrak{F}(X)} \sum_{x \in A} |h(x)| g(x) \leq \sup_{A \in \mathfrak{F}(X)} \sum_{x \in A} |f(x)| g(x) < \infty.$$

We conclude that $h \in \overline{\mathcal{L}}(\ell_g)$. But $\{f \neq h\} \subset \{g = 0\}$. Therefore, by Proposition 3.35, $f \in \mathcal{L}^1(\ell_g)$.

Finally, the formula for $\int_{\ell_g} f$ follows easily from the appropriate formula for $\overline{\ell}_g(h)$. ☐

This example motivates our choice of integral. Compare statements (a) and (c) of Proposition 3.35 to statements (b) and (d) respectively. Compare Theorem 3.15 to Theorem 3.36 as well. Countability is assumed for the assertions about $(X, \overline{\mathcal{L}}(\ell_g), \overline{\ell_g})$. This assumption is not needed for the corresponding propositions about $(X, \mathcal{L}^1(\ell_g), \int_{\ell_g})$. Countability is irrelevant to the problem of extension. It is simply a consequence of Daniell's procedure, indicating a weakness in the classical construction (which stops after introducing the closure).

The situation encountered here is not unique to the space $(X, \mathcal{F}(X), \ell_g)$. In general, due to the nature of Daniell's procedure, artificial conditions, irrelevant to the problem at hand, appear in the description of the closure. These conditions disappear with the use of the integral presented in this book.

There is another question which we pursue. The classical Daniell construction has proven to be a very useful tool for the treatment of the Stieltjes functionals. What changes with the introduction of $(X, \mathcal{L}^1(\ell), \int_\ell)$?

(b) σ-finite Daniell spaces and Stieltjes integrals A Daniell space $(X, \mathcal{L}, \ell)$ is called σ-**finite** if there is a sequence $(A_n)_{n \in \mathbb{N}}$ in $\mathfrak{R}(\mathcal{L})$ for which $X = \bigcup_{n \in \mathbb{N}} A_n$.

Theorem 3.37 *If $(X, \mathcal{L}, \ell)$ is a σ-finite Daniell space, then*

$$(X, \overline{\mathcal{L}}(\ell), \overline{\ell}) = \left(X, \mathcal{L}^1(\ell), \int_\ell \right).$$

Proof. The inequality $\preccurlyeq$ always holds by Theorem 3.29(c). Take $f \in \mathcal{L}^1(\ell)$. By definition, there is a $g \in \overline{\mathcal{L}}(\ell)$ such that $f = g$ ℓ-a.e. By Theorem 3.22(b), $f \in \overline{\mathcal{L}}(\ell)$ whenever $\mathfrak{N}(\ell) \subset \overline{\mathfrak{N}}(\ell)$. But if $A \in \mathfrak{N}(\ell)$, then $A \cap B \in \overline{\mathfrak{N}}(\ell)$ for every $B \in \mathfrak{R}(\ell)$. In particular, this holds for $B \in \mathfrak{R}(\mathcal{L})$. Thus, if $(B_n)_{n \in \mathbb{N}}$ is a sequence in $\mathfrak{R}(\mathcal{L})$ such that $X = \bigcup_{n \in \mathbb{N}} B_n$, then

$$A = \bigcup_{n \in \mathbb{N}} (A \cap B_n) \in \overline{\mathfrak{N}}(\ell),$$

which proves the theorem. $\qquad\qquad\square$

Examples of σ-finite Daniell spaces are the spaces $(X, \mathcal{L}(\mathfrak{I}), \ell_{\mu_g})$, which are associated to Stieltjes measures μ_g on the ring of sets $\mathfrak{I}$ of interval forms on $\mathbb{R}$. The integrals related to these spaces are precisely the closures given by the Daniell process. The new definition has no effect on these particularly important classical examples. However, Example (a) shows that, in general, $\mathcal{L}^1(\ell)$ and $\overline{\mathcal{L}}(\ell)$ do not coincide.

Exercises

1. Show that a Daniell space $(X, \mathcal{L}, \ell)$ is σ-finite if and only if $\overline{\mathcal{L}}(\ell)$ contains a strictly positive element.

2. By Exercise 2(b) of Section 2.4, every positive linear functional ℓ on $c_0(X)$ is null-continuous and satisfies $\ell(f) = \sum_{x \in X} f(x) g(x)$ for some $g \in \ell^1(X)_+$. Show that the following hold.

(a) $c_0(X) \subset \overline{\mathcal{L}}(\ell_g) \subset \mathcal{L}^1(\ell_g)$.

(b) $(X, \mathcal{L}^1(\ell), \int_\ell) = (X, \mathcal{L}^1(\ell_g), \int_{\ell_g})$.

(c) $\mathcal{N}(\ell) = \mathcal{N}(\ell_g)$.

(d) $\mathfrak{N}(\ell) = \mathfrak{N}(\ell_g)$.

3. Let X be an uncountable set and consider the positive linear functional
   $\ell : c_f(X) \to \mathbb{R}$, $\alpha e_X + h \mapsto \alpha$, with $\alpha \in \mathbb{R}$ and $h \in \mathcal{F}(X)$ (cf. Exercise 3 of Section 2.4). Prove the following.

   (a) $\mathcal{L}^1(\ell) = \{\alpha e_X + h \,|\, \alpha \in \mathbb{R},\ h \in \mathbb{R}^X,\ \{h \neq 0\}\ \text{countable}\}$.

   (b) $\mathcal{N}(\ell) = \{f \in \mathbb{R}^X \,|\, \{f \neq 0\}\ \text{is countable}\}$.

   (c) $\mathfrak{N}(\ell) = \{A \subset X \,|\, A\ \text{is countable}\}$.

3.8 The induction principle

An important task of integration theory is the investigation of properties of integrable functions. Often, the original space $(X, \mathcal{L}, \ell)$ has certain properties which we would like $(X, \mathcal{L}^1(\ell), \int_\ell)$ to retain. We shall see numerous examples of this. Following the pattern of the construction procedure, we can prove the properties of $\mathcal{L}^1(\ell)$ and $\int_\ell$ by first examining $(\mathcal{L}^\uparrow, \ell^\uparrow)$ and $(\mathcal{L}^\downarrow, \ell^\downarrow)$, then $(\overline{\mathcal{L}}(\ell), \overline{\ell})$, and finally $(\mathcal{L}^1(\ell), \int_\ell)$. This procedure can usually be abbreviated. With this in mind, we prove two theorems.

Theorem 3.38 *Let $P(\,\cdot\,)$ be a statement about functions from $\overline{\mathcal{L}}(\ell)$ and let the following conditions be satisfied.*

(i) *$P(f)$ is true for each $f \in \mathcal{L}$.*

(ii) *If $(f_n)_{n \in \mathbb{N}}$ is a monotone sequence in $\overline{\mathcal{L}}(\ell)$ with $\lim_{n \to \infty} \overline{\ell}(f_n) \notin \{\infty, -\infty\}$ and if $P(f_n)$ is true for every $n \in \mathbb{N}$, then $P(\lim_{n \to \infty} f_n)$ is true.*

(iii) *If $f, g \in \overline{\mathcal{L}}(\ell)$, $f \leq g$, $\overline{\ell}(f) = \overline{\ell}(g)$, and if both $P(f)$ and $P(g)$ are true, then $P(h)$ is true for every $h \in \overline{\mathcal{L}}(\ell)$, $f \leq h \leq g$.*

Then $P(f)$ is true for every $f \in \overline{\mathcal{L}}(\ell)$.

Proof. Let $\mathcal{F}$ be the set of all $f \in \overline{\mathcal{L}}(\ell)$ for which $P(f)$ is true. By (i), $\mathcal{L} \subset \mathcal{F}$. Applying (ii),

$$\{f \in \mathcal{L}^\uparrow \,|\, \ell^\uparrow(f) < \infty\} \subset \mathcal{F}$$

and

$$\{f \in \mathcal{L}^\downarrow \,|\, \ell^\downarrow(f) > -\infty\} \subset \mathcal{F}.$$

Now take $f \in \overline{\mathcal{L}}(\ell)$. For each $n \in \mathbb{N}$ there is a $1/n$-bracket (f_n', f_n'') of f, and we may choose these brackets in such a way that $(f_n')_{n \in \mathbb{N}}$ is increasing and $(f_n'')_{n \in \mathbb{N}}$ is decreasing. By (ii),

$$f' := \bigvee_{n \in \mathbb{N}} f_n' \in \mathcal{F} \quad \text{and} \quad f'' := \bigwedge_{n \in \mathbb{N}} f_n'' \in \mathcal{F}.$$

Furthermore, $f' \le f \le f''$, and $\overline{\ell}(f') = \overline{\ell}(f'') = \overline{\ell}(f)$. Hence, by (iii), $f \in \mathcal{F}$. Thus $\overline{\mathcal{L}}(\ell) \subset \mathcal{F}$. $\qquad\square$

We call Theorem 3.38 the **induction principle** for $\overline{\mathcal{L}}(\ell)$. It is easy to formulate a corresponding principle for $\mathcal{L}^1(\ell)$.

Theorem 3.39 *Let $P(\cdot)$ be a statement about ℓ-integrable functions and let the following conditions be satisfied.*

(i) *$P(f)$ is true for each $f \in \mathcal{L}$.*

(ii) *If $(f_n)_{n \in \mathbb{N}}$ is a monotone sequence in $\overline{\mathcal{L}}(\ell)$ with $\lim_{n \to \infty} \overline{\ell}(f_n) \notin \{\infty, -\infty\}$ and if $P(f_n)$ is true for each $n \in \mathbb{N}$, then $P(\lim_{n \to \infty} f_n)$ is true.*

(iii) *Let $P(f)$ be true for $f \in \overline{\mathcal{L}}(\ell)$. Then $P(g)$ is true for every $g \in \mathcal{L}^1(\ell)$ with $g = f$ ℓ-a.e.*

Then $P(f)$ is true for each $f \in \mathcal{L}^1(\ell)$.

Proof. Take $f, g \in \overline{\mathcal{L}}(\ell)$ with $f \le g$ and $\overline{\ell}(f) = \overline{\ell}(g)$. Suppose that both $P(f)$ and $P(g)$ are true. If $h \in \overline{\mathcal{L}}(\ell)$ and $f \le h \le g$, then $f = h$ ℓ-a.e. (Theorem 3.22) and hence $P(h)$ is true. By Theorem 3.38, $P(f)$ is true for every $f \in \overline{\mathcal{L}}(\ell)$. But for each $f \in \mathcal{L}^1(\ell)$ there is a $g \in \overline{\mathcal{L}}(\ell)$ with $f = g$ ℓ-a.e. Then, by (iii), $P(f)$ is true whenever $f \in \mathcal{L}^1(\ell)$. $\qquad\square$

Our first application is the following theorem.

Theorem 3.40 *Let $(X, \mathcal{L}_1, \ell_1)$ and $(X, \mathcal{L}_2, \ell_2)$ be Daniell spaces satisfying*

$$(X, \mathcal{L}_1, \ell_1) \preccurlyeq (X, \mathcal{L}_2, \ell_2).$$

Then

$$(X, \overline{\mathcal{L}}(\ell_1), \overline{\ell_1}) \preccurlyeq (X, \overline{\mathcal{L}}(\ell_2), \overline{\ell_2}).$$

Proof. Let $\mathcal{F}$ be the set of all $f \in \overline{\mathcal{L}}(\ell_1)$ such that $f \in \overline{\mathcal{L}}(\ell_2)$ and $\overline{\ell_2}(f) = \overline{\ell_1}(f)$. By hypothesis, $\mathcal{L}_1 \subset \mathcal{F}$. Let $(f_n)_{n \in \mathbb{N}}$ be a monotone sequence in $\mathcal{F}$ such that $\lim_{n \to \infty} \overline{\ell_1}(f_n) \notin \{\infty, -\infty\}$. Then $\lim_{n \to \infty} \overline{\ell_2}(f_n) \notin \{\infty, -\infty\}$. By the monotone convergence theorem, $\lim_{n \to \infty} f_n \in \overline{\mathcal{L}}(\ell_2)$ and

$$\overline{\ell_2}\Big(\lim_{n \to \infty} f_n \Big) = \lim_{n \to \infty} \overline{\ell_2}(f_n) = \lim_{n \to \infty} \overline{\ell_1}(f_n) = \overline{\ell_1}\Big(\lim_{n \to \infty} f_n \Big).$$

Thus, $\lim_{n \to \infty} f_n \in \mathcal{F}$. Finally, given $f, g \in \mathcal{F}$ with $f \le g$ and $\overline{\ell_1}(f) = \overline{\ell_1}(g)$, we have that $\overline{\ell_2}(f) = \overline{\ell_2}(g)$, and hence $h \in \overline{\mathcal{L}}(\ell_2)$ for every $h \in \overline{\mathcal{L}}(\ell_1)$ with $f \le h \le g$. Furthermore,

$$\overline{\ell_2}(h) = \overline{\ell_2}(f) = \overline{\ell_1}(f) = \overline{\ell_1}(h).$$

We conclude that $h \in \mathcal{F}$.

The hypotheses of Theorem 3.38 are therefore satisfied, where $P(f)$ is the proposition $f \in \mathcal{F}$, and we have $\overline{\mathcal{L}}(\ell_1) \subset \mathcal{F}$. $\qquad\square$

The proof of the next theorem is another useful application of the induction principle.

Theorem 3.41 *Let $(X, \mathcal{L}_1, \ell_1)$ and $(X, \mathcal{L}_2, \ell_2)$ be Daniell spaces satisfying*

$$(X, \mathcal{L}_1, \ell_1) \preccurlyeq (X, \overline{\mathcal{L}}(\ell_2), \overline{\ell}_2) \quad \text{and} \quad (X, \mathcal{L}_2, \ell_2) \preccurlyeq (X, \overline{\mathcal{L}}(\ell_1), \overline{\ell}_1).$$

Then

$$(X, \overline{\mathcal{L}}(\ell_1), \overline{\ell}_1) = (X, \overline{\mathcal{L}}(\ell_2), \overline{\ell}_2)$$

and

$$\left(X, \mathcal{L}^1(\ell_1), \int_{\ell_1}\right) = \left(X, \mathcal{L}^1(\ell_2), \int_{\ell_2}\right).$$

Proof. For $f \in \overline{\mathcal{L}}(\ell_1)$, let $P(f)$ be the proposition that $f \in \overline{\mathcal{L}}(\ell_2)$ and $\overline{\ell}_1(f) = \overline{\ell}_2(f)$. Then condition (i) of Theorem 3.38 is satisfied by hypothesis, condition (ii) follows from the monotone convergence theorem for $\overline{\mathcal{L}}(\ell_2)$, and condition (iii) follows from the sandwich principle for $\overline{\mathcal{L}}(\ell_2)$. Given $f \in \overline{\mathcal{L}}(\ell_1)$, Theorem 3.38 implies that $P(f)$ is true, i.e. $(X, \overline{\mathcal{L}}(\ell_1), \overline{\ell}_1) \preccurlyeq (X, \overline{\mathcal{L}}(\ell_2), \overline{\ell}_2)$. The converse follows analogously. Hence

$$(X, \overline{\mathcal{L}}(\ell_1), \overline{\ell}_1) = (X, \overline{\mathcal{L}}(\ell_2), \overline{\ell}_2).$$

The second statement follows easily from this and the definition of the integral. It is enough to note that the ℓ_1-null sets coincide with the ℓ_2-null sets, since $\mathfrak{R}(\ell_1) = \mathfrak{R}(\ell_2)$. ∎

These first two applications of the induction principle have been of a somewhat theoretical nature. The next application is a 'practical' result, namely the **translation invariance of Lebesgue measure**. As usual, we let λ denote Lebesgue measure on $\mathfrak{J}$, the ring of interval forms on $\mathbb{R}$.

Theorem 3.42 *Given $f \in \overline{\mathbb{R}}^{\mathbb{R}}$ and $\beta \in \mathbb{R}$, define*

$$f_\beta : \mathbb{R} \longrightarrow \overline{\mathbb{R}}, \quad x \longmapsto f(x + \beta)$$

(the translate of f by β). Then for every $f \in \overline{\mathbb{R}}^{\mathbb{R}}$,

$$f \in \mathcal{L}^1(\ell_\lambda) \iff f_\beta \in \mathcal{L}^1(\ell_\lambda), \quad \text{and in this case} \quad \int_{\ell_\lambda} f = \int_{\ell_\lambda} f_\beta.$$

Proof. Note that by Theorem 3.37, $\mathcal{L}^1(\ell_\lambda) = \overline{\mathcal{L}}(\ell_\lambda)$. For $f \in \overline{\mathcal{L}}(\ell_\lambda)$, let $P(f)$ be the proposition that $f_\beta \in \overline{\mathcal{L}}(\ell_\lambda)$ and $\overline{\ell}_\lambda(f) = \overline{\ell}_\lambda(f_\beta)$. Translating an interval does not change its Lebesgue measure. Consequently, $P(f)$ is true for every $f \in \mathcal{L}(\mathfrak{J})$. Conditions (ii) and (iii) of Theorem 3.38 follow from the monotone convergence theorem and the sandwich principle. Thus, by Theorem 3.38, $f_\beta \in \overline{\mathcal{L}}(\ell_\lambda)$ and $\overline{\ell}_\lambda(f) = \overline{\ell}_\lambda(f_\beta)$ for every $f \in \overline{\mathcal{L}}(\ell_\lambda)$. The converse follows analogously. ∎

Exercise 4 below extends this theorem. A further generalization for n-dimensional Lebesgue measure is presented in Section 5.3.

Exercises

1. Let $(X, \mathcal{L}, \ell)$ be a Daniell space. Prove the following.

   (a) $\left(X, \overline{\mathcal{L}}(\overline{\ell}|_{\overline{\mathcal{L}}(\ell) \cap \mathbb{R}^X}), \overline{\ell}|_{\overline{\mathcal{L}}(\ell) \cap \mathbb{R}^X}\right) = (X, \overline{\mathcal{L}}(\ell), \overline{\ell})$.

   (b) $\left(X, \mathcal{L}^1(\int_\ell |_{\mathcal{L}^1(\ell) \cap \mathbb{R}^X}), \int_{(\int_\ell |_{\mathcal{L}^1(\ell) \cap \mathbb{R}^X})}\right) = (X, \mathcal{L}^1(\ell), \int_\ell)$.

2. Let $(X, \mathcal{L}, \ell)$ be a Daniell space and suppose that $f \in \mathbb{R}^X$ is bounded. Prove the following.

   (a) If $fg \in \overline{\mathcal{L}}(\ell)$ for every $g \in \mathcal{L}$, then $fg \in \overline{\mathcal{L}}(\ell)$ for every $g \in \overline{\mathcal{L}}(\ell)$ and
   $$|\overline{\ell}(fg)| \leq \overline{\ell}(|fg|) \leq \overline{\ell}(|g|) \sup_{x \in X} |f(x)|.$$

   (b) If $fg \in \mathcal{L}^1(\ell)$ for every $g \in \mathcal{L}$, then $fg \in \mathcal{L}^1(\ell)$ for every $g \in \mathcal{L}^1(\ell)$ and
   $$\left| \int_\ell (fg) \right| \leq \int_\ell |fg| \leq \int_\ell (|g|) \sup_{x \in X} |f(x)|.$$

3. Let $(X, \mathcal{L}, \ell)$ be a Daniell space and take $f \in \overline{\mathbb{R}}_+^X$. Prove the following.

   (a) If $f \wedge g \in \overline{\mathcal{L}}(\ell)$ for every $g \in \mathcal{L}_+$, then $f \wedge g \in \overline{\mathcal{L}}(\ell)$ for every $g \in \overline{\mathcal{L}}(\ell)$.

   (b) If $f \wedge g \in \mathcal{L}^1(\ell)$ for every $g \in \mathcal{L}_+$, then $f \wedge g \in \mathcal{L}^1(\ell)$ for every $g \in \mathcal{L}^1(\ell)$.

4. For $\alpha, \beta \in \mathbb{R}$, $\alpha \neq 0$, define $\varphi : \mathbb{R} \to \mathbb{R}$, $x \mapsto \alpha x + \beta$, and let ℓ_λ be the functional on $\mathcal{L} := \mathcal{L}(\mathfrak{I})$ generated by Lebesgue measure on $\mathfrak{I}$. Show that for $f \in \overline{\mathbb{R}}^{\mathbb{R}}$
   $$f \circ \varphi \in \mathcal{L}^1(\ell_\lambda) \iff f \in \mathcal{L}^1(\ell_\lambda), \quad \text{and in this case} \int_{\ell_\lambda} f \circ \varphi = \frac{1}{|\alpha|} \int_{\ell_\lambda} f.$$

   For $\alpha = 1$, this is precisely the translation invariance of Lebesgue measure.

3.9 Functionals on $\mathbb{R}^X$

Since the results of this section are not used again, this section may be omitted on first reading.

We characterize null-continuous positive linear functionals on $\mathbb{R}^X$. We shall see that these functionals have a rather special form. Why is this of interest? Suppose that $\mathbb{R}^X \subset \mathcal{L}^1(\ell)$. Then, by Theorem 3.29(b), (the restriction of) $\int_\ell$ is a null-continuous positive linear functional on $\mathbb{R}^X$ and thus of the particular form alluded to. As a consequence, the space $\mathcal{L}^1(\ell)$ can contain all of $\mathbb{R}^X$ only for special Daniell spaces $(X, \mathcal{L}, \ell)$. In other words, the extension procedure described in this chapter *cannot*, in general, render every function in $\mathbb{R}^X$ integrable – there are limitations which lie in the nature of things. We have already seen this for Lebesgue measure λ on $\mathbb{R}$, for example. There the function $e_{\mathbb{R}}$ is not ℓ_λ-integrable.

We need some preliminaries before presenting our characterization. We need some properties of ultrafilters.

The non-empty subset $\mathfrak{F}$ of $\mathfrak{P}(X)$ is called a **filterbase** (on X) if

(i) $\emptyset \notin \mathfrak{F}$,

(ii) $A, B \in \mathfrak{F} \implies \exists C \in \mathfrak{F}, C \subset A \cap B$.

A filterbase $\mathfrak{F}$ on X is called a **filter** (on X) if, in addition,

(iii) $A \in \mathfrak{F}, A \subset B \subset X \implies B \in \mathfrak{F}$.

For example, if A is a non-empty subset of X, then $\{A\}$ is a filterbase on X, while the set of all supersets of A is a filter on X.

Proposition 3.43 *Let $\mathfrak{F}$ be a filter on X. Then $X \in \mathfrak{F}$, and $\bigcap_{\iota \in I} A_\iota \in \mathfrak{F}$ for every non-empty finite family $(A_\iota)_{\iota \in I}$ in $\mathfrak{F}$.*

Proof. The first claim is obvious, and the second follows by complete induction on the number of elements in I. $\qquad\qquad\square$

If $\mathfrak{F}, \mathfrak{G}$ are filters on X, then $\mathfrak{F}$ is called **coarser** than $\mathfrak{G}$, and $\mathfrak{G}$ is called **finer** than $\mathfrak{F}$, whenever $\mathfrak{F} \subset \mathfrak{G}$.

The following result shows that filterbases always generate filters.

Proposition 3.44 *If $\mathfrak{F}$ is a filterbase on X, then*

$$\mathfrak{G}_{\mathfrak{F}} := \{B \subset X \mid B \supset A \text{ for some } A \in \mathfrak{F}\}$$

is the coarsest filter on X containing $\mathfrak{F}$.

Proof. Clearly $\emptyset \notin \mathfrak{G}_{\mathfrak{F}}$. Take $B_1, B_2 \in \mathfrak{G}_{\mathfrak{F}}$. There are $A_1, A_2 \in \mathfrak{F}$ with $A_1 \subset B_1$ and $A_2 \subset B_2$. We can find an $A \in \mathfrak{F}$ contained in $A_1 \cap A_2$, and hence contained in $B_1 \cap B_2$. Thus $B_1 \cap B_2 \in \mathfrak{G}_{\mathfrak{F}}$. That (iii) is satisfied for $\mathfrak{G}_{\mathfrak{F}}$ follows by a similar argument. Hence $\mathfrak{G}_{\mathfrak{F}}$ is a filter on X containing $\mathfrak{F}$. That it is the coarsest filter with this property, is obvious. $\qquad\square$

We call $\mathfrak{G}_{\mathfrak{F}}$ the **filter generated by** $\mathfrak{F}$ and we say that $\mathfrak{F}$ is a **base** for $\mathfrak{G}_{\mathfrak{F}}$. For example, the filter generated by $\{A\}$ is the set of all supersets of A.

Note that in general there are many filters containing a given filterbase.

Those filters which are maximal with respect to the inclusion relation on the set of all filters on X are called **ultrafilters** (on X). (Recall that an element w of an ordered set Z is called **maximal** if there are no strictly larger elements in Z, i.e. if $z \geq w$ for $z \in Z$ then $z = w$.) Hence a filter is an ultrafilter on X if there are no strictly finer filters on X. A simple example of an ultrafilter is the filter generated by the filterbase $\{\{x\}\}$, for fixed $x \in X$, i.e. the set $\{A \subset X \mid x \in A\}$. Ultrafilters of this type are called **trivial**.

For a general assertion about the existence of ultrafilters, we need Zorn's lemma, which is logically equivalent to the axiom of choice. We shall not delve into a discussion of this logical dependence, but take the lemma for granted.

Theorem 3.45 (Zorn's Lemma) *Let* Z *be a non-empty inductively ordered set (i.e. an ordered set in which every totally ordered subset has an upper bound) and take* $z \in Z$. *Then there is a maximal element* w *of* Z *with* $z \le w$.

Theorem 3.46 *Let* $\mathfrak{F}$ *be a filter on* X. *Then there is an ultrafilter on* X *which is finer than* $\mathfrak{F}$.

Proof. To apply Zorn's lemma, it is sufficient to show that the set of all filters on X is inductively ordered by inclusion. Let Φ be a totally ordered set of filters on X. If $\Phi = \emptyset$, then every filter on X, e.g. the filter $\{X\}$, is an upper bound for Φ. Now suppose that $\Phi \neq \emptyset$. If we can show that $\mathfrak{F} := \bigcup\{\mathfrak{G} \mid \mathfrak{G} \in \Phi\}$ is a filter, then the claim will be proven. Obviously $\mathfrak{F} \neq \emptyset$ and $\emptyset \notin \mathfrak{F}$. To verify (ii), take $A, B \in \mathfrak{F}$. There are $\mathfrak{G}, \mathfrak{H} \in \Phi$ such that $A \in \mathfrak{G}$, $B \in \mathfrak{H}$. Since Φ is totally ordered, we have $\mathfrak{G} \subset \mathfrak{H}$ or $\mathfrak{H} \subset \mathfrak{G}$. In the first case $A \cap B \in \mathfrak{H}$ and in the second $A \cap B \in \mathfrak{G}$. Hence in both cases $A \cap B \in \mathfrak{F}$. The proof of (iii) is trivial. $\qquad\square$

How can we decide whether a given filter is an ultrafilter? The verification of its maximality with respect to inclusion is often cumbersome. The following criterion, however, is very useful.

Theorem 3.47 *Let* $\mathfrak{F}$ *be a filter on* X. *Then the following are equivalent.*

(a) $\mathfrak{F}$ *is an ultrafilter.*

(b) *If* $A \subset X$ *and* $A \cap B \neq \emptyset$ *whenever* $B \in \mathfrak{F}$, *then* $A \in \mathfrak{F}$.

(c) *If* $A \cup B \in \mathfrak{F}$ *for subsets* A, B *of* X, *then* $A \in \mathfrak{F}$ *or* $B \in \mathfrak{F}$.

(d) *If* $A \subset X$, *then* $A \in \mathfrak{F}$ *or* $X \setminus A \in \mathfrak{F}$.

Proof. (a)$\Rightarrow$(b). If A is as described, then $\{A \cap B \mid B \in \mathfrak{F}\}$ is clearly a filterbase on X. In view of Proposition 3.44, there is a filter $\mathfrak{G}$ containing all sets of the form $A \cap B$ ($B \in \mathfrak{F}$). Hence, by (iii), $\mathfrak{G}$ contains A and every $B \in \mathfrak{F}$. But $\mathfrak{F}$ is an ultrafilter. Hence $\mathfrak{G} = \mathfrak{F}$. Thus $A \in \mathfrak{F}$.

(b)$\Rightarrow$(c). Assume that $A \cup B \in \mathfrak{F}$, but $A \notin \mathfrak{F}$ and $B \notin \mathfrak{F}$. By (b), there are $A_1, B_1 \in \mathfrak{F}$ with $A_1 \cap A = \emptyset$ and $B_1 \cap B = \emptyset$. Thus

$$\emptyset = \big(A \cap (A_1 \cap B_1)\big) \cup \big(B \cap (A_1 \cap B_1)\big) = (A \cup B) \cap (A_1 \cap B_1) \in \mathfrak{F},$$

which is a contradiction.

(c)$\Rightarrow$(d) is obvious, since $X \in \mathfrak{F}$.

(d)$\Rightarrow$(a). Let $\mathfrak{G}$ be a filter on X with $\mathfrak{F} \subset \mathfrak{G}$. Take $A \in \mathfrak{G}$. Then $X \setminus A \notin \mathfrak{G}$ and hence $X \setminus A \notin \mathfrak{F}$. By (d), $A \in \mathfrak{F}$. Therefore $\mathfrak{F} = \mathfrak{G}$. $\qquad\square$

Let Γ denote the set of all ultrafilters on X.

Proposition 3.48 *Given $\mathfrak{F} \in \Gamma$, choose $A_{\mathfrak{F}} \in \mathfrak{F}$. Then there is a finite subset Γ_0 of Γ with $X = \bigcup_{\mathfrak{F} \in \Gamma_0} A_{\mathfrak{F}}$.*

Proof. Otherwise,

$$\mathfrak{G} := \left\{ X \setminus \bigcup_{\mathfrak{F} \in \Gamma_0} A_{\mathfrak{F}} \,\middle|\, \Gamma_0 \text{ is a finite subset of } \Gamma \right\}$$

is a filterbase on X. Then there is an ultrafilter $\mathfrak{F}_0$ with $\mathfrak{G} \subset \mathfrak{F}_0$ (Proposition 3.44 and Theorem 3.46), from which there follows the contradiction $X \setminus A_{\mathfrak{F}_0} \in \mathfrak{G} \subset \mathfrak{F}_0$. $\square$

We now turn our attention to a special class of ultrafilters. We call an ultrafilter $\mathfrak{F}$ **δ-stable** if $\bigcap_{n \in \mathbb{N}} A_n \in \mathfrak{F}$ for every sequence $(A_n)_{n \in \mathbb{N}}$ in $\mathfrak{F}$. Note that every trivial ultrafilter is δ-stable. But the existence of non-trivial δ-stable ultrafilters cannot be proved within the standard axioms of set theory! An additional axiom is required to guarantee their existence. In any case, non-trivial δ-stable ultrafilters can exist only on 'big' sets X. Surprisingly enough, this existence problem is closely related to measure theory, as we shall see in the exercises.

Proposition 3.49 *Let $(\mathfrak{F}_n)_{n \in \mathbb{N}}$ be a sequence of distinct δ-stable ultrafilters on X. Then there is a disjoint sequence $(C_n)_{n \in \mathbb{N}}$ in $\mathfrak{P}(X)$ such that $C_n \in \mathfrak{F}_n$ for every $n \in \mathbb{N}$.*

Proof. Take $n \in \mathbb{N}$. For each $m \neq n$, there is, by the ultrafilter property of $\mathfrak{F}_m$, a set $A_m \in \mathfrak{F}_m$ not belonging to $\mathfrak{F}_n$. In view of Theorem 3.47 (a)$\Rightarrow$(d), we have $B_n := \bigcap_{m \in \mathbb{N} \setminus \{n\}} (X \setminus A_m) \in \mathfrak{F}_n$, but $B_n \notin \mathfrak{F}_m$ for every $m \neq n$. Put $C_n := B_n \setminus \bigcup_{m < n} B_m$. By Theorem 3.47 (a)$\Rightarrow$(c), $C_n \in \mathfrak{F}_n$. $\square$

Note that the preceding proposition is trivial if every $\mathfrak{F}_n$ is trivial.

The next result leads us close to the connection between the δ-stable ultrafilters on X and the functionals on $\mathbb{R}^X$.

Theorem 3.50 *Let $\mathfrak{F}$ be an ultrafilter on X. Then the following assertions are equivalent.*

(a) $\mathfrak{F}$ is δ-stable.

(b) For each $f \in \mathbb{R}^X$ there is an $\alpha_{f,\mathfrak{F}} \in \mathbb{R}$ such that $\{f = \alpha_{f,\mathfrak{F}}\} \in \mathfrak{F}$.

Moreover, if these equivalent conditions are satisfied, then $\alpha_{f,\mathfrak{F}}$ is uniquely determined and the map

$$\mathbb{R}^X \longrightarrow \mathbb{R}, \quad f \longmapsto \alpha_{f,\mathfrak{F}}$$

is a null-continuous positive linear functional on $\mathbb{R}^X$.

Proof. (a)$\Rightarrow$(b). Put $A := \{\alpha \in \overline{\mathbb{R}} \mid \{f \geq \alpha\} \in \mathfrak{F}\}$. Since $X \in \mathfrak{F}$, we have that $-\infty \in A$. Let $\alpha_0 := \sup A$. Take a sequence $(\alpha_n)_{n \in \mathbb{N}}$ in A with $\alpha_n \uparrow \alpha_0$. Then

$$\{f \geq \alpha_0\} = \bigcap_{n \in \mathbb{N}} \{f \geq \alpha_n\} \in \mathfrak{F}.$$

Similarly, if $B := \{\beta \in \overline{\mathbb{R}} \mid \{f \leq \beta\} \in \mathfrak{F}\}$ and $\beta_0 := \inf B$, then $\{f \leq \beta_0\} \in \mathfrak{F}$. In view of

$$\{\alpha_0 \leq f \leq \beta_0\} = \{\alpha_0 \leq f\} \cap \{f \leq \beta_0\} \in \mathfrak{F},$$

we see that $\alpha_0 \leq \beta_0$. Assume that $\alpha_0 < \beta_0$. Then there is a $\gamma \in \mathbb{R}$ with $\alpha_0 < \gamma < \beta_0$. By Theorem 3.47 (a)$\Rightarrow$(c), we have $\{\alpha_0 \leq f < \gamma\} \in \mathfrak{F}$ or $\{\gamma \leq f \leq \beta_0\} \in \mathfrak{F}$. In the first case, we obtain a contradiction by noting that $\{f \leq \gamma\} \in \mathfrak{F}$ and thus $\gamma \in B$. In the second case, we obtain the contradiction $\gamma \in A$ by a similar argument. Hence $\alpha_0 = \beta_0$, and thus $\{f = \alpha_0\} \in \mathfrak{F}$.

(b)$\Rightarrow$(a). Let $(A_n)_{n \in \mathbb{N}}$ be a sequence in $\mathfrak{F}$. To show that $\bigcap_{n \in \mathbb{N}} A_n \in \mathfrak{F}$, we may suppose that $(A_n)_{n \in \mathbb{N}}$ is decreasing and that $A_n \setminus A_{n+1} \neq \emptyset$ whenever $n \in \mathbb{N}$. Put

$$f : X \longrightarrow \mathbb{R}, \quad x \longmapsto \begin{cases} n & \text{if } x \in A_n \setminus A_{n+1} \text{ for some } n \in \mathbb{N}, \\ 0 & \text{if } x \in \bigcap_{n \in \mathbb{N}} A_n, \\ -1 & \text{otherwise.} \end{cases}$$

If $\alpha_{f,\mathfrak{F}} = n$ for some $n \in \mathbb{N}$, then $\{f = \alpha_{f,\mathfrak{F}}\} \cap A_{n+1} = \emptyset$, which is impossible. Since $X \setminus A_1 \notin \mathfrak{F}$, $\alpha_{f,\mathfrak{F}} = -1$ is also impossible. Thus $\alpha_{f,\mathfrak{F}} = 0$, i.e. $\bigcap_{n \in \mathbb{N}} A_n \in \mathfrak{F}$.

Of course, if (a) or (b) is satisfied, then the number $\alpha_{f,\mathfrak{F}}$ is unique, since for $\alpha \neq \beta$ we have $\{f = \alpha\} \cap \{f = \beta\} = \emptyset$. The functional

$$\varphi : \mathbb{R}^X \longrightarrow \mathbb{R}, \quad f \longmapsto \alpha_{f,\mathfrak{F}}$$

is clearly positive. That φ is linear follows from the relation

$$\{\beta_1 f_1 + \beta_2 f_2 = \beta_1 \alpha_{f_1,\mathfrak{F}} + \beta_2 \alpha_{f_2,\mathfrak{F}}\} \supset \{f_1 = \alpha_{f_1,\mathfrak{F}}\} \cap \{f_2 = \alpha_{f_2,\mathfrak{F}}\},$$

for arbitrary $f_1, f_2 \in \mathbb{R}^X$ and $\beta_1, \beta_2 \in \mathbb{R}$. To prove that φ is also null-continuous, let $f_n \downarrow 0$. Then

$$C := \bigcap_{n \in \mathbb{N}} \{f_n = \alpha_{f_n,\mathfrak{F}}\} \in \mathfrak{F}.$$

There is some $x \in C$. Then $\inf_{n \in \mathbb{N}} \alpha_{f_n,\mathfrak{F}} = \inf_{n \in \mathbb{N}} f_n(x) = 0$. $\square$

For the rest of this section, let φ be a positive linear functional on $\mathbb{R}^X$. Letting Φ denote the set of all δ-stable ultrafilters on X, we associate the mapping

$$g : \Phi \longrightarrow \mathbb{R}, \quad \mathfrak{F} \longmapsto \inf_{A \in \mathfrak{F}} \varphi(e_A)$$

to φ.

Proposition 3.51 *The set* $\{g \neq 0\}$ *is finite.*

Proof. Otherwise, there would be a sequence $(\mathfrak{F}_n)_{n \in \mathbb{N}}$ of distinct elements of Φ with $g(\mathfrak{F}_n) > 0$ for every $n \in \mathbb{N}$. Take a disjoint sequence $(C_n)_{n \in \mathbb{N}}$ in $\mathfrak{P}(X)$ with $C_n \in \mathfrak{F}_n$ for every $n \in \mathbb{N}$ (Proposition 3.49). Put $f := \sum_{n=1}^{\infty} \frac{1}{g(\mathfrak{F}_n)} e_{C_n}$. Then, for each $m \in \mathbb{N}$,

$$\varphi(f) \geq \varphi\left(\sum_{n=1}^{m} \frac{1}{g(\mathfrak{F}_n)} e_{C_n} \right) = \sum_{n=1}^{m} \frac{1}{g(\mathfrak{F}_n)} \varphi(e_{C_n}) \geq m.$$

Hence $\varphi(f) = \infty$, which is a contradiction.   □

In view of the preceding proposition, we can associate another map to φ, namely the map

$$\psi : \mathbb{R}^X \longrightarrow \mathbb{R}, \quad f \longmapsto \sum_{\mathfrak{F} \in \Phi} g(\mathfrak{F}) \alpha_{f,\mathfrak{F}}.$$

Proposition 3.52 ψ *is a null-continuous positive linear functional on* $\mathbb{R}^X$ *and* $\varphi - \psi$ *is a positive linear functional on* $\mathbb{R}^X$.

Proof. The first claim follows from the fact that $f \mapsto \alpha_{f,\mathfrak{F}}$ is null-continuous, positive and linear (Theorem 3.50). This immediately implies that $\varphi - \psi$ is linear. To show that $\varphi - \psi$ is positive, take $f \in \mathbb{R}_+^X$. We can write $\{g \neq 0\} = \{\mathfrak{F}_1, \ldots, \mathfrak{F}_n\}$, for distinct elements $\mathfrak{F}_k$ of Φ. By Theorem 3.50 (a)$\Rightarrow$(b) and Proposition 3.49, we may select pairwise disjoint sets $A_k \in \mathfrak{F}_k$ with $A_k \subset \{f = \alpha_{f,\mathfrak{F}_k}\}$ for each $k \leq n$. We have $f \geq \sum_{k=1}^{n} \alpha_{f,\mathfrak{F}_k} e_{A_k}$ and thus, since φ is positive and linear,

$$\varphi(f) \geq \sum_{k=1}^{n} \alpha_{f,\mathfrak{F}_k} \varphi(e_{A_k}) \geq \sum_{k=1}^{n} \alpha_{f,\mathfrak{F}_k} g(\mathfrak{F}_k) = \psi(f).$$

Since f was arbitrary, $\varphi - \psi$ is positive.   □

Proposition 3.53 *For each* $\mathfrak{F} \in \Gamma$ *there is an* $A \in \mathfrak{F}$ *with* $(\varphi - \psi)(e_A) = 0$.

Proof. We first consider the case $\mathfrak{F} \in \Phi$. Take $B \in \mathfrak{F}$ such that $B \notin \mathfrak{G}$ whenever $\mathfrak{G} \in \{g \neq 0\}$, $\mathfrak{G} \neq \mathfrak{F}$. There is a sequence $(C_n)_{n \in \mathbb{N}}$ in $\mathfrak{F}$ with $g(\mathfrak{F}) = \inf_{n \in \mathbb{N}} \varphi(e_{C_n})$. We have $C := \bigcap_{n \in \mathbb{N}} C_n \in \mathfrak{F}$ and, since φ is positive, $g(\mathfrak{F}) \leq \varphi(e_C) \leq \inf_{n \in \mathbb{N}} \varphi(e_{C_n}) \leq g(\mathfrak{F})$, whence $\varphi(e_C) = g(\mathfrak{F})$. Put $A := B \cap C$. Then $\varphi(e_A) = g(\mathfrak{F})$. Clearly $\alpha_{e_A,\mathfrak{F}} = 1$ and $\alpha_{e_A,\mathfrak{G}} = 0$ for every $\mathfrak{G} \in \{g \neq 0\}$, $\mathfrak{G} \neq \mathfrak{F}$. We conclude that $\psi(e_A) = g(\mathfrak{F}) = \varphi(e_A)$.

Now suppose that $\mathfrak{F} \in \Gamma \setminus \Phi$. If B is as in the first case, then $\alpha_{e_B,\mathfrak{G}} = 0$ for every $\mathfrak{G} \in \{g \neq 0\}$ and hence $\psi(e_B) = 0$. Since $\mathfrak{F} \notin \Phi$, there is a decreasing sequence $(C_n)_{n \in \mathbb{N}}$ in $\mathfrak{F}$ with $C := \bigcap_{n \in \mathbb{N}} C_n \notin \mathfrak{F}$. For $n \in \mathbb{N}$ put $A_n := C_n \setminus C$. Then $A_n \in \mathfrak{F}$ (Theorem 3.47 (a)$\Rightarrow$(c)). Assume that $\varphi(e_{A_n}) > 0$ for every $n \in \mathbb{N}$. Then, in view of $\bigcap_{n \in \mathbb{N}} A_n = \emptyset$, the function

$f := \sum_{n=1}^{\infty} \frac{1}{\varphi(e_{A_n})} e_{A_n}$ is real-valued. Thus for each $m \in \mathbb{N}$

$$\varphi(f) \geq \varphi\left(\sum_{n=1}^{m} \frac{1}{\varphi(e_{A_n})} e_{A_n}\right) = \sum_{n=1}^{m} \frac{1}{\varphi(e_{A_n})} \varphi(e_{A_n}) = m,$$

which is impossible. Hence there is a $k \in \mathbb{N}$ with $\varphi(e_{A_k}) = 0$, and for $A := A_k \cap B$ we obtain $\varphi(e_A) = \psi(e_A) = 0$, because φ and ψ are positive linear functionals. $\qquad\qquad \square$

Corollary 3.54 *If* $f \in \mathbb{R}_+^X$ *and* $\mathfrak{F} \in \Gamma$, *then there is an* $A_{\mathfrak{F}} \in \mathfrak{F}$ *with* $(\varphi - \psi)(fe_{A_{\mathfrak{F}}}) = 0$.

Proof. Take $A \in \mathfrak{F}$ with $(\varphi - \psi)(e_A) = 0$ (Proposition 3.53). Given $n \in \mathbb{N}$, we have $nfe_A \leq f^2 + n^2 e_A$ and hence

$$n(\varphi - \psi)(fe_A) \leq (\varphi - \psi)(f^2) + n^2(\varphi - \psi)(e_A) = (\varphi - \psi)(f^2),$$

from which the assertion follows. $\qquad\qquad \square$

We have now completed the preliminaries to the result heralded above, which fully characterizes the positive linear functionals on $\mathbb{R}^X$.

Theorem 3.55 *For each positive linear functional* φ *on* $\mathbb{R}^X$, *there is a map* $g : \Phi \to \mathbb{R}_+$ *with* $\{g \neq 0\}$ *finite such that*

$$\varphi(f) = \sum_{\mathfrak{F} \in \Phi} g(\mathfrak{F}) \alpha_{f, \mathfrak{F}}$$

for every $f \in \mathbb{R}^X$. *Moreover,* φ *is automatically null-continuous.*

Proof. Take $f \in \mathbb{R}_+^X$. Given $\mathfrak{F} \in \Gamma$, choose $A_{\mathfrak{F}} \in \mathfrak{F}$ with $(\varphi - \psi)(fe_{A_{\mathfrak{F}}}) = 0$ (Corollary 3.54). By Proposition 3.48, there is a finite subset Γ_0 of Γ such that $X = \bigcup_{\mathfrak{F} \in \Gamma_0} A_{\mathfrak{F}}$. Then

$$0 \leq (\varphi - \psi)(f) \leq (\varphi - \psi)\left(\sum_{\mathfrak{F} \in \Gamma_0} fe_{A_{\mathfrak{F}}}\right) = 0.$$

For $f \in \mathbb{R}^X$, we obtain

$$(\varphi - \psi)(f) = (\varphi - \psi)(f^+) - (\varphi - \psi)(f^-) = 0.$$

Hence $\varphi = \psi$. The null-continuity of φ is a consequence of Proposition 3.52. $\qquad\qquad \square$

Exercises

1. When is $\mathfrak{P}(X) \setminus \{\emptyset\}$ a filter?

2. Verify that the set of all neighbourhoods of a point in a topological space is a filter. When is it an ultrafilter?

3. Let $(x_n)_{n \in \mathbb{N}}$ be a sequence in X. Show that

$$\{\{x_m \mid m \geq n\} \mid n \in \mathbb{N}\}$$

is a filterbase on X and describe the filter it generates. Can you find explicitly a finer ultrafilter?

4. Let $\mathfrak{F}, \mathfrak{G}$ be filterbases on X such that $A \cap B \neq \emptyset$ for all $A \in \mathfrak{F}$ and $B \in \mathfrak{G}$. Show that there is an ultrafilter $\mathfrak{H}$ with $\mathfrak{F} \subset \mathfrak{H}$ and $\mathfrak{G} \subset \mathfrak{H}$.

5. Show that there are no non-trivial δ-stable ultrafilters on the countable set X.

6. Let $\mathfrak{F}$ be an ultrafilter. Put

$$\mu : \mathfrak{P}(X) \longrightarrow \mathbb{R}_+, \quad A \longmapsto \begin{cases} 1 & \text{if } A \in \mathfrak{F}, \\ 0 & \text{if } A \notin \mathfrak{F}. \end{cases}$$

Prove the following.

(a) μ is a positive content.
(b) μ is a positive measure if and only if $\mathfrak{F}$ is δ-stable.
(c) If μ is a positive measure and $f \in \mathbb{R}^X$, then $f \in \mathcal{L}^1(\ell_\mu)$ and $\int_{\ell_\mu} f = \alpha_{f,\mathfrak{F}}$.

7. Prove that the following are equivalent.

(a) Every δ-stable ultrafilter on X is trivial.
(b) There is no positive measure μ on $\mathfrak{R} := \mathfrak{P}(X)$ which takes precisely two values (say 0 and α with $\alpha > 0$) such that $\mu(\{x\}) = 0$ for every $x \in X$.

8. Show that the map g in Theorem 3.55 is uniquely determined.

9. Describe explicitly the map g in Theorem 3.55 for the functional ℓ_{δ_x} generated by the Dirac measure δ_x.

10. Let φ be a positive linear functional on $\mathbb{R}^X$. Describe $\mathcal{L}^1(\varphi)$, using the notation of Theorem 3.55.

11. Let $(X, \mathcal{L}, \ell)$ be a Daniell space. Show that the following are equivalent.

(a) $\mathcal{L}^1(\ell) = \overline{\mathbb{R}}^X$.
(b) $X \in \mathfrak{N}(\ell)$.
(c) $\ell = 0$.

3.10 Summary

Looking back over the construction of the integral, we can distinguish two stages, beginning with a Daniell space:

$$(X, \mathcal{L}, \ell) \xrightarrow{\text{1st stage}} (X, \overline{\mathcal{L}}(\ell), \overline{\ell}) \xrightarrow{\text{2nd stage}} (X, \mathcal{L}^1(\ell), \int_\ell).$$

The fundamental theorems on convergence, the monotone convergence theorem and the Lebesgue convergence theorem, are obtained during the first stage and are not improved during the second stage. This might create the impression that the second stage is of little importance. We have seen in the examples, however, that it is precisely in this second stage that any of the artificial conditions which may have appeared during the first stage are removed. Only in this manner do we obtain a satisfactory notion of the integral; this will be confirmed repeatedly in the following chapters.

There are other, more fundamental reasons for choosing $(X, \mathcal{L}^1(\ell), \int_\ell)$ as our integral. They are explained in detail in [CW], Chapter 4. We do not discuss these at this point.

The closure $(X, \overline{\mathcal{L}}(\ell), \overline{\ell})$ will still be useful to us when formulating certain propositions, but more especially in proofs.

4

Measure and integral

Until now our discussion has been general. We now turn to the single most important special case, namely that of functionals which are derived from a positive measure space $(X, \mathfrak{R}, \mu)$. We have seen that for each such measure space $(X, \mathfrak{R}, \mu)$ there is an associated Daniell space $(X, \mathcal{L}(\mathfrak{R}), \ell_\mu)$. We shall see that a by-product of the extension theory for Daniell spaces described in the previous chapter is an extension theory for positive measure spaces.

We note that the 'classical' approach to measure theory, due to Carathéodory, follows another path: given a positive measure space $(X, \mathfrak{R}, \mu)$, one extends μ to an 'outer measure' μ^* on $\mathfrak{P}(X)$ and then defines the set $\mathfrak{L}(\mu)$ of μ-integrable sets by a 'measurability condition' in terms of μ^*. The integral with respect to μ is then defined in terms of approximations by means of $\mathfrak{L}(\mu)$-step functions. This path can be followed to construct the same extension of μ and the same integral with respect to μ as we do. But the availability of the more general Daniell approach has led us to prefer to derive integration with respect to a measure as a special case of integration with respect to a null-continuous functional.

Let $(X, \mathfrak{R}, \mu)$ be a positive measure space. For convenience we shall use the following notation:

$$\overline{\mathcal{L}}(\mu) := \overline{\mathcal{L}}(\ell_\mu), \qquad \overline{\mathcal{N}}(\mu) := \overline{\mathcal{N}}(\ell_\mu), \qquad \overline{\mathfrak{N}}(\mu) := \overline{\mathfrak{N}}(\ell_\mu),$$
$$\mathcal{L}^1(\mu) := \mathcal{L}^1(\ell_\mu), \quad \mathcal{N}(\mu) := \mathcal{N}(\ell_\mu), \quad \mathfrak{N}(\mu) := \mathfrak{N}(\ell_\mu).$$

We write μ-a.e. for ℓ_μ-a.e. The elements of $\mathcal{L}^1(\mu)$ are called μ-**integrable functions**. The elements of $\mathcal{N}(\mu)$ are called μ-**null functions** and the elements of $\mathfrak{N}(\mu)$ μ-**null sets**. Given $f \in \mathcal{L}^1(\mu)$, define

$$\int f(x) \, d\mu(x) := \int f \, d\mu := \int_{\ell_\mu} f.$$

$\int f \, d\mu$ is called the μ-**integral of** f. In general, we shall write $\int f \, d\mu$ for the integral. But occasionally, especially when we are integrating with respect to two or more measures on different spaces at the same time – as, for example, in the case of product measures – it is convenient to distinguish explicitly the provenance of the integration. In such a case we shall write $\int f(x)\,d\mu(x)$ to indicate that f is considered as a function of $x \in X$ and μ is a measure on the same space X. If, for example, we are given a function f of two variables on a space $X \times Y$, it should be clear that $\int f(x, y)\,d\nu(y)$ indicates that ν is a measure on Y, that we are integrating with respect to the variable $y \in Y$ and that x is kept constant during this integration.

Given a set X, let $\mathfrak{R}$ and $\mathfrak{S}$ be subsets of $\mathfrak{P}(X)$ and $\mu : \mathfrak{R} \to \mathbb{R}$ and $\nu : \mathfrak{S} \to \mathbb{R}$ mappings. Define – as we did for functionals –

$$(X, \mathfrak{R}, \mu) \preccurlyeq (X, \mathfrak{S}, \nu)$$

if and only if $\mathfrak{R} \subset \mathfrak{S}$ and $\nu|_{\mathfrak{R}} = \mu$. In such a case $(X, \mathfrak{S}, \nu)$ is said to be an **extension** of $(X, \mathfrak{R}, \mu)$.

4.1 Extensions of positive measure spaces

Given a positive measure space $(X, \mathfrak{R}, \mu)$, define

$$\mathfrak{L}(\mu) := \{A \subset X \,|\, e_A \in \mathcal{L}^1(\mu)\}.$$

The elements of $\mathfrak{L}(\mu)$ are said to be **μ-integrable**. Given $A \in \mathfrak{L}(\mu)$, define

$$\mu^X(A) := \int e_A\,d\mu.$$

We next show that $(X, \mathfrak{L}(\mu), \mu^X)$ is an extension of $(X, \mathfrak{R}, \mu)$ which bears the same relationship to the original space $(X, \mathfrak{R}, \mu)$ as the space $(X, \mathcal{L}^1(\mu), \int \cdot\,d\mu)$ does to $(X, \mathcal{L}(\mathfrak{R}), \ell_\mu)$. The following results clearly indicate that we have reached one of the main aims of measure theory, namely extending a measure μ on a (possibly small) ring of sets with possibly rather poor properties to a measure (namely μ^X) on a larger class of sets (namely $\mathfrak{L}(\mu)$) with far better properties.

In Theorems 4.1–4.4, $(X, \mathfrak{R}, \mu)$ always denotes a positive measure space.

Theorem 4.1

(a) $(X, \mathfrak{L}(\mu), \mu^X)$ *is a positive measure space.*

(b) $(X, \mathfrak{R}, \mu) \preccurlyeq (X, \mathfrak{L}(\mu), \mu^X)$.

(c) $\mathfrak{L}(\mu)$ *is a δ-ring.*

(d) $\mathfrak{N}(\mu) = \{A \in \mathfrak{L}(\mu) \,|\, \mu^X(A) = 0\}$.

Proof. These properties are elementary consequences of those of $\mathcal{L}^1(\mu)$ and $\int \cdot\,d\mu$.

Take $A, B \in \mathfrak{L}(\mu)$. Then

$$e_{A \cup B} = e_A \vee e_B \in \mathcal{L}^1(\mu), \quad \text{so } A \cup B \in \mathfrak{L}(\mu).$$
$$e_{A \cap B} = e_A \wedge e_B \in \mathcal{L}^1(\mu), \quad \text{so } A \cap B \in \mathfrak{L}(\mu).$$
$$e_{A \setminus B} = e_A - e_{A \cap B} \in \mathcal{L}^1(\mu), \quad \text{so } A \setminus B \in \mathfrak{L}(\mu).$$

Moreover, $\emptyset \in \Re \subset \mathfrak{L}(\mu)$. If A and B are disjoint, then $e_{A \cup B} = e_A + e_B$, so that

$$\mu^X(A \cup B) = \int e_{A \cup B} d\mu = \int e_A d\mu + \int e_B d\mu = \mu^X(A) + \mu^X(B).$$

$A \subset B$ implies that $e_A \leq e_B$ and therefore

$$\mu^X(A) = \int e_A d\mu \leq \int e_B d\mu = \mu^X(B).$$

The null-continuity of μ^X follows from Theorem 4.2(a) whose proof follows (without, of course, making use of the present theorem).

If $(A_n)_{n \in \mathbb{N}}$ is an arbitrary sequence in $\mathfrak{L}(\mu)$, then – once again in anticipation of Theorem 4.2(a) –

$$\bigcap_{n \in \mathbb{N}} A_n = \downarrow \bigcap_{m \leq n} A_m \in \mathfrak{L}(\mu).$$

Thus $\mathfrak{L}(\mu)$ is a δ-ring. If $A \in \Re$, then $e_A \in L(\Re) \subset L^1(\mu)$ and

$$\mu^X(A) = \int e_A d\mu = \ell_\mu(e_A) = \mu(A).$$

(d) is a consequence of Theorem 3.30(c). $\qquad \square$

The next result corresponds to the monotone convergence theorem.

Theorem 4.2

(a) *Let* $(A_n)_{n \in \mathbb{N}}$ *be a decreasing sequence in* $\mathfrak{L}(\mu)$. *Then*

$$\bigcap_{n \in \mathbb{N}} A_n \in \mathfrak{L}(\mu) \quad and \quad \mu^X\left(\bigcap_{n \in \mathbb{N}} A_n\right) = \downarrow \mu^X(A_n).$$

(b) *If* $(A_n)_{n \in \mathbb{N}}$ *is an increasing sequence in* $\mathfrak{L}(\mu)$ *with*

$$\sup_{n \in \mathbb{N}} \mu^X(A_n) < \infty,$$

then

$$\bigcup_{n \in \mathbb{N}} A_n \in \mathfrak{L}(\mu) \quad and \quad \mu^X\left(\bigcup_{n \in \mathbb{N}} A_n\right) = \uparrow \mu^X(A_n).$$

Proof. (a) $\left(e_{A_n}\right)_{n \in \mathbb{N}}$ is a decreasing sequence in $L^1(\mu)$ and

$$e_{\bigcap_{n \in \mathbb{N}} A_n} = \downarrow e_{A_n} \geq 0.$$

By the monotone convergence theorem $\bigcap_{n \in \mathbb{N}} A_n \in \mathfrak{L}(\mu)$ and

$$\mu^X\left(\bigcap_{n \in \mathbb{N}} A_n\right) = \downarrow \int e_{A_n} d\mu = \downarrow \mu^X(A_n).$$

A similar argument proves (b). $\qquad \square$

Our next theorem is the analogue of the Lebesgue convergence theorem for $\mathfrak{L}(\mu)$.

Theorem 4.3 *Let $(A_n)_{n \in \mathbb{N}}$ be a sequence in $\mathfrak{L}(\mu)$. Suppose that $A :=$ $\lim_{n \to \infty} A_n$ in the complete lattice $\mathfrak{P}(X)$ and assume that there is a $B \in \mathfrak{L}(\mu)$ with $A_n \subset B$ for every $n \in \mathbb{N}$. Then*

$$A \in \mathfrak{L}(\mu) \quad and \quad \mu^X(A) = \lim_{n \to \infty} \mu^X(A_n).$$

Proof. The assertion follows of course from the Lebesgue convergence theorem. For given $n \in \mathbb{N}$, $e_{A_n} \in \mathcal{L}^1(\mu)$ and $e_A = \lim_{n \to \infty} e_{A_n}$. Moreover, $e_B \in \mathcal{L}^1(\mu)$ and $e_{A_n} \leq e_B$ for each $n \in \mathbb{N}$. Thus the hypotheses of the Lebesgue convergence theorem are fulfilled and so it follows that $e_A \in \mathcal{L}^1(\mu)$. But then $A \in \mathfrak{L}(\mu)$ and

$$\mu^X(A) = \int e_A d\mu = \lim_{n \to \infty} \int e_{A_n} d\mu = \lim_{n \to \infty} \mu^X(A_n).$$

$\square$

The next property is also of major importance.

Theorem 4.4 *Let $(A_n)_{n \in \mathbb{N}}$ be a sequence in $\mathfrak{L}(\mu)$ with $\sum_{n \in \mathbb{N}} \mu^X(A_n)$ $< \infty$. Then*

$$\bigcup_{n \in \mathbb{N}} A_n \in \mathfrak{L}(\mu) \quad and \quad \mu^X \left(\bigcup_{n \in \mathbb{N}} A_n \right) \leq \sum_{n \in \mathbb{N}} \mu^X(A_n).$$

If, in addition, the A_n are pairwise disjoint, then

$$\mu^X \left(\bigcup_{n \in \mathbb{N}} A_n \right) = \sum_{n \in \mathbb{N}} \mu^X(A_n).$$

Proof. Given $n \in \mathbb{N}$, define

$$B_n := \bigcup_{k \leq n} A_k \setminus \bigcup_{k < n} A_k.$$

$(B_n)_{n \in \mathbb{N}}$ is a disjoint sequence in $\mathfrak{L}(\mu)$ and

$$\bigcup_{n \in \mathbb{N}} A_n = \bigcup_{n \in \mathbb{N}} B_n = \uparrow \bigcup_{k \leq n} B_k.$$

But for each $n \in \mathbb{N}$, $B_n \subset A_n$ so that

$$\sup_{n \in \mathbb{N}} \mu^X \left(\bigcup_{k \leq n} B_k \right) = \sup_{n \in \mathbb{N}} \sum_{k \leq n} \mu^X(B_k) \leq \sum_{n \in \mathbb{N}} \mu^X(A_n) < \infty.$$

Theorem 4.2(b) implies that $\bigcup_{n \in \mathbb{N}} A_n = \bigcup_{n \in \mathbb{N}} B_n \in \mathfrak{L}(\mu)$ and

$$\mu^X \left(\bigcup_{n \in \mathbb{N}} A_n \right) = \uparrow \mu^X \left(\bigcup_{k \leq n} B_k \right) = \uparrow \sum_{k \leq n} \mu^X(B_k)$$

$$\leq \uparrow \sum_{k \leq n} \mu^X(A_k) = \sum_{n \in \mathbb{N}} \mu^X(A_n).$$

If the A_n are pairwise disjoint, then $B_n = A_n$ for each $n \in \mathbb{N}$ and therefore

$$\mu^* \left(\bigcup_{n \in \mathbb{N}} A_n \right) = \uparrow \sum_{k \leq n} \mu^*(B_k) = \uparrow \sum_{k \leq n} \mu^X(A_k) = \sum_{n \in \mathbb{N}} \mu^*(A_n).$$

□

This has important consequences, as the following corollary shows.

Corollary 4.5

(a) *If μ is a positive content on the ring of sets $\mathfrak{R} \subset \mathfrak{P}(X)$, then the following are equivalent.*

(a1) *$(X, \mathfrak{R}, \mu)$ is a positive measure space.*

(a2) *If $(A_n)_{n \in \mathbb{N}}$ is a disjoint sequence in $\mathfrak{R}$ for which $\bigcup_{n \in \mathbb{N}} A_n \in \mathfrak{R}$, then*

$$\mu \left(\bigcup_{n \in \mathbb{N}} A_n \right) = \sum_{n \in \mathbb{N}} \mu(A_n).$$

(b) *Let $(X, \mathfrak{R}, \mu)$ be a positive measure space. Take $A \in \mathfrak{R}$ and let $(A_n)_{n \in \mathbb{N}}$ be a sequence in $\mathfrak{R}$ such that $A \subset \bigcup_{n \in \mathbb{N}} A_n$. Then*

$$\mu(A) \leq \sum_{n \in \mathbb{N}} \mu(A_n).$$

Proof. (a1)⇒(a2) follows immediately from Theorem 4.4.

(a2)⇒(a1). Let $(A_n)_{n \in \mathbb{N}}$ be a decreasing sequence in $\mathfrak{R}$ with $\downarrow A_n = \emptyset$. Given $n \in \mathbb{N}$, put $B_n := A_n \setminus A_{n+1}$. Then $(B_n)_{n \in \mathbb{N}}$ is a disjoint sequence in $\mathfrak{R}$ and $\bigcup_{n \in \mathbb{N}} B_n = A_1 \in \mathfrak{R}$. It follows that

$$\mu(A_1) = \sum_{n \in \mathbb{N}} \mu(B_n) = \sum_{n \in \mathbb{N}} \left(\mu(A_n) - \mu(A_{n+1}) \right)$$

$$= \sup_{n \in \mathbb{N}} \sum_{k=1}^{n} \left(\mu(A_k) - \mu(A_{k+1}) \right) = \sup_{n \in \mathbb{N}} \left(\mu(A_1) - \mu(A_{n+1}) \right)$$

$$= \mu(A_1) - \downarrow \mu(A_n).$$

Thus $\downarrow \mu(A_n) = 0$ and so $(X, \mathfrak{R}, \mu)$ is a positive measure space.

(b) We may assume that $\sum_{n \in \mathbb{N}} \mu(A_n) < \infty$. Then, by Theorem 4.4,

$$\mu(A) \leq \sum_{n \in \mathbb{N}} \mu(A \cap A_n) \leq \sum_{n \in \mathbb{N}} \mu(A_n).$$

□

The property of positive measure spaces described in (a2) is called **σ-additivity**. Positive measure spaces are often defined by σ-additivity instead of null-continuity. Corollary 4.5 shows that both approaches lead to the same concept of a measure space.

We call property (b) **σ-subadditivity**. Every positive measure is σ-subadditive.

The next proposition contains criteria for a function or a set to be null which shall prove useful in several places later in our investigations.

Proposition 4.6 *Let $(X, \mathfrak{R}, \mu)$ be a positive measure space. Then*

(a) For each $f \in \overline{\mathbb{R}}^X$, the following are equivalent.

(a1) $f \in \mathcal{N}(\mu)$.

(a2) $fe_A \in \overline{\mathcal{N}}(\mu)$ for every $A \in \mathfrak{R}$.

(b) For each $B \subset X$, the following are equivalent.

(b1) $B \in \mathfrak{N}(\mu)$.

(b2) $B \cap A \in \overline{\mathfrak{N}}(\mu)$ for every $A \in \mathfrak{R}$.

Proof. (a1)$\Rightarrow$(a2) is trivial.

(a2)$\Rightarrow$(a1). Take $A \in \mathfrak{R}(\ell_\mu)$. There is a $g \in \overline{\mathcal{L}}(\mu)_+$ with $A = \{g > 0\}$. Furthermore, there is a $g'' \in \mathcal{L}(\mathfrak{R})^\uparrow$, $g'' \geq g$, and there is an increasing sequence $(g_n)_{n \in \mathbb{N}}$ in $\mathcal{L}(\mathfrak{R})$ for which $g'' = \uparrow g_n$. We have

$$A \subset \{g'' > 0\} = \uparrow \{g_n > 0\},$$

and $\{g_n > 0\} \in \mathfrak{R}$ for every $n \in \mathbb{N}$. Then

$$|f|e_A \leq \uparrow |f|e_{\{g_n > 0\}},$$

and, by hypothesis, $fe_{\{g_n > 0\}} \in \overline{\mathcal{N}}(\mu)$ for every $n \in \mathbb{N}$. Therefore $|f|e_{\{g_n > 0\}} \in \overline{\mathcal{N}}(\mu)$ for every $n \in \mathbb{N}$. We conclude that $\uparrow |f|e_{\{g_n > 0\}} \in \overline{\mathcal{N}}(\mu)$. It follows that $|f|e_A \in \overline{\mathcal{N}}(\mu)$ and so, finally, $fe_A \in \overline{\mathcal{N}}(\mu)$.

(b) follows from (a). $\square$

We have achieved a substantial simplification by needing only consider the behaviour of sets in $\mathfrak{R}$ to determine whether a function or set is null. The preceding proposition shows clearly that our concept of null set is a 'local' concept. For a set to be a null set it suffices that it is 'locally' a null set, i.e. that its intersection with every element of $\mathfrak{R}$ is a null set.

Our next proposition provides another convenient characterization of μ-null sets. Observe that this characterization does not depend on any extension of μ.

Proposition 4.7 *Let $(X, \mathfrak{R}, \mu)$ be a positive measure space. Then for each $B \subset X$ the following are equivalent.*

(a) $B \in \mathfrak{N}(\mu)$.

(b) Given $A \in \mathfrak{R}$ and $\varepsilon > 0$, there is a sequence $(A_n)_{n \in \mathbb{N}}$ in $\mathfrak{R}$ such that $A \cap B \subset \bigcup_{n \in \mathbb{N}} A_n$ and $\sum_{n \in \mathbb{N}} \mu(A_n) < \varepsilon$.

Proof. (a)⇒(b). Take $A \in \mathfrak{R}$ and $\varepsilon > 0$. Since $A \cap B \in \overline{\mathfrak{N}}(\mu)$, we see that $e_{A \cap B} \in \overline{\mathcal{L}}(\mu)$ and $\overline{\ell_\mu}(e_{A \cap B}) = 0$. Thus there is a $g \in \mathcal{L}(\mathfrak{R})^\uparrow$ with $g \geq e_{A \cap B}$ and $\ell_\mu^\uparrow(g) < \frac{\varepsilon}{3}$. Moreover, there is a sequence $(g_n)_{n \in \mathbb{N}}$ in $\mathcal{L}(\mathfrak{R})_+$ with $g_n \uparrow g$. Given $n \in \mathbb{N}$, put $B_n := \{g_n \geq \frac{1}{2}\}$. Then $(B_n)_{n \in \mathbb{N}}$ is an increasing sequence in $\mathfrak{R}$ such that $A \cap B \subset \bigcup_{n \in \mathbb{N}} B_n$ and, since $e_{B_n} \leq 2g_n \leq 2g$, we see that $\mu(B_n) \leq 2 \int g d\mu < \frac{2}{3}\varepsilon$ for every $n \in \mathbb{N}$.

Define $A_1 := B_1$ and $A_n := B_n \setminus B_{n-1}$ whenever $n > 1$. Then $\bigcup_{n \in \mathbb{N}} A_n = \bigcup_{n \in \mathbb{N}} B_n$. Moreover, since $\mu(B_n) = \sum_{k=1}^n \mu(A_k)$ for every $n \in \mathbb{N}$,

$$\sum_{n \in \mathbb{N}} \mu(A_n) = \sup_{n \in \mathbb{N}} \mu(B_n) \leq \frac{2}{3}\varepsilon < \varepsilon.$$

(b)⇒(a). Take $A \in \mathfrak{R}$ and $\varepsilon > 0$. Let $(A_n)_{n \in \mathbb{N}}$ be a sequence satisfying condition (b). Then $(0, e_{\bigcup_{n \in \mathbb{N}} A_n})$ is an ε-bracket for $e_{A \cap B}$. Since ε is arbitrary, it follows that $e_{A \cap B} \in \overline{\mathcal{N}}(\mu)$. But A is also arbitrary and so, by Proposition 4.6(b), $B \in \mathfrak{N}(\mu)$. □

The spaces $\mathcal{L}^1(\mu)$ of integrable functions derived from positive measure spaces have an important property known as the **Stone property**, which is described in the next theorem.

Theorem 4.8 *Let $(X, \mathfrak{R}, \mu)$ be a positive measure space. Then $f \wedge \alpha \in \mathcal{L}^1(\mu)$ for all $f \in \mathcal{L}^1(\mu)$ and $\alpha \in \mathbb{R}$, $\alpha > 0$.*

Proof. We apply the induction principle. Let $\mathcal{F}$ be the set of all $f \in \mathcal{L}^1(\mu)$ with the property described. By Proposition 2.16, $\mathcal{L}(\mathfrak{R}) \subset \mathcal{F}$. Take $\alpha > 0$ and let $(f_n)_{n \in \mathbb{N}}$ be a monotone sequence in $\mathcal{F}$ with $\lim_{n \to \infty} \int f_n d\mu \notin \{\infty, -\infty\}$. Then $(f_n \wedge \alpha)_{n \in \mathbb{N}}$ is a monotone sequence in $\mathcal{L}^1(\mu)$. If $(f_n)_{n \in \mathbb{N}}$ is increasing, then $\int f_n \wedge \alpha d\mu \leq \int f_n d\mu$ for every $n \in \mathbb{N}$ and hence $\uparrow \int f_n \wedge \alpha d\mu < \infty$. If $(f_n)_{n \in \mathbb{N}}$ is decreasing, then

$$\int f_n \wedge \alpha d\mu \geq \int f_n d\mu - \int |f_1| d\mu,$$

so that $\downarrow \int f_n \wedge \alpha d\mu > -\infty$. The monotone convergence theorem implies that $\lim_{n \to \infty}(f_n \wedge \alpha) \in \mathcal{L}^1(\mu)$. Thus $\lim_{n \to \infty} f_n \in \mathcal{F}$, since

$$\lim_{n \to \infty}(f_n \wedge \alpha) = \left(\lim_{n \to \infty} f_n\right) \wedge \alpha.$$

Finally, if $f \in \mathcal{F}$, $g \in \mathcal{L}^1(\mu)$ and $g = f$ μ-a.e., then $f \wedge \alpha = g \wedge \alpha$ μ-a.e. Hence $g \wedge \alpha \in \mathcal{L}^1(\mu)$ and $g \in \mathcal{F}$. Theorem 3.39 now implies that $\mathcal{F} = \mathcal{L}^1(\mu)$. □

Corollary 4.9 *Let $(X, \mathfrak{R}, \mu)$ be a positive measure space. Then*

$$\{f > \alpha\}, \ \{f \geq \alpha\} \ and \ \{f = \alpha\}$$

belong to $\mathfrak{L}(\mu)$ for all $f \in \mathcal{L}^1(\mu)$ and $\alpha \in \mathbb{R}$, $\alpha > 0$.

Proof. We may suppose that $f(x) > -\infty$ for every $x \in X$. The corollary then follows from Theorem 4.8, the monotone convergence theorem and the following relationships:

$$e_{\{f>\alpha\}} = \uparrow \left(n(f - f \wedge \alpha) \wedge 1 \right) \quad \text{and} \quad e_{\{f>\alpha\}} \le (1/\alpha)|f|;$$
$$e_{\{f\ge\alpha\}} = \downarrow e_{\{f>\alpha-1/n\}};$$
$$e_{\{f=\alpha\}} = e_{\{f\ge\alpha\}} - e_{\{f>\alpha\}}.$$

$\square$

Our next proposition describes the approximation of (arbitrary) positive functions by positive step functions. We strongly recommend that the reader sketch the first three or four members of the sequence $(f_n)_{n\in\mathbb{N}}$ mentioned in this proposition.

Proposition 4.10 *Take $f \in \overline{\mathbb{R}}_+^X$, and for each $n \in \mathbb{N}$ define*

$$f_n := \frac{1}{2^n} \sum_{k=1}^{n2^n} e_{\{f\ge k/2^n\}}.$$

Then $0 \le f_n \uparrow f$.

Proof. Take $x \in X$. If $f(x) = \infty$, then $f_n(x) = n$ for every $n \in \mathbb{N}$. It follows that $f_n(x) \uparrow f(x)$.

Now suppose that $f(x) < \infty$. Take $n > f(x)$. There is a unique $k \in \mathbb{N}$ with $k/2^n \le f(x) < (k+1)/2^n$. Then $f_n(x) = k/2^n$ and hence $0 \le f(x) - f_n(x) < 1/2^n$. It follows that $f(x) = \lim_{n\to\infty} f_n(x)$. To show that $(f_n(x))_{n\in\mathbb{N}}$ is an increasing sequence, take $n \in \mathbb{N}$. If $f(x) \ge k/2^n$ for some k with $1 \le k \le n2^n$, then $f(x) \ge 2k/2^{n+1}$ and $1 \le 2k \le n2^{n+1}$. Thus there are at least twice as many j for which $f(x) \ge j/2^{n+1}$ and $1 \le j \le n2^{n+1}$ as there are k for which $f(x) \ge k/2^n$ and $1 \le k \le n2^n$. It follows that

$$f_n(x) = \frac{1}{2^n} \sum_{k=1}^{n2^n} e_{\{f\ge k/2^n\}} \le \frac{1}{2} \cdot \frac{1}{2^n} \sum_{j=1}^{n2^{n+1}} e_{\{f\ge j/2^{n+1}\}} \le f_{n+1}(x).$$

$\square$

Restricting attention to integrable functions and using the decomposition $f = f^+ - f^-$, the next corollary follows immediately from Proposition 4.10 and Corollary 4.9.

Corollary 4.11 *Let $(X, \mathfrak{R}, \mu)$ be a positive measure space and take $f \in \mathcal{L}^1(\mu)$. Then there is a sequence $(f_n)_{n\in\mathbb{N}}$ of $\mathfrak{L}(\mu)$-step functions converging pointwise to f. Moreover, if $f \ge 0$, then the sequence can be chosen as in Proposition 4.10.*

We saw in Theorem 4.8 that the spaces $\mathcal{L}^1(\mu)$ enjoy the Stone property. It is natural to ask whether *every* space $\mathcal{L}^1(\ell)$ derived from a Daniell space $(X, \mathcal{L}, \ell)$ possesses this property. A simple example, contained in the exercises, shows that this is not the case. The next question is how we can characterize those spaces $(X, \mathcal{L}^1(\ell), \int_\ell)$ for which $\mathcal{L}^1(\ell)$ has the Stone property. Our next theorem provides the answer: if $\mathcal{L}^1(\ell)$ has the Stone property, then there is a measure μ which generates the same integral as ℓ. In other words, of all the spaces $\mathcal{L}^1(\ell)$, it is the spaces $\mathcal{L}^1(\mu)$ derived from positive measures which exhibit the Stone property.

Theorem 4.12 *Let $(X, \mathcal{L}, \ell)$ be a Daniell space such that $f \wedge \alpha \in \mathcal{L}^1(\ell)$ whenever $f \in \mathcal{L}^1(\ell)$ and $\alpha > 0$. Define*

$$\mathfrak{R} := \{A \subset X \mid e_A \in \mathcal{L}^1(\ell)\},$$

$$\mu : \mathfrak{R} \longrightarrow \mathbb{R}, \quad A \longmapsto \int_\ell e_A.$$

Then:

(a) $(X, \mathfrak{R}, \mu)$ is a positive measure space, and $\mathfrak{R}$ is a δ-ring.

(b) $(X, \mathcal{L}^1(\mu), \int \cdot \, d\mu) = (X, \mathcal{L}^1(\ell), \int_\ell)$.

Proof. (a) This can be proved similarly to Theorem 4.1.

(b) Since the crucial step is to show that every μ-null set is also an ℓ-null set, we begin by proving this. Take $B \in \mathfrak{N}(\mu)$ and $A \in \mathfrak{R}$. We first show that $A \cap B \in \mathfrak{N}(\ell)$. Given $k \in \mathbb{N}$, Proposition 4.7 ensures that there is a sequence $(B_{kn})_{n\in\mathbb{N}}$ in $\mathfrak{R}$ such that

$$A \cap B \subset \bigcup_{n\in\mathbb{N}} B_{kn} \subset A \quad \text{and} \quad \sum_{n\in\mathbb{N}} \mu(B_{kn}) < \frac{1}{k}.$$

Then $\sum_{n\in\mathbb{N}} \int_\ell e_{B_{kn}} < \frac{1}{k}$. By Theorem 3.34,

$$\sum_{n\in\mathbb{N}} e_{B_{kn}} \in \mathcal{L}^1(\ell) \quad \text{and} \quad \int_\ell \left(\sum_{n\in\mathbb{N}} e_{B_{kn}} \right) < \frac{1}{k}.$$

Put $B_k := \bigcup_{n\in\mathbb{N}} B_{kn}$. Then, by the Stone property,

$$e_{B_k} = \left(\sum_{n\in\mathbb{N}} e_{B_{kn}} \right) \wedge 1 \in \mathcal{L}^1(\ell) \quad \text{and} \quad \int_\ell e_{B_k} < \frac{1}{k}.$$

Put $f := \lim_{k\to\infty} e_{B_k}$. Then the Lebesgue convergence theorem (Theorem 3.33) implies that $f \in \mathcal{N}(\ell)$. Since $0 \leq e_{A\cap B} \leq f$, it follows that $A \cap B \in \mathfrak{N}(\ell)$.

Now take $A \in \mathfrak{R}(\ell)$. Then $A = \{h > 0\}$ for some $h \in \overline{\mathcal{L}}(\ell)_+$. Given $m \in \mathbb{N}$, put $A_m := \{h \geq 1/m\}$. Then, arguing as in Corollary 4.9, we see that $A_m \in \mathfrak{R}$. Thus $A_m \cap B \in \mathfrak{N}(\ell)$ for every $m \in \mathbb{N}$, by what we have already proved. Since $A = \bigcup_{m\in\mathbb{N}} A_m$, we conclude that $A \cap B \in \mathfrak{N}(\ell)$ as well. Hence, by Theorem 3.30(f), $A \cap B \in \overline{\mathfrak{N}}(\ell)$. Since A was arbitrary, it follows that $B \in \mathfrak{N}(\ell)$.

We now prove assertion (b). For '$\preccurlyeq$', we use the induction principle (Theorem 3.39). Put

$$\mathcal{F} := \left\{ f \in \mathcal{L}^1(\mu) \,\middle|\, f \in \mathcal{L}^1(\ell) \text{ and } \int f \, d\mu = \int_\ell f \right\}.$$

Then $\mathcal{L}(\mathfrak{R}) \subset \mathcal{F}$. By the monotone convergence theorem, condition (ii) of Theorem 3.39 is fulfilled. Finally, take $f \in \overline{\mathcal{L}}(\mu) \cap \mathcal{F}$ and $g \in \mathcal{L}^1(\mu)$ with $g = f$ μ-a.e. By the first part, $g = f$ ℓ-a.e. Hence $g \in \mathcal{F}$ (Theorem 3.30(b)). Theorem 3.39 now implies that $\mathcal{F} = \mathcal{L}^1(\mu)$.

To complete the proof of (b) we must show that $\mathcal{L}^1(\ell) \subset \mathcal{L}^1(\mu)$. Take $f \in \mathcal{L}^1(\ell)_+$. By the Stone property the set $\{f \geq \alpha\}$ belongs to $\mathfrak{R}$ for each $\alpha > 0$. Thus the functions f_n defined in Proposition 4.10 are μ-integrable and

$$\sup_{n \in \mathbb{N}} \int f_n \, d\mu = \sup_{n \in \mathbb{N}} \int_\ell f_n \leq \int_\ell f < \infty.$$

The monotone convergence theorem implies that $f \in \mathcal{L}^1(\mu)$. Since $f = f^+ - f^-$ for any $f \in \mathcal{L}^1(\ell)$, we see that $\mathcal{L}^1(\ell) \subset \mathcal{L}^1(\mu)$. □

It follows from the definition of the integral that for each μ-integrable set A there is a function $f \in \overline{\mathcal{L}}(\mu)$ with $e_A = f$ μ-a.e. The next proposition asserts that this f can be chosen to be a characteristic function as well.

Proposition 4.13 *Let $(X, \mathfrak{R}, \mu)$ be a positive measure space. Take $A \subset X$. Then the following are equivalent.*

(a) $A \in \mathfrak{L}(\mu)$.

(b) There is a $B \subset A$ with $e_B \in \overline{\mathcal{L}}(\mu)$ and $A \setminus B \in \mathfrak{N}(\mu)$.

Proof. Take $A \in \mathfrak{L}(\mu)$. Then there is an $f \in \overline{\mathcal{L}}(\mu)_+$ such that $e_A = f$ μ-a.e. Put $B := \{f = 1\} \cap A$. By Corollary 4.9, $B \in \mathfrak{L}(\mu)$. Hence $e_B \in \mathcal{L}^1(\mu)$ and $\{e_B \neq 0\} \subset \{f \neq 0\}$. Thus by Theorem 3.30(f), $e_B \in \overline{\mathcal{L}}(\mu)$. Clearly $A \setminus B \in \mathfrak{N}(\mu)$, completing the proof that (a)$\Rightarrow$(b). The converse is trivial. □

Theorem 4.25 below is another approximation theorem for integrable sets. It is formulated in terms of the original ring of sets $\mathfrak{R}$.

In order to simplify the statement of another important property, define for each positive measure space $(X, \mathfrak{R}, \mu)$,

$$\mathfrak{M}(\mu) := \{A \subset X \mid A \cap B \in \mathfrak{L}(\mu) \text{ for all } B \in \mathfrak{L}(\mu)\}.$$

The sets in $\mathfrak{M}(\mu)$ are called **μ-measurable**.

Proposition 4.14

(a) $\mathfrak{M}(\mu)$ is a σ-algebra on X.

(b) $\mathfrak{L}(\mu) \subset \mathfrak{M}(\mu)$.

Proof. (a) follows from the definition of $\mathfrak{M}(\mu)$ since, according to Theorem 4.1(c), $\mathfrak{L}(\mu)$ is a δ-ring.

(b) follows from the fact that $\mathfrak{L}(\mu)$ is a ring of sets. □

Theorem 4.15 *Let $(X, \mathfrak{R}, \mu)$ be a positive measure space. Take $f \in \mathcal{L}^1(\mu)$ and $A \in \mathfrak{M}(\mu)$. Then $fe_A \in \mathcal{L}^1(\mu)$.*

Proof. Take $A \in \mathfrak{M}(\mu)$ and put

$$\mathcal{F} := \{f \in \mathcal{L}^1(\mu) \mid fe_A \in \mathcal{L}^1(\mu)\}.$$

Assume that $f = \sum_{\iota \in I} \alpha_\iota e_{A_\iota} \in \mathcal{L}(\mathfrak{R})$. Note that $A_\iota \in \mathfrak{R} \subset \mathfrak{L}(\mu)$, so that $A_\iota \cap A \in \mathfrak{L}(\mu)$ and $e_{A_\iota \cap A} \in \mathcal{L}^1(\mu)$ for all $\iota \in I$. Then

$$fe_A = \sum_{\iota \in I} \alpha_\iota e_{A_\iota \cap A} \in \mathcal{L}^1(\mu)$$

and hence $\mathcal{L}(\mathfrak{R}) \subset \mathcal{F}$.

Let $(f_n)_{n \in \mathbb{N}}$ be a monotone sequence in $\mathcal{F}$ such that $\lim_{n \to \infty} \int f_n d\mu$ is finite. Then $(f_n e_A)_{n \in \mathbb{N}}$ is a monotone sequence in $\mathcal{L}^1(\mu)$. If $(f_n)_{n \in \mathbb{N}}$ is increasing, then for each $n \in \mathbb{N}$, $\int f_n e_A \, d\mu \leq \int |f_1| \, d\mu + \int f_n d\mu$ and hence $\uparrow \int f_n e_A \, d\mu < \infty$. If $(f_n)_{n \in \mathbb{N}}$ is decreasing, then given $n \in \mathbb{N}$, $\int f_n e_A \, d\mu \geq \int -|f_1| \, d\mu + \int f_n d\mu$ and hence $\downarrow \int f_n e_A \, d\mu > -\infty$. By the monotone convergence theorem,

$$\left(\lim_{n \to \infty} f_n \right) e_A = \lim_{n \to \infty} (f_n e_A) \in \mathcal{L}^1(\mu).$$

Thus $\lim_{n \to \infty} f_n \in \mathcal{F}$.

Given $f \in \mathcal{F}$ and $g \in \mathcal{L}^1(\mu)$, if $f = g$ μ-a.e., then $fe_A = ge_A$ μ-a.e. It follows that $ge_A \in \mathcal{L}^1(\mu)$. Thus g is also in $\mathcal{F}$. The induction principle (Theorem 3.39) now implies that $\mathcal{F} = \mathcal{L}^1(\mu)$. □

Given $f \in \mathcal{L}^1(\mu)$ and $A \in \mathfrak{M}(\mu)$ we define

$$\int_A f \, d\mu := \int fe_A d\mu.$$

$\int_A f \, d\mu$ is called the **integral of f on A**.

We now turn to the problem of finding sufficient conditions for two positive measure spaces to generate the same integral.

Theorem 4.16 *Let $\mathfrak{R}$ be a ring of sets. Let $(X, \mathfrak{S}_1, \mu_1)$ and $(X, \mathfrak{S}_2, \mu_2)$ be positive measure spaces satisfying one of the two conditions below.*

(a) $\mathfrak{R} \subset \mathfrak{S}_1 \subset \mathfrak{R}_\delta$ and $\mathfrak{R} \subset \mathfrak{S}_2 \subset \mathfrak{R}_\delta$ and $\mu_1|_\mathfrak{R} = \mu_2|_\mathfrak{R}$.

(b) $(X, \mathfrak{S}_1, \mu_1) \preccurlyeq \left(X, (\mathfrak{S}_2)_\delta, {\mu_2}^X \big|_{(\mathfrak{S}_2)_\delta} \right)$ and
   $(X, \mathfrak{S}_2, \mu_2) \preccurlyeq \left(X, (\mathfrak{S}_1)_\delta, {\mu_1}^X \big|_{(\mathfrak{S}_1)_\delta} \right)$.

Then

$$\left(X, \mathcal{L}^1(\mu_1), \int \cdot \, d\mu_1 \right) = \left(X, \mathcal{L}^1(\mu_2), \int \cdot \, d\mu_2 \right)$$

and

$$(X, \mathfrak{L}(\mu_1), {\mu_1}^X) = (X, \mathfrak{L}(\mu_2), {\mu_2}^X).$$

Proof. (a) Define $\ell_1 := \ell_{\mu_1}|_{\mathcal{L}(\mathfrak{R})}$ and $\ell_2 := \ell_{\mu_2}|_{\mathcal{L}(\mathfrak{R})}$. It follows from the hypotheses that

$$(X, \mathcal{L}(\mathfrak{R}), \ell_1) \preccurlyeq (X, \mathcal{L}(\mathfrak{S}_2), \ell_{\mu_2})$$

and

$$(X, \mathcal{L}(\mathfrak{R}), \ell_2) \preccurlyeq (X, \mathcal{L}(\mathfrak{S}_1), \ell_{\mu_1}).$$

Then, using Theorem 3.40,

$$(X, \overline{\mathcal{L}}(\ell_1), \overline{\ell_1}) \preccurlyeq (X, \overline{\mathcal{L}}(\mu_2), \overline{\ell_{\mu_2}})$$

and

$$(X, \overline{\mathcal{L}}(\ell_2), \overline{\ell_2}) \preccurlyeq (X, \overline{\mathcal{L}}(\mu_1), \overline{\ell_{\mu_1}}).$$

Note that

$$\mathfrak{R}_\delta \subset \{A \subset X \,|\, e_A \in \overline{\mathcal{L}}(\ell_1)\} \cap \{A \subset X \,|\, e_A \in \overline{\mathcal{L}}(\ell_2)\}.$$

Thus

$$(X, \mathcal{L}(\mathfrak{S}_1), \ell_{\mu_1}) \preccurlyeq (X, \overline{\mathcal{L}}(\mu_2), \overline{\ell_{\mu_2}})$$

and

$$(X, \mathcal{L}(\mathfrak{S}_2), \ell_{\mu_2}) \preccurlyeq (X, \overline{\mathcal{L}}(\mu_1), \overline{\ell_{\mu_1}}).$$

The claim now follows by Theorem 3.41.

(b) $\{A \subset X \,|\, e_A \in \overline{\mathcal{L}}(\mu_2)\}$ is a δ-ring which contains $\mathfrak{S}_2$ and hence also $(\mathfrak{S}_2)_\delta$. It follows that

$$(X, \mathcal{L}(\mathfrak{S}_1), \ell_{\mu_1}) \preccurlyeq (X, \overline{\mathcal{L}}(\mu_2), \overline{\ell_{\mu_2}}).$$

Similarly,

$$(X, \mathcal{L}(\mathfrak{S}_2), \ell_{\mu_2}) \preccurlyeq (X, \overline{\mathcal{L}}(\mu_1), \overline{\ell_{\mu_1}}).$$

The claim now follows by Theorem 3.41 again. $\square$

We commenced with a positive measure space $(X, \mathfrak{R}, \mu)$ and obtained the integral $(X, \mathcal{L}^1(\mu), \int \cdot \, d\mu)$. But $(X, \mathfrak{L}(\mu), \mu^X)$ is also a positive measure space. What happens when we apply our extension procedure to this space? Do we obtain a further extension of $(X, \mathcal{L}^1(\mu), \int \cdot \, d\mu)$? Fortunately not, as the following theorem asserts. We say 'fortunately', because otherwise our integral $(X, \mathcal{L}^1(\mu), \int \cdot \, d\mu)$ would still not have been 'complete'.

Theorem 4.17 *Let $(X, \mathfrak{R}, \mu)$ be a positive measure space. Then μ and μ^X generate the same integral on X, that is*

$$\left(X, \mathcal{L}^1(\mu), \int \cdot \, d\mu\right) = \left(X, \mathcal{L}^1(\mu^X), \int \cdot \, d(\mu^X)\right).$$

Proof. The assertion follows immediately by applying Theorem 4.12 to the Daniell space $(X, \mathcal{L}(\mathfrak{R}), \ell_\mu)$. $\qquad\qquad\qquad\qquad\qquad\qquad\qquad\qquad\qquad\qquad\quad$ □

The positive measure space $(X, \mathfrak{R}, \mu)$ is called **σ-finite** if there is a sequence $(A_n)_{n\in\mathbb{N}}$ in $\mathfrak{R}$ with $X = \bigcup_{n\in\mathbb{N}} A_n$. If $(X, \mathfrak{R}, \mu)$ is σ-finite, then clearly $(X, \mathcal{L}(\mathfrak{R}), \ell_\mu)$ is also σ-finite. This, together with Theorem 3.37, immediately establishes our next theorem.

Theorem 4.18 *For each σ-finite positive measure space $(X, \mathfrak{R}, \mu)$,*

$$\left(X, \mathcal{L}^1(\mu), \int \cdot\, d\mu\right) = (X, \overline{\mathcal{L}}(\mu), \overline{\ell_\mu}).$$

A particular application of this theorem is to Stieltjes measures.

Some comments on the restriction of positive measures are in order here. The reader has undoubtedly met functions which are defined only on intervals of $\mathbb{R}$ – or more generally on such subsets of $\mathbb{R}^n$ as closed n-dimensional rectangular prisms – but can nevertheless be integrated. But our discussion has considered Stieltjes measures, for example, only on all of $\mathbb{R}$ (see Section 2.4). We now show that this is sufficient. The theory of integration on measurable subsets follows satisfactorily from the general theory by means of the operation of 'restriction'.

We consider a positive measure space $(X, \mathfrak{R}, \mu)$. Let A be a μ-measurable subset of X. For each $f \in \overline{\mathbb{R}}^X$, let $f|_A$ denote, as usual, the restriction of f to A. We define

$$\mathcal{L}(\mu)|_A := \{B \in \mathcal{L}(\mu) \mid B \subset A\},$$
$$\mathcal{L}^1(\mu)|_A := \{f|_A \mid f \in \mathcal{L}^1(\mu)\},$$
$$\mu^X|_A(B) := \mu^X(B) \quad \text{whenever } B \in \mathcal{L}(\mu)|_A,$$

and

$$\ell_A(f|_A) := \int f e_A d\mu \quad \text{whenever } f \in \mathcal{L}^1(\mu).$$

(Note that the last definition is justified by Theorem 4.15.)

$(A, \mathcal{L}(\mu)|_A, \mu^X|_A)$ is called the **restriction** of $(X, \mathcal{L}(\mu), \mu^X)$ to A. Theorem 4.19 describes integration with respect to such restriction: the $\mu^X|_A$-integrable functions are precisely the restrictions to A of the μ-integrable functions. We thus obtain a very satisfactory concept of integration for the restriction to arbitrary measurable sets. Of course the restrictions to the n-dimensional rectangular prisms mentioned above are an example.

Theorem 4.19 *Let A be a μ-measurable subset of X, where $(X, \mathfrak{R}, \mu)$ is a positive measure space. Then*

$$\left(A, \mathcal{L}^1(\mu)|_A, \ell_A\right) = \left(A, \mathcal{L}^1(\mu^X|_A), \int \cdot\, d(\mu^X|_A)\right).$$

Proof. Let $\mathcal{F}$ be the set of all $f \in \mathcal{L}^1(\mu^X|_A)$ such that $f \in \mathcal{L}^1(\mu)|_A$ and

$$\ell_A(f) = \int f\,d(\mu^X|_A).$$

It is immediate from the definition that $\mathcal{F}$ contains every $\mathfrak{L}(\mu)|_A$-step function. It is also easy to see that if $(f_n)_{n\in\mathbb{N}}$ is a monotone sequence in $\mathcal{F}$ for which $\left(\int f_n d(\mu^X|_A)\right)_{n\in\mathbb{N}}$ is bounded, then its limit is also in $\mathcal{F}$. Finally,

$$\mathfrak{N}(\mu^X|_A) = \{B \in \mathfrak{N}(\mu) \mid B \subset A\}.$$

Since '$\supset$' is trivial, it only remains to prove the opposite inclusion. Take $B \in \mathfrak{N}(\mu^X|_A)$ and $C \in \mathfrak{R}$. Then $B \cap C \subset A \cap C \in \mathfrak{L}(\mu)|_A$ and hence $B \cap C \in \overline{\mathfrak{N}}(\mu^X|_A)$ by Theorem 3.30(f). Consequently, given $\varepsilon > 0$, there is an $f \in \mathcal{L}\big(\mathfrak{L}(\mu)|_A\big)^{\uparrow}$ such that

$$e_{B\cap C} \le f \quad\text{and}\quad \int f\,d(\mu^X|_A) < \varepsilon.$$

By the argument above, $f \in \mathcal{L}^1(\mu)|_A$ and $\ell_A(f) < \varepsilon$, which implies that $\mu^X(B \cap C) < \varepsilon$. Since ε is arbitrary, $B \cap C \in \overline{\mathfrak{N}}(\mu)$. Hence, by Proposition 4.6(b), $B \in \mathfrak{N}(\mu)$.

The induction principle (Theorem 3.39) implies that $\mathcal{F} = \mathcal{L}^1(\mu^X|_A)$.

It remains to show that $\mathcal{L}^1(\mu)|_A \subset \mathcal{L}^1(\mu^X|_A)$. This also follows by the induction principle: let $\mathcal{F}$ be the set of all $f \in \mathcal{L}^1(\mu)$ for which $f|_A \in \mathcal{L}^1(\mu^X|_A)$ and

$$\int f|_A\,d(\mu^X|_A) = \int_A f\,d\mu,$$

bearing in mind the above characterization of $\mu^X|_A$-null sets. □

An important application of the restriction concept is Lebesgue measure λ on $\mathfrak{J}$ restricted to a λ-measurable subset A of $\mathbb{R}$. We put $\lambda_A := \lambda^{\mathbb{R}}|_A$ and call λ_A **Lebesgue measure on A**. By Theorem 4.19 a function $f \in \overline{\mathbb{R}}^A$ is Lebesgue integrable on A (i.e. with respect to λ_A) if and only if it is the restriction to A of a Lebesgue integrable function $g \in \overline{\mathbb{R}}^{\mathbb{R}}$, and in this case $\int f\,d\lambda_A = \int g e_A d\lambda$, which is as it should be.

We can now describe the relationship of the Riemann integral to the Lebesgue integral, which is that the Daniell extension procedure applied to the Riemann integral yields precisely the Lebesgue integral! In the following discussion $A := [a, b]$ is a compact interval in $\mathbb{R}$, $\mathcal{R}$ is the set of Riemann integrable functions on A and, for each $f \in \mathcal{R}$, $\ell_R(f) := \int_a^b f(x)\,dx$ is the Riemann integral of f (see the introduction to Chapter 3).

Theorem 4.20

(a) $(A, \mathcal{R}, \ell_R)$ is a Daniell space.

(b) If f is a Riemann integrable function on A, then f is also Lebesgue *integrable on A and*

$$\int_a^b f(x)\,dx = \int f\,d\lambda_A.$$

(c) $(A, \overline{\mathcal{L}}(\ell_R), \overline{\ell_R}) = (A, \mathcal{L}^1(\lambda_A), \int \cdot\,d\lambda_A).$

Proof. (a) It is easy to see (and presumably well known to the reader from elementary calculus) that $\mathcal{R}$ is a vector lattice of functions and that ℓ_R is a positive linear functional on $\mathcal{R}$. We know already that (the restriction of) ℓ_R is null-continuous on the space $C(A)$ of continuous functions on A (Section 2.4, Example (a)). This result will be used to prove null-continuity in the general case.

We begin with a preliminary observation. Each positive Riemann integrable function f can be approximated from below by a positive lower Darboux function $f_*(3)$ (see the introduction to Chapter 3) and, being a step function, $f_*(3)$ can be approximated from below by a positive continuous function g. (Possibly the easiest way to construct such a g is to replace $f_*(3)$ by a linear function in a neighbourhood of each point of discontinuity of $f_*(3)$.) Of course, given any $\varepsilon' > 0$, the function g can be chosen to satisfy $\int_a^b (f(x) - g(x))\,dx < \varepsilon'$.

Now let $(f_n)_{n\in\mathbb{N}}$ be a sequence in $\mathcal{R}$ such that $0 = \downarrow f_n$, and take $\varepsilon > 0$. By the above considerations, given $n \in \mathbb{N}$ there is a continuous function g_n on A such that $0 \le g_n \le f_n$ and $\ell_R(f_n) - \ell_R(g_n) < \varepsilon/2^n$. Given $n \in \mathbb{N}$, put $h_n := \bigwedge_{m\le n} g_m$. Then each h_n is continuous and $0 = \downarrow h_n$, which implies that $0 = \downarrow \overline{\ell}_R(h_n)$. But

$$f_n - h_n = \bigwedge_{m\le n} f_m - \bigwedge_{m\le n} g_m \le \sum_{m=1}^n (f_m - g_m)$$

and so

$$\ell_R(f_n) - \ell_R(h_n) \le \sum_{m=1}^n \ell_R(f_m - g_m) < \sum_{m=1}^n \frac{\varepsilon}{2^m} < \varepsilon$$

for every $n \in \mathbb{N}$. Thus $\inf_{n\in\mathbb{N}} \ell_R(f_n) \le \varepsilon$ and, since ε was arbitrary, we conclude that $0 = \downarrow \ell_R(f_n)$.

(b) λ_A is defined on the ring of sets $\mathfrak{R} := \mathfrak{L}(\lambda^{\mathbb{R}})|_A$. Since every subinterval of A belongs to $\mathfrak{R}$, the upper and lower Darboux functions of f belong to $\mathcal{L}(\mathfrak{R})$ and can hence be used to produce ε-brackets for f. Thus $f \in \overline{\mathcal{L}}(\lambda_A)$ and since ℓ_R and ℓ_{λ_A} coincide on the characteristic functions of intervals, it follows easily that

$$\int_a^b f(x)\,dx = \overline{\ell_{\lambda_A}}(f) = \int f\,d\lambda_A$$

holds as well.

(c) We use the induction principle to prove '$\succeq$'. First, note that by Theorem 4.18 $\mathcal{L}^1(\lambda) = \overline{\mathcal{L}}(\lambda)$. Let $\mathcal{F}$ denote the set of all $f \in \overline{\mathcal{L}}(\lambda)$ for which $f|_A \in \overline{\mathcal{L}}(\ell_R)$ and $\int f e_A d\lambda = \overline{\ell_R}(f|_A)$. Clearly $\mathcal{L}(\mathfrak{I}) \subset \mathcal{F}$. Let $(f_n)_{n\in\mathbb{N}}$ be a monotone sequence in $\mathcal{F}$ with $\lim_{n\to\infty} \overline{\ell_\lambda}(f_n) \notin \{\infty, -\infty\}$. Then $f := \lim_{n\to\infty} f_n \in \overline{\mathcal{L}}(\lambda)$ and thus $f e_A \in \overline{\mathcal{L}}(\lambda)$. Hence

$$\lim_{n\to\infty} \overline{\ell_R}(f_n|_A) = \lim_{n\to\infty} \int f_n e_A d\lambda = \int f e_A d\lambda \notin \{\infty, -\infty\}.$$

Applying the monotone convergence theorem to $\overline{\mathcal{L}}(\ell_R)$, it follows that

$$f|_A \in \overline{\mathcal{L}}(\ell_R) \quad \text{and} \quad \overline{\ell_R}(f|_A) = \lim_{n\to\infty} \overline{\ell_R}(f_n|_A) = \int f e_A d\lambda,$$

i.e. that $f \in \mathcal{F}$.

Now take $f, g \in \mathcal{F}$ with $f \le g$ and $\overline{\ell_\lambda}(f) = \overline{\ell_\lambda}(g)$. Then $\overline{\ell_\lambda}(f e_A) \le \overline{\ell_\lambda}(g e_A)$ and $\overline{\ell_\lambda}(f e_{\mathbb{R}\backslash A}) \le \overline{\ell_\lambda}(g e_{\mathbb{R}\backslash A})$. But $\overline{\ell_\lambda}(f e_A) + \overline{\ell_\lambda}(f e_{\mathbb{R}\backslash A}) = \overline{\ell_\lambda}(g e_A) + \overline{\ell_\lambda}(g e_{\mathbb{R}\backslash A})$ as well. Thus

$$\overline{\ell_R}(f|_A) = \int f e_A d\lambda = \int g e_A d\lambda = \overline{\ell_R}(g|_A).$$

Hence, by Theorem 3.22(c), every $h \in \overline{\mathcal{L}}(\lambda)$ satisfying $f \le h \le g$ belongs to $\mathcal{F}$.

Theorem 3.38 implies that $\mathcal{F} = \overline{\mathcal{L}}(\lambda)$ (where $P(f)$ is of course the proposition $f \in \mathcal{F}$). Take $g \in \mathcal{L}^1(\lambda_A)$. By Theorem 4.19 there is an $f \in \mathcal{L}^1(\lambda) = \overline{\mathcal{L}}(\lambda)$ such that $g = f|_A$. Since $f \in \mathcal{F}$, we see that $g \in \overline{\mathcal{L}}(\ell_R)$ and (again using Theorem 4.19) that

$$\int g \, d\lambda_A = \int f e_A \, d\lambda = \overline{\ell_R}(g).$$

We have thus proved '$\succeq$'.

We use the induction principle for the proof of '$\preceq$' as well. This time let $\mathcal{F}$ be the set of all $f \in \overline{\mathcal{L}}(\ell_R)$ for which $f \in \mathcal{L}^1(\lambda_A)$ and $\overline{\ell_R}(f) = \int f \, d\lambda_A$. We leave it as an exercise for the reader to complete the details. $\square$

The disadvantages of the Riemann integral are thus overcome with the introduction of the Lebesgue integral. Since the Lebesgue integral is the Daniell extension of the Riemann integral, it exhibits all the favourable properties enjoyed by any Daniell integral.

There is an elegant characterization of Riemann integrable functions: a bounded function f on $[a, b]$ is Riemann integrable on $[a, b]$ if and only if $\{x \in [a, b] \mid f$ is discontinuous at $x\}$ is a Lebesgue null set. The exercises contain hints as to how this statement may be proved.

Exercises

1. Find a positive measure space $(X, \mathfrak{R}, \mu)$ and an $f \in \mathcal{L}^1(\mu)$ such that $f^2 \notin \mathcal{L}^1(\mu)$.

2. Show that the sequence $(f_n)_{n \in \mathbb{N}}$ defined in Proposition 4.10 converges uniformly to $f \in \overline{\mathbb{R}}_+^X$ on every subset A of X for which $\sup_{x \in A} f(x) < \infty$.

3. Determine $\mathfrak{M}(\delta_x)$ for the Dirac measure δ_x (cf. Exercise 5 of Section 2.4).

4. Let X be uncountable. Define

$$\mathfrak{R} := \{A \subset X \,|\, A \text{ is countable or } X \setminus A \text{ is countable}\}$$

and

$$\mu : \mathfrak{R} \longrightarrow \mathbb{R}, \quad A \longmapsto \begin{cases} 0 & \text{if } A \text{ is countable} \\ 1 & \text{if } X \setminus A \text{ is countable.} \end{cases}$$

Show that $\mathfrak{M}(\mu) = \mathfrak{L}(\mu) = \mathfrak{R}$.

5. Consider $X :=]0,1[$, $f := id_X$, $\mathcal{L} := \{\alpha f \,|\, \alpha \in \mathbb{R}\}$. Define $\ell : \mathcal{L} \to \mathbb{R}$, $\alpha f \mapsto \alpha$. Verify the following statements.

   (a) $(X, \mathcal{L}, \ell)$ is a Daniell space.
   (b) $(X, \mathcal{L}^1(\ell), \int_\ell) = (X, \mathcal{L}, \ell)$.
   (c) $\mathcal{L}^1(\ell)$ does not have the Stone property. Hence there is no measure μ for which $(X, \mathcal{L}^1(\ell), \int_\ell) = (X, \mathcal{L}^1(\mu), \int \cdot d\mu)$.

6. Let $(X, \mathfrak{R}, \mu)$ be a positive measure space. Prove that the following are equivalent.

   (a) $\mu = 0$.
   (b) $X \in \mathfrak{N}(\mu)$.
   (c) $\overline{\mathbb{R}}^X = \mathcal{N}(\mu)$.
   (d) $\overline{\mathbb{R}}^X = \mathcal{L}^1(\mu)$.
   (e) $\int f d\mu = 0$ whenever $f \in \mathcal{L}^1(\mu)$.

7. Define

$$\mathfrak{R} := \{A \subset \mathbb{R} \,|\, A \text{ is countable or } \mathbb{R} \setminus A \text{ is countable}\},$$

$$\mu : \mathfrak{R} \longrightarrow \mathbb{R}, \quad A \longmapsto \begin{cases} 0 & \text{if } A \text{ is countable} \\ 1 & \text{if } \mathbb{R} \setminus A \text{ is countable,} \end{cases}$$

$$\mathfrak{S} := \{A \subset \mathbb{R} \,|\, A \text{ is finite or } \mathbb{R} \setminus A \text{ is finite}\}, \qquad \nu := \mu|_{\mathfrak{S}},$$

$$\mathfrak{S}_{\mathbb{R}\setminus\mathbb{Q}} := \{A \subset \mathbb{R} \setminus \mathbb{Q} \,|\, A \in \mathfrak{S}\}, \qquad \nu_{\mathbb{R}\setminus\mathbb{Q}} := \nu|_{\mathfrak{S}_{\mathbb{R}\setminus\mathbb{Q}}}.$$

Prove the following statements.

   (a) $\mathfrak{R}$ is a σ-algebra on $\mathbb{R}$.
   (b) $\mathfrak{S}$ is a ring of sets and $\mathfrak{S}_{\mathbb{R}\setminus\mathbb{Q}}$ is a δ-ring.
   (c) μ, ν and $\nu_{\mathbb{R}\setminus\mathbb{Q}}$ are positive measures.
   (d) μ and ν generate the same integral on $\mathbb{R}$.
   (e) $e_{\mathbb{R}\setminus\mathbb{Q}}^{\mathbb{R}} \in \mathcal{L}^1(\nu)$, $e_{\mathbb{R}\setminus\mathbb{Q}}^{\mathbb{R}\setminus\mathbb{Q}} \in \mathcal{L}^1(\nu_{\mathbb{R}\setminus\mathbb{Q}})$, and $\int e_{\mathbb{R}\setminus\mathbb{Q}}^{\mathbb{R}} d\nu = 1$, $\int e_{\mathbb{R}\setminus\mathbb{Q}}^{\mathbb{R}\setminus\mathbb{Q}} d\nu_{\mathbb{R}\setminus\mathbb{Q}} = 0$.

8. Let $(X, \mathfrak{R}, \mu)$ be a positive measure space and take $f \in \mathcal{L}^1(\mu)_+$. Prove the following.

   (a) If $\beta_n \downarrow 0$, then $\int f \wedge \beta_n d\mu \downarrow 0$.

   (b) Define $\alpha := \sup_{x \in X} f(x) \in \overline{\mathbb{R}}$ and suppose that $0 \le \alpha_n \uparrow \alpha$. Then $\int f d\mu = \uparrow \int f \wedge \alpha_n d\mu$.

9. Let $(X, \mathfrak{R}, \mu)$ be a positive measure space and take $f \in \mathcal{L}^1(\mu)$, $\alpha > 0$. Show that

$$\mu^X(\{f \ge \alpha\}) \le \frac{1}{\alpha} \int |f| d\mu.$$

10. Let $(X, \mathfrak{R}, \mu)$ be a positive measure space and take $f \in \mathcal{L}^1(\mu)$, $\varepsilon > 0$. Show that there is a $\delta > 0$ such that $\int_A |f| d\mu < \varepsilon$ whenever $A \in \mathcal{L}(\mu)$ satisfies $\mu^X(A) < \delta$.
    Hint: First consider the case where f is bounded.

11. Let $(X, \mathfrak{R}, \mu)$ be a positive measure space. Prove that μ is bounded if and only if $X \in \mathcal{L}(\mu)$ and that in this case

$$\mu^X(X) = \sup_{A \in \mathfrak{R}} \mu(A).$$

12. Let $(X, \mathfrak{R}, \mu)$ be a positive measure space. Take $f \in \mathcal{L}^1(\mu)$ and $A \in \mathcal{L}(\mu)$. Prove the following.

    (a) There are a sequence $(A_n)_{n \in \mathbb{N}}$ in $\mathfrak{R}$ and a set $B \in \mathfrak{N}(\mu)$ such that $\{f \ne 0\} \subset \bigcup_{n \in \mathbb{N}} A_n \cup B$.

    (b) $A \subset \bigcup_{n \in \mathbb{N}} A_n \cup B$ for some suitable sequence $(A_n)_{n \in \mathbb{N}}$ in $\mathfrak{R}$ and some suitable set $B \in \mathfrak{N}(\mu)$.

    (c) If $f \ge 0$, then $\int f d\mu = \sup_{B \in \mathfrak{R}} \int_B f d\mu$.

    (d) $\int f d\mu = \sup_{B \in \mathfrak{R}} \int_B f d\mu + \inf_{B \in \mathfrak{R}} \int_B f d\mu$.

    (e) $\mu^X(A) = \sup_{B \in \mathfrak{R}} \mu^X(A \cap B)$.

13. Let $\mathfrak{R}$ be a δ-ring and μ a positive measure on $\mathfrak{R}$. Show that:

    (a) $\{f \ge \alpha\} \in \mathfrak{R}$ whenever $f \in \mathcal{L}(\mathfrak{R})^{\downarrow}$ and $\alpha > 0$;

    (b) $\int e_A d\mu = \sup\{\mu(B) \mid B \in \mathfrak{R}, B \subset A\}$ for any $A \in \mathcal{L}(\mu)$;

    (c) (b) does not hold for general rings of sets.

    Hint for (b): First prove the statement for $A \in \{C \mid e_C \in \overline{\mathcal{L}}(\mu)\}$ by applying (a).

14. Let $(X, \mathfrak{R}, \mu)$ be a positive measure space. Show that there is a uniquely determined positive measure ν on $\mathfrak{R}_\delta$ with $\nu|_{\mathfrak{R}} = \mu$ and that, moreover, $\nu = \mu^X|_{\mathfrak{R}_\delta}$.

15. Let $(X, \mathfrak{R}, \mu)$ be a positive measure space. Take $A \in \mathfrak{M}(\mu)$, $f \in \mathcal{L}^1(\mu^X|_A)$. Define

$$\tilde{f} : X \longrightarrow \overline{\mathbb{R}}, \quad x \longmapsto \begin{cases} f(x) & \text{if } x \in A \\ 0 & \text{otherwise.} \end{cases}$$

Show that $\tilde{f} \in \mathcal{L}^1(\mu)$.

16. Determine which of the following functions are Lebesgue integrable:

$$\frac{1}{x^2} \sin \frac{1}{x} \qquad \text{on an interval } [a, \infty[, \text{ with } a > 0;$$

$$\frac{1}{x^2} \sin \frac{1}{x} \qquad \text{on } [0, \infty[;$$

$$\frac{x \log x}{(1 + x^2)^3} \qquad \text{on } [0, \infty[;$$

$$\frac{x^{3/2}}{1 + x^2} \qquad \text{on } [0, \infty[;$$

$$\frac{\sin x}{x} \qquad \text{on } [0, \infty[.$$

17. Let f be a bounded function on the interval $[a, b]$. Show that f is Riemann integrable on $[a, b]$ if and only if $\{x \in [a, b] \mid f$ is discontinuous at $x\}$ is a Lebesgue null set.

Hints: We use the notation from the introduction to Chapter 3.

'$\Rightarrow$'. Given $n \in \mathbb{N}$, let $\mathfrak{Z}_n$ be a subdivision of $[a, b]$ with

$$\int_a^b \left(f^*(\mathfrak{Z}_n) - f_*(\mathfrak{Z}_n) \right)(x)\, dx < \frac{1}{n}.$$

Define $g := \bigvee_{n \in \mathbb{N}} f_*(\mathfrak{Z}_n)$ and $h := \bigwedge_{n \in \mathbb{N}} f^*(\mathfrak{Z}_n)$. Show that $\{h > g\} \in \mathfrak{N}(\lambda)$. Prove that f is continuous at every $x \in [a, b]$ which is not a division point of any $\mathfrak{Z}_n$ and for which $g(x) = f(x) = h(x)$.

'$\Leftarrow$'. Let $(\mathfrak{Z}_n)_{n \in \mathbb{N}}$ be a sequence of subdivisions of $[a, b]$ such that $\mathfrak{Z}_{n+1}$ is obtained from $\mathfrak{Z}_n$ by dividing each partitioning interval of $\mathfrak{Z}_n$ into equal halves. Then for each $n \in \mathbb{N}$,

$$f_*(\mathfrak{Z}_n) \leq f_*(\mathfrak{Z}_{n+1}) \leq f \leq f^*(\mathfrak{Z}_{n+1}) \leq f^*(\mathfrak{Z}_n).$$

Define g and h as above. Take $x \in [a, b]$ at which f is continuous and show that $g(x) = h(x)$. Thus $g = h$ λ-a.e. on $[a, b]$. Prove that

$$\lim_{n \to \infty} \int f_*(\mathfrak{Z}_n)\, d\lambda_{[a,b]} = \lim_{n \to \infty} \int f^*(\mathfrak{Z}_n)\, d\lambda_{[a,b]}$$

and conclude that f is Riemann integrable on $[a, b]$.

18. Show that $A \subset \mathbb{R}$ is a Lebesgue null set if and only if, given any $\varepsilon > 0$, there is a sequence $([a_n, b_n])_{n \in \mathbb{N}}$ of non-empty intervals in $\mathbb{R}$ such that

$$A \subset \bigcup_{n \in \mathbb{N}} [a_n, b_n] \quad \text{and} \quad \sum_{n \in \mathbb{N}} (b_n - a_n) < \varepsilon.$$

19. Let $f : I \to \mathbb{R}$ be a Lipschitz continuous function on the interval $I \subset \mathbb{R}$ – i.e. there is an $L \in \mathbb{R}$ such that

$$|f(x) - f(y)| \le L|x - y|$$

for all $x, y \in I$. Show that $f(N)$ is a Lebesgue null set whenever $N \subset I$ is. Is the same true for arbitrary continuous functions? (Consider the function f defined in Exercise 4 of Section 3.3.)

4.2 Examples

(a) We first consider the example of a positive measure μ^g on the ring of sets $\mathfrak{F}(X)$ which consists of the finite subsets of a set X. (This was introduced in Section 2.4.) Recall that $g \in \mathbb{R}_+^X$ and

$$\mu^g(A) := \sum_{x \in A} g(x), \quad \text{where } A \in \mathfrak{F}(X).$$

We have already established that $\ell_{\mu^g} = \ell_g$ for the functional ℓ_g on $\mathcal{F}(X)$. (This was described in detail in Sections 2.4, 3.5 and 3.7.) Theorem 3.36 allows us easily to deduce the next theorem.

Theorem 4.21

(a) $\mathfrak{L}(\mu^g) = \{ A \subset X \mid \sup_{B \in \mathfrak{F}(X)} \sum_{x \in A \cap B} g(x) < \infty \}$.

(b) Given $A \in \mathfrak{L}(\mu^g)$,

$$(\mu^g)^X(A) = \sup_{B \in \mathfrak{F}(X)} \sum_{x \in A \cap B} g(x).$$

(c) $\mathfrak{M}(\mu) = \mathfrak{P}(X)$.

(b) **Stieltjes measures** The positive measure spaces $(\mathbb{R}, \mathfrak{I}, \mu_g)$ were introduced in Section 2.4. Recall that $\mathfrak{I}$ denoted the ring of sets of interval forms on $\mathbb{R}$ and μ_g the Stieltjes measure corresponding to the increasing, left continuous function g on $\mathbb{R}$. We prove:

Proposition 4.22 $\mathfrak{L}(\mu_g)$ *contains*

(a) *all bounded open sets of $\mathbb{R}$ and*

(b) *all compact subsets of $\mathbb{R}$.*

Proof. Each open set of $\mathbb{R}$ is the union of a countable family of pairwise disjoint open intervals. If $]\alpha, \beta[$ $(\alpha < \beta)$ is such an interval, then

$$]\alpha, \beta[= \uparrow [\alpha + 1/n, \beta[.$$

Theorem 4.2(b) implies that $]\alpha, \beta[\in \mathfrak{L}(\mu_g)$. Using Theorem 4.4, we find that each bounded open set must be contained in $\mathfrak{L}(\mu_g)$. We obtain (b) simply by noting that each compact subset of $\mathbb{R}$ is the set-theoretical difference between two bounded open sets. $\qquad\square$

'Borel sets' form an important class of sets. Let X be a topological space. Then the **Borel sets of X** are the elements of the σ-algebra on X generated by the open sets. We write $\mathfrak{B}(X)$ for the set of all Borel sets of X.

When $X = \mathbb{R}$, we have the following theorem.

Theorem 4.23 *Every bounded Borel set of $\mathbb{R}$ is contained in $\mathfrak{L}(\mu_g)$.*

Proof. Define

$$\mathfrak{S} := \left\{ A \in \mathfrak{B}(\mathbb{R}) \,\middle|\, A \cap]\alpha, \beta[\in \mathfrak{L}(\mu_g) \text{ for all } \alpha, \beta \in \mathbb{R} \text{ with } \alpha \leq \beta \right\}.$$

By Proposition 4.22(a), $\mathfrak{S}$ contains every open subset of $\mathbb{R}$. Since

$$(A \cup B) \cap]\alpha, \beta[= (A \cap]\alpha, \beta[) \cup (B \cap]\alpha, \beta[),$$
$$(A \setminus B) \cap]\alpha, \beta[= (A \cap]\alpha, \beta[) \setminus (B \cap]\alpha, \beta[)$$

and

$$\left(\bigcap_{n \in \mathbb{N}} A_n \right) \cap]\alpha, \beta[= \bigcap_{n \in \mathbb{N}} (A_n \cap]\alpha, \beta[),$$

it follows easily from Theorem 4.1(c) that $\mathfrak{S}$ is a δ-ring. Moreover, since $\mathfrak{S}$ contains the open set $\mathbb{R}$, $\mathfrak{S}$ is a σ-algebra on $\mathbb{R}$. But as we have seen, $\mathfrak{S}$ contains every open subset of $\mathbb{R}$. Hence $\mathfrak{B}(\mathbb{R}) \subset \mathfrak{S}$.

If A is a bounded Borel set, then there are $\alpha, \beta \in \mathbb{R}$, $\alpha \leq \beta$, such that $A \subset]\alpha, \beta[$. Hence

$$A = A \cap]\alpha, \beta[\in \mathfrak{L}(\mu_g).$$

$\qquad\square$

Theorem 4.23 indicates how extensive $\mathfrak{L}(\mu_g)$ is. However, we should also mention that the space $\mathfrak{L}(\mu_g)$ does not consist only of Borel sets. An example is given in the exercises. On the other hand, an obvious question is whether there are bounded sets not contained in $\mathfrak{L}(\mu_g)$. Indeed there are such 'bad' sets! We next present one example, due to Vitali, for Lebesgue measure λ. Note that such an example cannot be trivial, since by Theorem 4.23 all 'good' bounded sets are λ-integrable. The axiom of choice is used in the construction.

Theorem 4.24 $[0, 1]$ *has a subset which is not Lebesgue integrable.*

Proof. Define a relation, $\sim$, on $[0, 1]$ by

$$x \sim y :\iff x - y \in \mathbb{Q}.$$

This is clearly an equivalence relation.

Let A be a minimal complete set of representatives for the equivalence classes, so that A contains a unique representative of each equivalence class. Note that the axiom of choice ensures that there is such a set. We prove that A is not Lebesgue integrable. Given $q \in Q_1 := \mathbb{Q} \cap [-1, 1]$, put

$$A_q := \{a + q \mid a \in A\}.$$

Then the sets A_q are pairwise disjoint. To see this, take $q, r \in Q_1$ and $x \in A_q \cap A_r$. There are $a_1, a_2 \in A$ satisfying $x = a_1 + q$ and $x = a_2 + r$. It follows that $a_1 - a_2 = r - q \in \mathbb{Q}$. Thus $a_1 = a_2$, by the definition of A. Hence $q = r$ establishing our claim.

We have

$$[0, 1] \subset \bigcup_{q \in Q_1} A_q \subset [-1, 2]. \tag{1}$$

Since the latter inclusion is obvious, we only prove the first. Given $x \in [0, 1]$, there is some $a \in A$ for which $x \sim a$. Hence $x - a \in \mathbb{Q}$. But $x - a \in [-1, 1]$ as well. Finally $x \in A_{x-a}$.

Now assume that $A \in \mathfrak{L}(\lambda)$. Then, by the translation invariance of Lebesgue measure (Theorem 3.42), $A_q \in \mathfrak{L}(\lambda)$ and $\lambda^{\mathbb{R}}(A_q) = \lambda^{\mathbb{R}}(A)$ for every $q \in Q_1$. Hence, if A were a λ-null set, $\bigcup_{q \in Q_1} A_q$ would also be one, contradicting the first inclusion in (1). Thus $\alpha := \lambda^{\mathbb{R}}(A) > 0$. Fix $n \in \mathbb{N}$ with $n\alpha > 3$ and take n distinct elements $q_1, \dots, q_n$ of Q_1. Then the second inclusion in (1) implies that $\bigcup_{k=1}^{n} A_{q_k} \subset [-1, 2]$ and hence $n\alpha \le \lambda^{\mathbb{R}}([-1, 2]) = 3$. This is a contradiction. Thus A cannot be Lebesgue integrable. $\qquad\qquad\square$

Arguing similarly to the above, the reader will be able to show that every λ-integrable set with measure strictly greater than 0 contains a bounded subset which is not λ-integrable.

Exercises

1. Determine $\mathfrak{L}(\mu^{ex})$.

2. Let g be an increasing left continuous function on $\mathbb{R}$ and μ_g the Stieltjes measure generated by it. Verify the following.

    (a) If g is continuous, then $\{A \subset \mathbb{R} \mid A \text{ is countable}\} \subset \mathfrak{N}(\mu_g)$.

    (b) If g is strictly increasing and U is a non-empty open subset of $\mathbb{R}$, then $U \notin \mathfrak{N}(\mu_g)$.

    (c) If g is strictly increasing and if f_1, f_2 are continuous functions on $\mathbb{R}$, then $f_1 = f_2$ whenever $f_1 = f_2$ μ_g-a.e.

3. The **generalized Cantor set** is constructed by slightly modifying the construction presented in Section 3.3. Let $(\alpha_n)_{n \in \mathbb{N}}$ be a sequence in $]0, 1[$. We define the intervals I_{ni} by redefining their lengths as follows:

$\lambda^{\mathbb{R}}(I_{11}) := \alpha_1$ and $\lambda^{\mathbb{R}}(I_{n+1,i}) := \alpha_{n+1}\lambda^{\mathbb{R}}(J_{ni})$ for $n \geq 1$. We denote the generalized Cantor set by

$$C((\alpha_n)_{n\in\mathbb{N}}).$$

The Cantor set C described earlier is then given by $C = C((1/3)_{n\in\mathbb{N}})$.

Prove the following statements.

(a) $\lambda^{\mathbb{R}}(C((\alpha_n)_{n\in\mathbb{N}})) = 1 - \sum_{n\in\mathbb{N}} \alpha_n(1-\alpha_1)\ldots(1-\alpha_{n-1})$.

(b) For each $\varepsilon > 0$ there is a nowhere dense compact subset A of $[0,1]$ with $\lambda^{\mathbb{R}}(A) > 1 - \varepsilon$. Conclude from this that e_A cannot be Riemann integrable. However, note that e_C *is* Riemann integrable, where C is the classical Cantor set.

(c) There is a Lebesgue integrable function on $[0,1]$ such that no Riemann integrable function on $[0,1]$ is equal to it λ-a.e.

4. Let C be a generalized Cantor set with $\lambda^{\mathbb{R}}(C) > 0$. Define $X := \mathbb{R} \setminus C$ and $\mathfrak{R} := \{A \cap X \mid A \in \mathfrak{I}_0\}$. Prove the following.

(a) For each $B \in \mathfrak{R}$ there are uniquely determined $a_B, b_B \in \mathbb{R}$ with $B = [a_B, b_B[\cap X$. $\mathfrak{R}$ is a semi-ring of sets.

(b) The map $\mu : \mathfrak{R} \to \mathbb{R}$, $B \mapsto b_B - a_B$ is a null-continuous positive content on $\mathfrak{R}$.

(c) There is a disjoint sequence $(B_n)_{n\in\mathbb{N}}$ in $\mathfrak{R}$ for which $\bigcup_{n\in\mathbb{N}} B_n \in \mathfrak{R}$ but $\mu(\bigcup_{n\in\mathbb{N}} B_n) > \sum_{n\in\mathbb{N}} \mu(B_n)$. Hence Corollary 4.5 does not hold in general for semi-rings.

5. Let X be a topological space. Prove the following statements.

(a) $\mathfrak{B}(X)$ is the σ-ring generated by the open sets of X,
    is the δ-ring generated by the open sets of X,
    is the δ-ring generated by the closed sets of X,
    is the σ-ring generated by the closed sets of X,
    is the σ-algebra generated by the closed sets of X.

(b) If X is a Hausdorff space and

$$\mathfrak{K}(X) := \{K \subset X \mid K \text{ is compact}\},$$

then $\mathfrak{K}(X) \subset \mathfrak{B}(X)$. It follows that the σ-algebra generated by $\mathfrak{K}(X)$ is contained in $\mathfrak{B}(X)$. Show that this σ-algebra and $\mathfrak{B}(X)$ need not coincide.

(c) If X is a Hausdorff space and $\mathfrak{R} := \{A \subset X \mid A \text{ is relatively compact}\}$, then $\mathfrak{K}(X)_\delta = \mathfrak{B}(X) \cap \mathfrak{R}$.

Hint for (c)'$\supset$': Take $A \in \mathfrak{B}(X) \cap \mathfrak{R}$. Show that $\{B \subset X \mid \overline{A} \cap B \in \mathfrak{K}(X)_\delta\}$ is a δ-ring which contains the closed sets.

6. Let A be a dense set of $\mathbb{R}$ and define

$$\mathfrak{A} := \{]-\infty, x] \mid x \in A\}, \qquad \mathfrak{B} := \{]-\infty, x[\mid x \in A\},$$
$$\mathfrak{C} := \{[x, \infty[\mid x \in A\}, \qquad \mathfrak{D} := \{]x, \infty[\mid x \in A\}.$$

Verify that $\mathfrak{B}(\mathbb{R}) = \mathfrak{A}_\sigma = \mathfrak{B}_\sigma = \mathfrak{C}_\sigma = \mathfrak{D}_\sigma$.

7. Let X, Y be topological spaces and $f : X \to Y$ a continuous mapping. Show that $f^{-1}(A) \in \mathfrak{B}(X)$ whenever $A \in \mathfrak{B}(Y)$.

8. We construct a λ-integrable subset of $[0, 1]$ which is not a Borel set. Let f be the Cantor function defined in Exercise 4 of Section 3.3 and define $g(x) := x + f(x)$ for every $x \in [0, 1]$. Prove the following.

   (a) g is a strictly increasing homeomorphism from $[0, 1]$ onto $[0, 2]$.

   (b) $g([0, 1] \setminus C) \in \mathfrak{L}(\lambda)$ and $\lambda^{\mathbb{R}}\big(g([0, 1] \setminus C)\big) = 1$. (Here C denotes the Cantor set.)

   (c) $g(C) \in \mathfrak{L}(\lambda)$ and $\lambda^{\mathbb{R}}(g(C)) = 1$.

   Let A be a subset of $g(C)$ which is not λ-integrable. Show that

   (d) $g^{-1}(A) \in \mathfrak{N}(\lambda)$.

   (e) $g^{-1}(A)$ is not a Borel subset of $\mathbb{R}$. (Use Exercise 7.)

4.3 Locally integrable functions

In this section we consider a positive measure space $(X, \mathfrak{R}, \mu)$.

A function $f \in \overline{\mathbb{R}}^X$ is said to be **locally μ-integrable** if for every $A \in \mathfrak{R}$, $f e_A \in \mathcal{L}^1(\mu)$. We write $\mathcal{L}^1_{loc}(\mu)$ for the set of all locally μ-integrable functions in $\overline{\mathbb{R}}^X$.

Note that every constant function is locally μ-integrable (but in general not μ-integrable). The function $f \in \overline{\mathbb{R}}^{\mathbb{R}}$ defined by $f(x) := 0$ for $x \le 0$ and $f(x) := 1/x$ for $x > 0$ is not locally Lebesgue integrable.

Before investigating locally μ-integrable functions more closely we prove a remarkable theorem describing the approximation of μ-integrable sets using elements of the given ring of sets $\mathfrak{R}$.

Theorem 4.25 *Take $A \in \mathfrak{L}(\mu)$ and $\varepsilon > 0$.*

(a) There is a decreasing sequence $(B_k)_{k \in \mathbb{N}}$ in $\mathfrak{R}$ such that $\bigcap_{k \in \mathbb{N}} B_k \subset A$ and

$$\mu^X(A) - \mu^X\left(\bigcap_{k \in \mathbb{N}} B_k\right) < \varepsilon.$$

(b) There is a $B \in \mathfrak{R}$ with $\mu^X(A \triangle B) < \varepsilon$.

Proof. (a) First take $A \subset X$ with $e_A \in \overline{\mathcal{L}}(\mu)$. Then there is an increasing sequence $(f_n)_{n \in \mathbb{N}}$ in $\mathcal{L}(\mathfrak{R})^\downarrow$ for which $0 \le f_n \le 2e_A$ for every $n \in \mathbb{N}$ and

$\overline{\ell_\mu}(2e_A) = \uparrow \ell_\mu^+(f_n)$. Define $f := \uparrow f_n$. Then $0 \le f \le 2e_A$ and $2e_A = f$ μ-a.e. (Theorem 3.22(c)). It follows that $A \setminus \{f > 1\} \in \mathfrak{N}(\mu)$. But

$$\{f > 1\} = \uparrow \{f_n > 1\}.$$

Given $\varepsilon > 0$, choose $n \in \mathbb{N}$ such that

$$\mu^X(\{f > 1\}) - \mu^X(\{f_n > 1\}) < \varepsilon.$$

Since $f_n \in \mathcal{L}(\mathfrak{R})^\downarrow$, there is a decreasing sequence $(g_k)_{k\in\mathbb{N}}$ of $\mathfrak{R}$-step functions with $f_n = \downarrow g_k$. Put $B_k := \{g_k \ge 1\}$. Then $(B_k)_{k\in\mathbb{N}}$ is a decreasing sequence in $\mathfrak{R}$ (Proposition 2.16) and

$$\bigcap_{k\in\mathbb{N}} B_k = \{f_n \ge 1\}.$$

For arbitrary $A \in \mathcal{L}(\mu)$ the result follows from Proposition 4.13.
(b) follows from (a). □

We next establish some fundamental properties of $\mathcal{L}^1_{loc}(\mu)$.

Theorem 4.26

(a) $\mathcal{L}^1_{loc}(\mu) \cap \mathbb{R}^X$ is a vector lattice of functions.

(b) $\{|f| = \infty\} \in \mathfrak{N}(\mu)$ for every $f \in \mathcal{L}^1_{loc}(\mu)$.

(c) $f \in \overline{\mathbb{R}}^X$ is contained in $\mathcal{L}^1_{loc}(\mu)$ if and only if $f = g$ μ-a.e. for some $g \in \mathcal{L}^1_{loc}(\mu) \cap \mathbb{R}^X$.

(d) If $f \in \mathcal{L}^1_{loc}(\mu)$, then $fe_A \in \mathcal{L}^1(\mu)$ for all $A \in \mathfrak{R}_\delta$.

(e) $\mathcal{L}^1(\mu) \subset \mathcal{L}^1_{loc}(\mu)$.

Proof. (a) follows immediately from the definition of $\mathcal{L}^1_{loc}(\mu)$.

(b),(c) Take $f \in \mathcal{L}^1_{loc}(\mu)$ and put $A := \{|f| = \infty\}$. Then $fe_B \in \mathcal{L}^1(\mu)$ for every $B \in \mathfrak{R}$, so that $A \cap B \in \overline{\mathfrak{N}}(\mu)$. Consequently, by Proposition 4.6(b), $A \in \mathfrak{N}(\mu)$. Define $g := fe_{X\setminus A}$. Then $g \in \mathcal{L}^1_{loc}(\mu) \cap \mathbb{R}^X$ and $f = g$ μ-a.e. The converse is clear.

(d) It is easy to see that $\{A \subset X \mid fe_A \in \mathcal{L}^1(\mu)\}$ is a δ-ring. It contains $\mathfrak{R}$ and therefore $\mathfrak{R}_\delta$.

(e) follows from Theorem 4.15. □

The next theorem bears a certain resemblance to Proposition 4.6: it, too, characterizes global properties by local behaviour.

Theorem 4.27 *Take* $f, g \in \mathcal{L}^1_{loc}(\mu)$.

(a) If $\int_A f \, d\mu \ge 0$ for every $A \in \mathfrak{R}$, then $f \ge 0$ μ-a.e.

(b) If $\int_A f \, d\mu \ge \int_A g \, d\mu$ for every $A \in \mathfrak{R}$, then $f \ge g$ μ-a.e.

(c) If $\int_A f \, d\mu = 0$ for every $A \in \mathfrak{R}$, then $f = 0$ μ-a.e.

Proof. It is sufficient to prove (a), since (b) and (c) are easy consequences. Let $(A_n)_{n\in\mathbb{N}}$ be a decreasing sequence in $\mathfrak{R}$ and put $A := \downarrow A_n$. Then $fe_A = \lim_{n\to\infty} fe_{A_n}$ and for each $n \in \mathbb{N}$,

$$|fe_{A_n}| \leq |f|e_{A_1} \in \mathcal{L}^1(\mu).$$

The Lebesgue convergence theorem implies that

$$\int fe_A d\mu = \lim_{n\to\infty} \int fe_{A_n} d\mu \geq 0.$$

Now put $B := \{f < 0\}$. Given $A \in \mathfrak{R}$, $A \cap B = \{fe_A < 0\} \in \mathfrak{L}(\mu)$. To see this, put $C_n := \{f^-e_A > 1/n\}$ for $n \in \mathbb{N}$, note that

$$\{fe_A < 0\} = \bigcup_{n\in\mathbb{N}} C_n$$

and apply Corollary 4.9 and Theorem 4.2(b). Let $\mathfrak{S}$ be the set of all intersections of decreasing sequences in $\mathfrak{R}$. By Theorem 4.25, there is an increasing sequence $(B_n)_{n\in\mathbb{N}}$ in $\mathfrak{S}$ such that $B_n \subset A \cap B$ for all $n \in \mathbb{N}$ and

$$A \cap B \setminus \bigcup_{n\in\mathbb{N}} B_n \in \mathfrak{N}(\mu).$$

Then by the Lebesgue convergence theorem,

$$\int_{A\cap B} f\, d\mu = \int_{\bigcup_{n\in\mathbb{N}} B_n} f\, d\mu = \lim_{n\to\infty} \int_{B_n} f\, d\mu \geq 0.$$

Take $n \in \mathbb{N}$. It follows that

$$\int e_{C_n} d\mu \leq n \int f^-e_{A\cap B} d\mu = n \int -fe_{A\cap B} d\mu \leq 0.$$

Thus $A \cap B = \bigcup_{n\in\mathbb{N}} C_n$ is also in $\mathfrak{N}(\mu)$. Since A is arbitrary, $B \in \mathfrak{N}(\mu)$. □

A useful formula expressing the integral of a positive function f in terms of the 'local' integrals $\int_A f\, d\mu$ is a corollary to this.

Corollary 4.28 *Take $f \in \overline{\mathbb{R}}_+^X$. Then the following are equivalent.*

(a) $f \in \mathcal{L}^1(\mu)$.

(b) $f \in \mathcal{L}^1_{loc}(\mu)$ *and* $\sup_{A\in\mathfrak{R}} \int_A f\, d\mu < \infty$.

Moreover, if these equivalent conditions are satisfied, then

$$\int f\, d\mu = \sup_{A\in\mathfrak{R}} \int_A f\, d\mu.$$

Proof. We need only prove (b)$\Rightarrow$(a) and deduce the formula. Put

$$\alpha := \sup_{A\in\mathfrak{R}} \int_A f\, d\mu.$$

There is an increasing sequence $(A_n)_{n\in\mathbb{N}}$ in $\mathfrak{R}$ with $\alpha = \sup_{n\in\mathbb{N}}\int_{A_n} f \, d\mu$. Put

$$A_0 := \bigcup_{n\in\mathbb{N}} A_n.$$

Using the monotone convergence theorem, we conclude that $f e_{A_0} \in \mathcal{L}^1(\mu)$ and $\int_{A_0} f \, d\mu = \alpha$. Take $B \in \mathfrak{R}$. Then

$$\alpha \geq \sup_{n\in\mathbb{N}} \int_{B\cup A_n} f \, d\mu = \int_{B\cup A_0} f \, d\mu$$

$$= \int_{B\setminus A_0} f \, d\mu + \int_{A_0} f \, d\mu = \int_{B\setminus A_0} f \, d\mu + \alpha.$$

Thus $\int_B f e_{X\setminus A_0} d\mu = 0$. Since B is arbitrary, Theorem 4.27(c) implies that $f e_{X\setminus A_0} = 0$ μ-a.e. Hence $f = f e_{A_0} + f e_{X\setminus A_0} \in \mathcal{L}^1(\mu)$ and

$$\int f \, d\mu = \int_{A_0} f \, d\mu = \alpha.$$

$\square$

Note that the preceding proof also implies the following.

Proposition 4.29 *Given $f \in \mathcal{L}^1(\mu)$, there are a sequence $(A_n)_{n\in\mathbb{N}}$ in $\mathfrak{R}$ and a set $C \in \mathfrak{N}(\mu)$ with*

$$\{f \neq 0\} \subset \left(\bigcup_{n\in\mathbb{N}} A_n \right) \cup C.$$

The significance of locally μ-integrable functions will be discussed in detail in Chapter 8.

Exercises

1. Find examples of locally integrable functions which are not integrable. Do so for Lebesgue measure and other measures as well.

2. Let $(X, \mathfrak{R}, \mu)$ be a positive measure space. Prove each of the following statements.

   (a) Suppose that $X \setminus A \in \mathfrak{N}(\mu)$ for some $A \in \mathfrak{R}$. Then $\mathcal{L}^1_{loc}(\mu) = \mathcal{L}^1(\mu)$.
   (b) If $\mathfrak{R}$ is a σ-ring, then $\mathcal{L}^1_{loc}(\mu) = \mathcal{L}^1(\mu)$.
   (c) Let $\mathfrak{F}$ be an ultrafilter on $\mathbb{N}$ containing no finite set. Define

   $$\mu : \mathfrak{P}(\mathbb{N}) \setminus \mathfrak{F} \longrightarrow \mathbb{R}, \quad A \longmapsto \sum_{n\in A} \frac{1}{n^2}.$$

   Then μ is a positive measure and $\mathcal{L}^1_{loc}(\mu) = \mathcal{L}^1(\mu)$. Nevertheless, there is no $A \in \mathfrak{P}(\mathbb{N}) \setminus \mathfrak{F}$ for which $\mathbb{N} \setminus A \in \mathfrak{N}(\mu)$.
   (d) If $\mathcal{L}^1_{loc}(\mu) = \mathcal{L}^1(\mu)$, then μ is bounded.

(e) There is a bounded positive measure μ with $\mathcal{L}^1_{loc}(\mu) \neq \mathcal{L}^1(\mu)$.

Hint for (c): Take $f \in \mathbb{R}^{\mathbb{N}}_+$ with $\sum_{n \in \mathbb{N}} \frac{f(n)}{n^2} = \infty$. Then there is an $A \subset \mathbb{N}$ with

$$\sum_{n \in A} \frac{f(n)}{n^2} = \sum_{n \in \mathbb{N} \setminus A} \frac{f(n)}{n^2} = \infty.$$

3. Let $(X, \mathfrak{R}, \mu)$ be a positive measure space and take $f \in \mathcal{L}^1_{loc}(\mu)$.

 (a) Show that $f e_A \in \mathcal{L}^1_{loc}(\mu)$ whenever $A \in \mathfrak{M}(\mu)$.
 (b) Does $A \in \overline{\mathfrak{L}}(\mu)$ imply that $f e_A \in \mathcal{L}^1(\mu)$?

4. Let $(X, \mathfrak{R}, \mu)$ be a positive measure space and $\mathfrak{S}$ a semi-ring with $\mathfrak{R} = \mathfrak{S}_r$. Show that Theorem 4.27 remains true if the condition 'for every $A \in \mathfrak{R}$' is replaced by the weaker condition 'for every $A \in \mathfrak{S}$'.

5. Let $(X, \mathfrak{R}, \mu)$ be a positive measure space and $(f_n)_{n \in \mathbb{N}}$ a sequence in $\mathcal{L}^1_{loc}(\mu)$. Prove the following.

 (a) If $(f_n)_{n \in \mathbb{N}}$ is bounded above in $\mathcal{L}^1_{loc}(\mu)$, then $\bigvee_{n \in \mathbb{N}} f_n \in \mathcal{L}^1_{loc}(\mu)$.
 (b) If $(f_n)_{n \in \mathbb{N}}$ is bounded below in $\mathcal{L}^1_{loc}(\mu)$, then $\bigwedge_{n \in \mathbb{N}} f_n \in \mathcal{L}^1_{loc}(\mu)$.
 (c) If $(|f_n|)_{n \in \mathbb{N}}$ is bounded in $\mathcal{L}^1_{loc}(\mu)$ and if $(f_n)_{n \in \mathbb{N}}$ converges μ-a.e. to $f \in \overline{\mathbb{R}}^X$, then $f \in \mathcal{L}^1_{loc}(\mu)$.

4.4 μ-measurable functions

In Section 4.1 we defined the class of μ-measurable sets which extended the collection of μ-integrable sets. In this section we introduce the related class of μ-measurable functions which, as we shall see, contains the μ-integrable functions. Some aspects of measure theory are more adequately formulated in this setting than in the context of integrable functions. We present Egoroff's theorem and the concept of convergence in measure as examples.

Throughout this section $(X, \mathfrak{R}, \mu)$ is always a positive measure space. Define

$$\mathcal{M}(\mu) := \left\{ f \in \overline{\mathbb{R}}^X \mid \{f < \alpha\} \in \mathfrak{M}(\mu) \text{ for every } \alpha \in \mathbb{R} \right\}.$$

The functions in $\mathcal{M}(\mu)$ are said to be **μ-measurable**. Note that if μ is a Stieltjes measure, then every monotone function f on $\mathbb{R}$ is μ-measurable since every $\{f < \alpha\}$ ($\alpha \in \mathbb{R}$) is an interval of $\mathbb{R}$.

The fact that $\mathfrak{M}(\mu)$ is a σ-algebra on X (Proposition 4.14(a)) lies at the heart of the proofs of the following results.

Our first observation is immediate.

Proposition 4.30

 (a) Every constant function in $\overline{\mathbb{R}}^X$ is μ-measurable.
 (b) If $A \subset X$, then $e_A \in \mathcal{M}(\mu)$ if and only if $A \in \mathfrak{M}(\mu)$.

Proposition 4.31 *Take $f \in \overline{\mathbb{R}}^X$. Then the following are equivalent.*

(a) $f \in \mathcal{M}(\mu)$.

(b) $\{f \le \alpha\} \in \mathfrak{M}(\mu)$ *for every* $\alpha \in \mathbb{R}$.

(c) $\{f > \alpha\} \in \mathfrak{M}(\mu)$ *for every* $\alpha \in \mathbb{R}$.

(d) $\{f \ge \alpha\} \in \mathfrak{M}(\mu)$ *for every* $\alpha \in \mathbb{R}$.

Proof. (a)$\Rightarrow$(b) follows from the fact that $\{f \le \alpha\} = \bigcap_{n\in\mathbb{N}}\{f < \alpha + \frac{1}{n}\}$.

(b)$\Rightarrow$(c) follows from the fact that $\{f > \alpha\} = X \setminus \{f \le \alpha\}$.

(c)$\Rightarrow$(d) follows from the fact that $\{f \ge \alpha\} = \bigcap_{n\in\mathbb{N}}\{f > \alpha - \frac{1}{n}\}$.

(d)$\Rightarrow$(a) follows from the fact that $\{f < \alpha\} = X \setminus \{f \ge \alpha\}$. $\square$

Proposition 4.32

(a) *Take $f, g \in \mathcal{M}(\mu)$ and $\alpha \in \mathbb{R}$. Then the sets*

$$\{f < g + \alpha\}, \quad \{f \le g + \alpha\}, \quad \{f = g + \alpha\}, \quad \{f \ne g + \alpha\}$$

are all μ-measurable.

(b) *If $f \in \mathcal{M}(\mu)$, then the sets*

$$\{f = \infty\}, \quad \{f = -\infty\}, \quad \{x \in X \mid f(x) \in \mathbb{R}\}$$

are μ-measurable.

Proof. (a) For the first set note that

$$\{f < g + \alpha\} = \bigcup_{\beta\in\mathbb{Q}} (\{f < \beta\} \setminus \{g < \beta - \alpha\}).$$

The statement now follows. For the other sets note the following:

$$\{f \le g + \alpha\} = X \setminus \{g < f - \alpha\},$$
$$\{f = g + \alpha\} = \{f \le g + \alpha\} \setminus \{f < g + \alpha\},$$
$$\{f \ne g + \alpha\} = X \setminus \{f = g + \alpha\}.$$

(b) follows from Proposition 4.31 and the equalities

$$\{f = \infty\} = \bigcap_{n\in\mathbb{N}}\{f > n\}, \qquad \{f = -\infty\} = \bigcap_{n\in\mathbb{N}}\{f < -n\},$$
$$\{x \in X \mid f(x) \in \mathbb{R}\} = X \setminus (\{f = \infty\} \cup \{f = -\infty\}).$$

$\square$

The next two theorems describe the structure of $\mathcal{M}(\mu)$.

Theorem 4.33

(a) $\alpha f \in \mathcal{M}(\mu)$ *whenever $\alpha \in \overline{\mathbb{R}}$ and $f \in \mathcal{M}(\mu)$.*

(b) $f + g \in \mathcal{M}(\mu)$ *for all $f, g \in \mathcal{M}(\mu)$ for which $f + g$ is defined.*

(c) $f \in \overline{\mathbb{R}}^X$ *is μ-measurable if and only if f^+ and f^- are μ-measurable.*

(d) *If $f \in \mathcal{M}(\mu)$, then $|f| \in \mathcal{M}(\mu)$.*

(e) *If $f \in \mathcal{M}(\mu)$ and $g \in \overline{\mathbb{R}}^X, g = f$ μ-a.e., then $g \in \mathcal{M}(\mu)$.*

(f) *$f e_A \in \mathcal{M}(\mu)$ whenever $f \in \mathcal{M}(\mu)$ and $A \in \mathfrak{M}(\mu)$.*

Proof. (a) For $\alpha = 0$ the assertion is trivial. Take $\alpha \in \mathbb{R}$, $\alpha > 0$. Then for any $\beta \in \mathbb{R}$, we have that $\{\alpha f < \beta\} = \{f < \beta/\alpha\}$. If on the other hand $\alpha \in \mathbb{R}$, $\alpha < 0$, then $\{\alpha f < \beta\} = \{f > \beta/\alpha\}$. In view of Proposition 4.31, it follows in both cases that $\alpha f \in \mathcal{M}(\mu)$. Next, note that

$$\{\infty f < \beta\} = \begin{cases} \{f \le 0\} & \text{if } \beta > 0, \\ \{f < 0\} & \text{if } \beta \le 0. \end{cases}$$

By Proposition 4.31 again, ∞f is μ-measurable. Finally, since $\{-\infty f < \beta\} = \{\infty f > -\beta\}$, another appeal to Proposition 4.31 shows that $-\infty f \in \mathcal{M}(\mu)$.

(b) We have

$$\{f + g < \alpha\} = \{f < -g + \alpha\}$$

for every $\alpha \in \mathbb{R}$, so that (a) together with Proposition 4.32 proves the statement.

(c) Since

$$\{f^+ < \alpha\} = \begin{cases} \{f < \alpha\} & \text{if } \alpha > 0, \\ \emptyset & \text{if } \alpha \le 0, \end{cases}$$

we have that $f^+ \in \mathcal{M}(\mu)$ whenever f is μ-measurable. Then, by (a), $f^- = (-f)^+ \in \mathcal{M}(\mu)$. The converse follows from (a) and (b).

(d) follows from (b) and (c), by virtue of $|f| = f^+ + f^-$.

(e) Since $\{g < \alpha\}$ differs from $\{f < \alpha\}$ only by a μ-null set, the assertion is obvious.

(f) In view of (c) we need only consider the case $f \ge 0$. Then

$$\{f e_A < \alpha\} = \begin{cases} \{f < \alpha\} \cup (X \setminus A) & \text{if } \alpha > 0, \\ \emptyset & \text{if } \alpha \le 0. \end{cases}$$

$\square$

In contrast to the situation in $\mathcal{L}^1(\mu)$, the product of two μ-measurable functions is again μ-measurable. Since we make no use of this fact we leave its proof as an exercise.

Corollary 4.34 *Every $\mathfrak{M}(\mu)$-step function is μ-measurable.*

Proof. This follows from Proposition 4.30(b) and Theorem 4.33(a),(b). $\square$

We next show that $\mathcal{M}(\mu)$ is closed with respect to the relevant limit operations.

Theorem 4.35

(a) *Given a countable family $(f_\iota)_{\iota \in I}$ in $\mathcal{M}(\mu)$, the functions $\bigwedge_{\iota \in I} f_\iota$ and $\bigvee_{\iota \in I} f_\iota$ belong to $\mathcal{M}(\mu)$.*

(b) *For each sequence* $(f_n)_{n \in \mathbb{N}}$ *in* $\mathcal{M}(\mu)$ *the functions* $\limsup_{n \to \infty} f_n$ *and* $\liminf_{n \to \infty} f_n$ *are* μ*-measurable.*

(c) *If* $(f_n)_{n \in \mathbb{N}}$ *is a sequence in* $\mathcal{M}(\mu)$ *converging pointwise in* $\overline{\mathbb{R}}^X$, *then* $\lim_{n \to \infty} f_n \in \mathcal{M}(\mu)$.

Proof. (a) The first assertion follows from the fact that for each $\alpha \in \mathbb{R}$

$$\left\{ \bigwedge_{\iota \in I} f_\iota < \alpha \right\} = \bigcup_{\iota \in I} \{ f_\iota < \alpha \}$$

and the second is a consequence of

$$\left\{ \bigvee_{\iota \in I} f_\iota \le \alpha \right\} = \bigcap_{\iota \in I} \{ f_\iota \le \alpha \}$$

and Proposition 4.31.

(b) is immediate from (a) and (c) follows from (b). □

The following fact is useful.

Proposition 4.36 *If* $f \in \overline{\mathbb{R}}^X$ *is* μ*-measurable and* B *is a Borel set of* $\mathbb{R}$, *then* $f^{-1}(B) \in \mathfrak{M}(\mu)$.

Proof. Since $\mathfrak{M}(\mu)$ is a σ-algebra, the same is true of

$$\mathfrak{T} := \{ B \subset \mathbb{R} \mid f^{-1}(B) \in \mathfrak{M}(\mu) \}.$$

By hypothesis $\mathfrak{T}$ contains every set of the form $]-\infty, \alpha[$. We conclude successively that $\mathfrak{T}$ also contains every set of the form $[\beta, \alpha[$, every set of the form $]\gamma, \alpha[$ and every open subset of $\mathbb{R}$. Since $\mathfrak{B}(\mathbb{R})$ is the σ-algebra on $\mathbb{R}$ generated by the open subsets of $\mathbb{R}$, $\mathfrak{B}(\mathbb{R})$ is also contained in $\mathfrak{T}$. □

What is the relationship between measurability and integrability? The answer to this question is provided by the next two results.

Proposition 4.37 $\mathcal{L}^1_{loc}(\mu)$ *(and hence* $\mathcal{L}^1(\mu)$ *too) is contained in* $\mathcal{M}(\mu)$.

Proof. In view of Theorem 4.33(c), it is sufficient to consider an $f \in \mathcal{L}^1_{loc}(\mu)_+$. Given $\alpha \le 0$, $\{ f \ge \alpha \} = X \in \mathfrak{M}(\mu)$. Now take $\alpha > 0$ and $A \in \mathfrak{L}(\mu)$. By Proposition 4.29, there are a sequence $(A_n)_{n \in \mathbb{N}}$ in $\mathfrak{R}$ and a $B \in \mathfrak{N}(\mu)$ with $A \subset (\bigcup_{n \in \mathbb{N}} A_n) \cup B$. Take $n \in \mathbb{N}$. By Corollary 4.9

$$\{ f \ge \alpha \} \cap A_n = \{ f e_{A_n} \ge \alpha \} \in \mathfrak{L}(\mu).$$

Hence

$$\{ f \ge \alpha \} \cap A = \left(\bigcup_{n \in \mathbb{N}} (\{ f \ge \alpha \} \cap A_n \cap A) \right) \cup (\{ f \ge \alpha \} \cap B \cap A) \in \mathfrak{L}(\mu),$$

proving the proposition. □

The next theorem's content is – loosely speaking – that the only way a measurable function can fail to be integrable is by being 'too big': as soon as a measurable function satisfies a suitable boundedness condition, it is immediately integrable.

Theorem 4.38 *Take $f \in \overline{\mathbb{R}}^X$.*

(a) *If $f \in \mathcal{M}(\mu)$ and $|f| \leq g$ μ-a.e. for some $g \in \mathcal{L}^1(\mu)$, then $f \in \mathcal{L}^1(\mu)$.*

(b) *If $f \in \mathcal{M}(\mu)$ and $|f| \leq g$ μ-a.e. for some $g \in \mathcal{L}^1_{loc}(\mu)$, then $f \in \mathcal{L}^1_{loc}(\mu)$.*

Proof. (a) We need only consider positive f. Take the sequence $(f_n)_{n \in \mathbb{N}}$ defined in Proposition 4.10. Given $\alpha > 0$, the μ-measurable set $\{f \geq \alpha\}$ is contained (except perhaps for a μ-null set) in $\{g \geq \alpha\}$, which is μ-integrable by Corollary 4.9. We conclude that $f_n \in \mathcal{L}^1(\mu)$ and $\int f_n d\mu \leq \int g\, d\mu$ for every $n \in \mathbb{N}$. By the monotone convergence theorem, $f \in \mathcal{L}^1(\mu)$.

(b) follows from (a) and Theorem 4.33(f). $\square$

We next investigate convergence properties of sequences of measurable functions. We start with the observation that the sequence $(f_n)_{n \in \mathbb{N}}$ defined by $f_n(x) := x^n$ converges pointwise but not uniformly to 0 on the interval $[0, 1[$. However, given $\varepsilon > 0$, the convergence *is* uniform on $[0, 1 - \varepsilon[$. Egoroff's theorem (our next theorem) generalizes this fact considerably. First define

$$\mathcal{L}^0(\mu) := \{f \in \mathcal{M}(\mu) \mid \{|f| = \infty\} \in \mathfrak{N}(\mu)\}.$$

By Proposition 4.37 and Theorem 4.26(b), $\mathcal{L}^1_{loc}(\mu) \subset \mathcal{L}^0(\mu)$.

Theorem 4.39 (Egoroff) *Take $f \in \mathcal{L}^0(\mu)$ and a sequence $(f_n)_{n \in \mathbb{N}}$ in $\mathcal{L}^0(\mu)$ with*

$$f(x) = \lim_{n \to \infty} f_n(x) \quad \mu\text{-a.e.}$$

Given $A \in \mathfrak{L}(\mu)$ and $\varepsilon > 0$, there is a $B \in \mathfrak{L}(\mu)$, $B \subset A$, with

$$\mu^X(A \setminus B) < \varepsilon \quad \text{and} \quad f = \lim_{n \to \infty} f_n \text{ uniformly on } B.$$

Proof. Define

$$C := \left(\bigcap_{n \in \mathbb{N}} \{|f_n| < \infty\} \right) \cap \{|f| < \infty\} \cap \{f = \lim_{n \to \infty} f_n\}.$$

By hypothesis $X \setminus C \in \mathfrak{N}(\mu)$.

Take $m \in \mathbb{N}$. Given $n \in \mathbb{N}$, define

$$A_{mn} := A \cap \left\{x \in C \,\middle|\, |f(x) - f_n(x)| \geq \frac{1}{m}\right\},$$

$$B_{mn} := \bigcup_{k \geq n} A_{mk}.$$

Since the sets A and C and the functions f and f_n are μ-measurable, each of the sets A_{mn} and B_{mn} belongs to $\mathfrak{L}(\mu)$ (Proposition 4.31, Theorem 4.33). We show that $\bigcap_{n \in \mathbb{N}} B_{mn} = \emptyset$. Indeed, assume that $x \in \bigcap_{n \in \mathbb{N}} B_{mn}$. Then for each $n \in \mathbb{N}$ there is a $k \geq n$ with $x \in A_{mk}$, i.e. $|f(x) - f_k(x)| \geq 1/m$, which contradicts the fact that $f(x) = \lim_{n \to \infty} f_n(x)$. Now since $(B_{mn})_{n \in \mathbb{N}}$

is decreasing, we conclude that $\inf_{n \in \mathbb{N}} \mu^X(B_{mn}) = 0$. Hence there is an $n_m \in \mathbb{N}$ with

$$\mu^X(B_{mn_m}) < \frac{\varepsilon}{2^m}.$$

Put

$$B := (A \cap C) \setminus \bigcup_{m \in \mathbb{N}} B_{mn_m}.$$

Obviously $B \in \mathfrak{L}(\mu)$. Moreover,

$$\mu^X(A \setminus B) = \mu^X(A \setminus C) + \mu^X \left(\bigcup_{m \in \mathbb{N}} B_{mn_m} \right)$$

$$\leq 0 + \sum_{m \in \mathbb{N}} \mu(B_{mn_m}) < \sum_{m \in \mathbb{N}} \frac{\varepsilon}{2^m} = \varepsilon.$$

To prove that $(f_n)_{n \in \mathbb{N}}$ converges to f uniformly on B, choose $\varepsilon' > 0$ and $m \in \mathbb{N}$ with $1/m < \varepsilon'$. If $x \in B$, then

$$x \in (A \cap C) \setminus \bigcup_{k \geq n_m} A_{mk},$$

so that

$$|f(x) - f_k(x)| < \frac{1}{m} < \varepsilon'$$

for every $k \geq n_m$. $\qquad \square$

The example preceding Egoroff's theorem shows that $A \setminus B$ need not be an element of $\mathfrak{N}(\mu)$. Indeed, given $B \in \mathfrak{L}(\lambda)$ with $[0, 1[\setminus B \in \mathfrak{N}(\lambda)$, B contains points arbitrarily close to 1. This fact makes it easy to see that $(f_n)_{n \in \mathbb{N}}$ cannot converge uniformly to 0 on B.

Our next observation is preliminary to our study of another notion of convergence.

Proposition 4.40 *Take $f, g \in \mathcal{M}(\mu)$ and $\alpha > 0$. Then*

$$\{x \in X \,|\, |f(x) - g(x)| \text{ is defined and } \geq \alpha\} \in \mathfrak{M}(\mu).$$

Proof. Put

$$A := \{x \in X \,|\, |f(x) - g(x)| \text{ is defined}\}.$$

Then, by Proposition 4.32(b),

$$A = X \setminus (\{f = g = \infty\} \cup \{f = g = -\infty\}) \in \mathfrak{M}(\mu).$$

Hence, by Theorem 4.33(f), $f e_A$ and $g e_A$ are both in $\mathcal{M}(\mu)$. Moreover,

$$\{x \in X \,|\, |f(x) - g(x)| \text{ is defined and } \geq \alpha\} = \{|f e_A - g e_A| \geq \alpha\} \in \mathfrak{M}(\mu).$$

$$\square$$

Let $(f_n)_{n \in \mathbb{N}}$ be a sequence in $\mathcal{M}(\mu)$ and take $f \in \mathcal{M}(\mu)$. Then $(f_n)_{n \in \mathbb{N}}$ is said to **converge in measure** μ to f if, given any $A \in \mathfrak{R}$ and any $\alpha > 0$,

$$\lim_{n \to \infty} \mu^X (\{x \in A \,|\, |f_n(x) - f(x)| \text{ is defined and } \geq \alpha\}) = 0. \qquad (1)$$

(Note that by the preceding proposition the set in question is indeed in $\mathfrak{L}(\mu)$.) We write $f_n \overset{\mu}{\to} f$ to denote that $(f_n)_{n \in \mathbb{N}}$ converges in measure μ to f.

Condition (1) is of particular interest for 'big' sets A and small values of α. It is obvious that every subsequence of a sequence which converges in measure also converges in measure to the same limit. However, the limit function is not uniquely determined in general! For if $f_n \overset{\mu}{\to} f$ and $g = f$ μ-a.e., then $f_n \overset{\mu}{\to} g$. On the other hand, two limit functions of a measure-convergent sequence cannot differ on anything but a null set. This could be proved here immediately, but we prefer to derive it as a corollary to another theorem (see below).

For our first example take the sequence $(f_n)_{n \in \mathbb{N}}$, defined by $f_n(x) := x^n$. This sequence converges in Lebesgue measure to 0 on $X = [0, 1[$. Obviously every uniformly convergent sequence of measurable functions converges in measure. In general, however, the concept of convergence in measure – which is important in probability theory for example – is not as easy to handle. We therefore seek other tests for determining whether a sequence converges in measure. But first we show that if $f_n \overset{\mu}{\to} f$ and $A \in \mathfrak{L}(\mu)$, then the defining property (1) is automatically satisfied.

Proposition 4.41 *Let f and f_n ($n \in \mathbb{N}$) be μ-measurable functions on X such that $f_n \overset{\mu}{\to} f$. Take $A \in \mathfrak{L}(\mu)$ and $\alpha > 0$. Then (1) is satisfied.*

Proof. Choose $A \in \mathfrak{L}(\mu)$, $\alpha > 0$ and let $\varepsilon > 0$. By Corollary 4.28, there is a $B \in \mathfrak{R}$ with $\mu^X(A \setminus B) < \varepsilon/2$. Given $n \in \mathbb{N}$, define

$$C_n := \{x \in X \mid |f_n(x) - f(x)| \text{ is defined and } \geq \alpha\}.$$

Then there is an $n_0 \in \mathbb{N}$ such that $\mu^X(B \cap C_n) < \varepsilon/2$ whenever $n \geq n_0$. Hence

$$\mu^X(A \cap C_n) \leq \mu^X(A \setminus B) + \mu^X(B \cap C_n) < \varepsilon,$$

which implies the statement. $\square$

We now prove the first of our main results on convergence in measure. Note that

$$\mathfrak{L}(\mu)_\sigma = \Big\{ \bigcup_{n \in \mathbb{N}} A_n \,\Big|\, A_n \in \mathfrak{L}(\mu) \text{ for every } n \in \mathbb{N} \Big\}$$

(see Exercise 17 of Section 2.4).

Theorem 4.42 *Let f and f_n ($n \in \mathbb{N}$) be μ-measurable functions on X such that $f_n \overset{\mu}{\to} f$. Take a set $A \in \mathfrak{L}(\mu)_\sigma$. Then there is a subsequence of $(f_n)_{n \in \mathbb{N}}$ which converges pointwise μ-a.e. to f on A.*

Proof. As a first step, suppose that $A \in \mathfrak{L}(\mu)$. Then Proposition 4.41 enables us to construct recursively a subsequence $(f_{n_k})_{k \in \mathbb{N}}$ such that for every $k \in \mathbb{N}$

$$\mu^X(B_k) < \frac{1}{2^k},$$

where
$$B_k := \left\{ x \in A \,\middle|\, |f_{n_k}(x) - f(x)| \text{ is defined and } \geq \frac{1}{k} \right\}.$$

Define
$$C_m := \bigcup_{k \geq m} B_k \text{ for every } m \in \mathbb{N} \quad \text{and} \quad C := \bigcap_{m \in \mathbb{N}} C_m.$$

It then follows from
$$\mu^X(C) = \inf_{m \in \mathbb{N}} \mu^X(C_m) \leq \inf_{m \in \mathbb{N}} \sum_{k \geq m} \frac{1}{2^k} = \inf_{m \in \mathbb{N}} \frac{1}{2^{m-1}} = 0$$

that $C \in \mathfrak{N}(\mu)$. We next show that $(f_{n_k})_{k \in \mathbb{N}}$ converges to f pointwise on $A \backslash C$. So take $x \in A \backslash C$. Then there is a k_0 with $x \in A \backslash C_{k_0}$. We distinguish three cases.

Case 1: $f(x) = \infty$. Then the expression $|f_{n_k}(x) - f(x)|$ is not defined for any $k \geq k_0$. (Otherwise, $|f_{n_k}(x) - f(x)| = \infty$ since $f(x) = \infty$, which would contradict the definition of C_{k_0}.) But this can happen only if $f_{n_k}(x) = \infty$ for every $k \geq k_0$, i.e.
$$f(x) = \lim_{k \to \infty} f_{n_k}(x).$$

Case 2: $f(x) = -\infty$. The argument is similar to the one above.

Case 3: $f(x) \in \mathbb{R}$. Take $\varepsilon > 0$ and choose $k_\varepsilon \in \mathbb{N}$ with $k_\varepsilon > \max\{k_0, \frac{1}{\varepsilon}\}$. If $k \geq k_\varepsilon$, then $x \notin B_k$ and so
$$|f_{n_k}(x) - f(x)| < \frac{1}{k} < \varepsilon,$$

which again shows that
$$f(x) = \lim_{k \to \infty} f_{n_k}(x).$$

Now, for the second step, let $A = \bigcup_{n \in \mathbb{N}} A_n$ where $(A_n)_{n \in \mathbb{N}}$ is a sequence in $\mathfrak{L}(\mu)$. For each $k \in \mathbb{N}$
$$f_{0,k} := f_k.$$

Take $m \in \mathbb{N}$. We use the result of the first step to construct recursively a subsequence, $(f_{m,k})_{k \in \mathbb{N}}$, of $(f_{m-1,k})_{k \in \mathbb{N}}$ such that $(f_{m,k})_{k \in \mathbb{N}}$ converges μ-a.e. on $A_1 \cup \cdots \cup A_m$ to f. Then the diagonal sequence $(f_{k,k})_{k \in \mathbb{N}}$ will be the required subsequence of $(f_n)_{n \in \mathbb{N}}$. Indeed, if $x \in A$ and x does not belong to any of the countably many null sets generated in the recursive construction, then, fixing $m \in \mathbb{N}$ with $x \in A_m$,
$$f(x) = \lim_{k \to \infty} f_{k,k}(x),$$

due to the fact that $(f_{k,k})_{k \in \mathbb{N}}$ is, except for its first $m - 1$ terms, a subsequence of $(f_{m,k})_{k \in \mathbb{N}}$. $\qquad\square$

Corollary 4.43 *Let f and f_n ($n \in \mathbb{N}$) be μ-measurable functions on X such that $f_n \xrightarrow{\mu} f$. If $(X, \mathfrak{R}, \mu)$ is σ-finite, then there is a subsequence of $(f_n)_{n \in \mathbb{N}}$ which converges pointwise μ-a.e. on X to f.*

The result mentioned above concerning the limit of a sequence which converges in measure is an easy consequence of Theorem 4.42:

Corollary 4.44 *Suppose that f, g and f_n ($n \in \mathbb{N}$) are μ-measurable functions on X. If $f_n \xrightarrow{\mu} f$ and $f_n \xrightarrow{\mu} g$, then $f = g$ μ-a.e.*

Proof. Take $A \in \mathfrak{R}$. Then there is a subsequence of $(f_n)_{n \in \mathbb{N}}$ which converges μ-a.e. to f on A and this subsequence in turn has a subsequence converging μ-a.e. to g on A. It follows that $f = g$ μ-a.e. on A, i.e. $\{f \neq g\} \cap A \in \mathfrak{N}(\mu)$. Proposition 4.6(b) now implies that $\{f \neq g\} \in \mathfrak{N}(\mu)$. □

If each of the functions f_n ($n \in \mathbb{N}$) is integrable, then we can also find a subsequence converging almost everywhere on the whole space X. We formulate this result in slightly greater generality.

Corollary 4.45 *Let f and f_n ($n \in \mathbb{N}$) be μ-measurable functions on X and suppose that $f_n \xrightarrow{\mu} f$. If there is an $A \in \mathfrak{L}(\mu)_\sigma$ such that $\bigcup_{n \in \mathbb{N}} \{f_n \neq 0\} \subset A$ (as is the case by Proposition 4.29 when each f_n is μ-integrable), then there is a subsequence of $(f_n)_{n \in \mathbb{N}}$ which converges pointwise μ-a.e. to f on X.*

Proof. By Theorem 4.42 there is a subsequence, $(f_{n_k})_{k \in \mathbb{N}}$, of $(f_n)_{n \in \mathbb{N}}$ converging μ-a.e. to f on A. Obviously $f_n e_{X \setminus A} \xrightarrow{\mu} f e_{X \setminus A}$. But $f_n e_{X \setminus A} \xrightarrow{\mu} 0$ as well. Thus by Corollary 4.44 $f = 0$ μ-a.e. on $X \setminus A$. It follows that $(f_{n_k})_{k \in \mathbb{N}}$ converges μ-a.e. to f on X. □

$f_n \xrightarrow{\mu} f$ does *not* in general imply that the sequence $(f_n)_{n \in \mathbb{N}}$ itself converges to f μ-a.e. As an example, consider Lebesgue measure λ and the functions

$$f_n := e_{[j2^{-k}, (j+1)2^{-k}[} \qquad \text{where } n = 2^k + j \text{ and } 0 \leq j < 2^k.$$

The first terms of the sequence $(f_n)_{n \in \mathbb{N}}$ are

$$e_{[0,1[}, \ e_{[0,\frac{1}{2}[}, \ e_{[\frac{1}{2},1[}, \ e_{[0,\frac{1}{4}[}, \ e_{[\frac{1}{4},\frac{1}{2}[}, \ \cdots$$

Now $f_n \xrightarrow{\lambda} 0$, but $(f_n(x))_{n \in \mathbb{N}}$ does not converge for any $x \in [0,1[$. On the other hand the subsequence $\left(e_{[0,2^{-k}[}\right)_{k \in \mathbb{N}}$ converges λ-a.e. to 0 (but not at 0, and so not everywhere!).

We use Egoroff's theorem to derive a converse to Theorem 4.42 in the case of functions belonging to $\mathcal{L}^0(\mu)$. The next result states in particular that every sequence in $\mathcal{L}^0(\mu)$ which converges pointwise μ-a.e. to a function in $\mathcal{L}^0(\mu)$ also converges in measure μ to this function.

Theorem 4.46 *Consider f and f_n ($n \in \mathbb{N}$) in $\mathcal{L}^0(\mu)$. Suppose that every subsequence of $(f_n)_{n \in \mathbb{N}}$ has a subsequence which converges to f μ-a.e. Then $f_n \xrightarrow{\mu} f$.*

Proof. Assume the statement is false. Then we can find an $A \in \mathfrak{R}$ and numbers $\alpha > 0$ and $\varepsilon > 0$ such that for every $n_0 \in \mathbb{N}$ there is an $n \geq n_0$ satisfying

$$\mu^X\left(\{x \in A \,|\, |f_n(x) - f(x)| \text{ is defined and } \geq \alpha\}\right) > \varepsilon.$$

Hence we can choose a subsequence $(f_{n_m})_{m \in \mathbb{N}}$ of $(f_n)_{n \in \mathbb{N}}$ such that for every $m \in \mathbb{N}$

$$\mu^X\left(\{x \in A \,|\, |f_{n_m}(x) - f(x)| \text{ is defined and } \geq \alpha\}\right) > \varepsilon. \tag{2}$$

By hypothesis $(f_{n_m})_{m \in \mathbb{N}}$ has a subsequence $(f_{n_{m(k)}})_{k \in \mathbb{N}}$ which converges to f μ-a.e. Now apply Egoroff's theorem to find a set $B \in \mathfrak{L}(\mu)$ contained in A such that

$$\mu^X(A \setminus B) < \varepsilon \quad \text{and} \quad f = \lim_{k \to \infty} f_{n_{m(k)}} \text{ uniformly on } B.$$

Thus there is a $k_0 \in \mathbb{N}$ such that for any $k \geq k_0$ and every $x \in B$

$$\left|f_{n_{m(k)}}(x) - f(x)\right| < \alpha.$$

We infer that for every $k \geq k_0$

$$\mu^X\left(\{x \in A \,|\, |f_{n_{m(k)}}(x) - f(x)| \text{ is defined and } \geq \alpha\}\right) \leq \mu^X(A \setminus B) < \varepsilon,$$

which contradicts (2). $\qquad\qquad\qquad\qquad\qquad\qquad\qquad\qquad\square$

We combine Corollary 4.45 and Theorem 4.46 to obtain our next corollary.

Corollary 4.47 *Take f and f_n ($n \in \mathbb{N}$) in $\mathcal{L}^0(\mu)$ and suppose that*

$$\bigcup_{n \in \mathbb{N}} \{f_n \neq 0\} \subset A$$

for some $A \in \mathfrak{L}(\mu)_\sigma$. Then the following are equivalent.

(a) $f_n \overset{\mu}{\to} f$.

(b) *Every subsequence of $(f_n)_{n \in \mathbb{N}}$ has a subsequence which converges to f μ-a.e.*

Observe again that the hypothesis is satisfied whenever $f \in \mathcal{L}^0(\mu)$ and every f_n is μ-integrable.

We draw the reader's attention to the fact that several authors use a 'global' notion of convergence in measure, defining '$f_n \overset{\mu}{\to} f$ globally' if given any $\alpha > 0$

$$\lim_{n \to \infty} \mu^X\left(\{x \in X \,|\, |f_n(x) - f(x)| \text{ is defined and } \geq \alpha\}\right) = 0.$$

Since global convergence in measure clearly implies convergence in measure, our 'local' concept is more general. It can be shown that if $f_n \overset{\mu}{\to} f$ globally, then $(f_n)_{n \in \mathbb{N}}$ has a subsequence converging to f μ-a.e. But Theorem 4.46 and Corollary 4.47 would not be true for global convergence in measure.

We shall meet convergence in measure again in Chapter 6.

Exercises

1. Prove each of the following.

   (a) Let $(A_\iota)_{\iota\in I}$ be a countable disjoint family in $\mathfrak{M}(\mu)$ with $\bigcup_{\iota\in I} A_\iota = X$ and $(f_\iota)_{\iota\in I}$ a family in $\mathcal{M}(\mu)$. Then the function

   $$f : X \longrightarrow \overline{\mathbb{R}}, \qquad x \longmapsto f_\iota(x) \qquad (x \in A_\iota,\ \iota \in I)$$

   is μ-measurable.

   (b) For each $f \in \mathcal{M}(\mu)$, the function

   $$h : X \longrightarrow \mathbb{R}, \qquad x \longmapsto \begin{cases} f(x) & \text{if } f(x) \in \mathbb{R} \\ 0 & \text{if } f(x) \notin \mathbb{R} \end{cases}$$

   is μ-measurable.

2. Verify the following.

   (a) $fg \in \mathcal{M}(\mu)$ for all $f,g \in \mathcal{M}(\mu)$.
   (b) Given $f \in \mathcal{M}(\mu)$, the function

   $$g : X \longrightarrow \mathbb{R}, \qquad x \longmapsto \begin{cases} 1/f(x) & \text{if } f(x) \in \mathbb{R} \setminus \{0\} \\ 0 & \text{otherwise} \end{cases}$$

   is μ-measurable.

   Hint for (a): First assume that $f,g \geq 0$, and observe that

   $$\{fg < \alpha\} = \{f = 0\} \cup \bigcup_{\substack{\beta\in\mathbb{Q}\\ \beta>0}} \left(\{g < \beta\} \cap \left\{f < \frac{\alpha}{\beta}\right\}\right) \qquad \text{for } \alpha > 0.$$

   For arbitrary f,g, put $A := \{f \geq 0\}$, $B := \{f < 0\}$, $C := \{g \geq 0\}$, $D := \{g < 0\}$, and consider fg on the sets $A \cap C$, $A \cap D$, $B \cap C$ and $B \cap D$.

3. Take $f \in \overline{\mathbb{R}}^X$. Define the **upper integral** of f with respect to μ by

   $$\ell_\mu^*(f) := \inf\{\ell_\mu^\uparrow(g) \mid g \in \mathcal{L}(\mathfrak{R})^\uparrow,\ g \geq f\ \mu\text{-a.e.}\}$$

   (see also Exercises 2 and 3 in Section 6.1). Prove the following.

   (a) $f \in \mathcal{L}^1(\mu)$ if and only if $f \in \mathcal{M}(\mu)$ and $\ell_\mu^*(|f|) < \infty$.
   (b) $f \in \mathcal{N}(\mu)$ if and only if $\ell_\mu^*(|f|) = 0$.

4. Take $f \in \overline{\mathbb{R}}_+^X$. Show that the following are equivalent.

   (a) $f \in \mathcal{M}(\mu)$.
   (b) $fe_A \in \mathcal{M}(\mu)$ for every $A \in \mathfrak{L}(\mu)$.
   (c) $f \wedge \alpha e_A \in \mathcal{L}^1(\mu)$ for every $A \in \mathfrak{L}(\mu)$ and every $\alpha > 0$.
   (d) $f \wedge n e_A \in \mathcal{L}^1(\mu)$ for every $A \in \mathfrak{L}(\mu)$ and every $n \in \mathbb{N}$.

5. Let μ be bounded. Prove the following assertions.

   (a) $\mathcal{M}(\mu)_+ = \{f \in \overline{\mathbb{R}}_+^X \mid f \wedge n \in \mathcal{L}^1(\mu)$ for all $n \in \mathbb{N}\}$.

   (b) $\mathcal{M}(\mu) = \{f \in \overline{\mathbb{R}}^X \mid (f \wedge n) \vee (-n) \in \mathcal{L}^1(\mu)$ for all $n \in \mathbb{N}\}$.

6. Determine which functions are measurable with respect to Dirac measure δ_x and which are measurable with respect to the measure μ defined in Exercise 4 of Section 4.1.

7. Show that Egoroff's theorem does not apply to sets $A \in \mathfrak{M}(\mu)$. (Consider the sequence $(e_{[n-1,n[})_{n\in\mathbb{N}}$ and a set $B \in \mathfrak{M}(\lambda)$ with

$$\lambda^{\mathbb{R}}([0,\infty[\setminus B) < 1/2. \,)$$

8. Find an example showing that Theorem 4.46 does not apply to functions $f_n \in \mathcal{M}(\mu)$.

9. Prove the following.

   (a) If $f_n \overset{\mu}{\to} f$ globally, then $(f_n)_{n\in\mathbb{N}}$ has a subsequence converging to f μ-a.e.

   (b) Convergence in measure does not imply global convergence in measure, and Theorem 4.46 and Corollary 4.47 do not hold for global convergence in measure. (Consider the sequence $(e_{[n-1,n[})_{n\in\mathbb{N}}$.)

10. Determine which of the following sequences $(f_n)_{n\in\mathbb{N}}$ converge in Lebesgue measure on $\mathbb{R}$:

    (a) $f_{2n} = e_{[-n,n]}$, $f_{2n-1} = e_{[-\frac{1}{n},\frac{1}{n}]}$;

    (b) $f_n = \infty e_{[0,n]}$;

    (c) $f_n(t) = \begin{cases} t^{\frac{1-n}{n}} & \text{if } 0 < t < 1, \\ 0 & \text{otherwise.} \end{cases}$

11. Let μ be counting measure on $\mathbb{N}$. Determine whether the sequences $(e_{\mathbb{N}\setminus\{1,...,n\}})_{n\in\mathbb{N}}$ and $(\infty e_{\{1,...,n\}})_{n\in\mathbb{N}}$ converge in measure μ.

4.5 Product measures and Fubini's theorem

'Product measures' provide one important way to construct positive measures. This section is devoted to them. We begin by describing the construction in the case of two positive measure spaces $(X, \mathfrak{R}, \mu)$ and $(Y, \mathfrak{S}, \nu)$. Define

$$\mathfrak{R} \square \mathfrak{S} := \{A \times B \mid A \in \mathfrak{R},\ B \in \mathfrak{S}\}.$$

Write $\mathfrak{R} \otimes \mathfrak{S}$ for the ring of sets generated by $\mathfrak{R} \square \mathfrak{S}$. The elements of $\mathfrak{R} \square \mathfrak{S}$ are called **rectangles**. Given $A \times B \in \mathfrak{R} \square \mathfrak{S}$, define

$$(\mu \square \nu)(A \times B) := \mu(A)\nu(B).$$

This definition provides a natural starting point, for recall that the area of a rectangle in the Euclidean plane is also calculated as the product of the lengths of its sides. Our aim is, of course, to extend the mapping $\mu \square \nu$ to a measure on $\mathfrak{R} \otimes \mathfrak{S}$. We first record some preliminary observations.

Proposition 4.48

(a) $\mathfrak{R} \square \mathfrak{S}$ *is a semi-ring of sets.*

(b) *If* $(A_\iota \times B_\iota)_{\iota \in I}$ *is a finite disjoint family in* $\mathfrak{R} \square \mathfrak{S}$ *for which* $\bigcup_{\iota \in I}(A_\iota \times B_\iota) = A \times B \in \mathfrak{R} \square \mathfrak{S}$, *then*

$$(\mu \square \nu)(A \times B) = \sum_{\iota \in I}(\mu \square \nu)(A_\iota \times B_\iota).$$

Proof. (a) follows from the general facts that for any sets A, B, C and D whatsoever,

$$(A \times B) \cap (C \times D) = (A \cap C) \times (B \cap D)$$
$$(A \times B) \setminus (C \times D) = \big((A \setminus C) \times B\big) \cup \big((A \cap C) \times (B \setminus D)\big)$$
$$\big((A \setminus C) \times B\big) \cap \big((A \cap C) \times (B \setminus D)\big) = \emptyset.$$

(b) Take $x \in X$. Then

$$e_A(x)e_B = \sum_{\iota \in I} e_{A_\iota}(x)e_{B_\iota}.$$

Each of the functions in this equality is an $\mathfrak{S}$-step function. It follows from Proposition 2.23 that for every $x \in X$

$$e_A(x)\nu(B) = \sum_{\iota \in I} e_{A_\iota}(x)\nu(B_\iota).$$

Thus

$$e_A \nu(B) = \sum_{\iota \in I} e_{A_\iota}\nu(B_\iota).$$

This is an equality between $\mathfrak{R}$-step functions. By the same arguments,

$$\mu(A)\nu(B) = \sum_{\iota \in I} \mu(A_\iota)\nu(B_\iota),$$

which is what we sought to prove. $\square$

Proposition 4.48 ensures that the hypotheses of Theorem 2.27 are satisfied. We can therefore extend the mapping $\mu \square \nu$ to $\mathfrak{R} \otimes \mathfrak{S}$. We have the following theorem concerning this extension.

Theorem 4.49 *There is a unique positive measure* $\mu \otimes \nu$ *on* $\mathfrak{R} \otimes \mathfrak{S}$ *such that* $\mu \otimes \nu|_{\mathfrak{R} \square \mathfrak{S}} = \mu \square \nu$.

Proof. Theorem 2.27 asserts the existence and uniqueness of a positive content $\mu \otimes \nu$ on $\mathfrak{R} \otimes \mathfrak{S}$ such that $\mu \otimes \nu|_{\mathfrak{R} \square \mathfrak{S}} = \mu \square \nu$. This is given by

$$(\mu \otimes \nu)\left(\bigcup_{\iota \in I}(A_\iota \times B_\iota)\right) = \sum_{\iota \in I} \mu(A_\iota)\nu(B_\iota)$$

where $(A_\iota \times B_\iota)_{\iota \in I}$ are arbitrary finite disjoint families in $\mathfrak{R} \square \mathfrak{S}$.

Let $(A_n)_{n \in \mathbb{N}}$ be a sequence in $\mathfrak{R} \otimes \mathfrak{S}$ for which $\downarrow A_n = \emptyset$. Given $n \in \mathbb{N}$,

$$A_n = \bigcup_{\iota \in I_n}(A_{n\iota} \times B_{n\iota}),$$

or equivalently,

$$e_{A_n}(x,y) = \sum_{\iota \in I_n} e_{A_{n\iota}}(x)\, e_{B_{n\iota}}(y) \qquad ((x,y) \in X \times Y),$$

where $(A_{n\iota} \times B_{n\iota})_{\iota \in I_n}$ is a finite disjoint family in $\mathfrak{R} \square \mathfrak{S}$. Take $x \in X$. Then

$$\sum_{\iota \in I_n} e_{A_{n\iota}}(x)\, e_{B_{n\iota}} \downarrow 0$$

in $\mathcal{L}(\mathfrak{S})$, and it follows that

$$\sum_{\iota \in I_n} e_{A_{n\iota}}(x)\, \nu(B_{n\iota}) \downarrow 0.$$

Since this holds for all $x \in X$,

$$\sum_{\iota \in I_n} e_{A_{n\iota}} \nu(B_{n\iota}) \downarrow 0$$

in $\mathcal{L}(\mathfrak{R})$. Thus

$$(\mu \otimes \nu)(A_n) = \sum_{\iota \in I_n} \mu(A_{n\iota})\nu(B_{n\iota}) \downarrow 0,$$

which shows that $\mu \otimes \nu$ is null-continuous. $\qquad\qquad\qquad\square$

The positive measure space $(X \times Y, \mathfrak{R} \otimes \mathfrak{S}, \mu \otimes \nu)$ is the **product** of $(X, \mathfrak{R}, \mu)$ and $(Y, \mathfrak{S}, \nu)$. A number of questions arise about the relationship between the integral with respect to the product and the integrals with respect to the factors. The best known of these is indubitably the one addressed by Fubini's theorem.

Suppose we are given a $\mu \otimes \nu$-integrable function f of two variables on the product space $X \times Y$. What is its integral? Fubini's result provides a convenient method for evaluating this integral provided that $f \in \overline{\mathcal{L}}(\mu \otimes \nu)$, a method which should be familiar to the reader from the Riemann integral on $\mathbb{R}^2$. The theorem states that the integral can be evaluated by means of two successive integrations, one with respect to the measure ν, the other with respect to μ. In the first x is kept fixed and the partial function $f(x, \cdot)$ is integrated with respect to ν. This yields a value $g(x) = \int f(x, y) \, d\nu(y)$. Then for the second step consider the result of the first integral as a function of x, say g. Now integrate g with respect to μ. The result is the desired integral $\int f \, d(\mu \otimes \nu)$. Moreover, the same result is obtained by integrating first with respect to μ and then integrating the resulting function (of y) with respect to ν, i.e. the order of integration is immaterial.

Care is required when attempting to prove this. For example, one has to ensure that the partial function $f(x, \cdot)$ is ν-integrable. This is not true in general for every $x \in X$, but fortunately the subset of X on which it is not true is in $\overline{\mathfrak{N}}(\mu)$. (This statement is already a non-trivial part of the theorem!) It therefore has no bearing upon the subsequent integration with respect to μ. But as a consequence we are faced with integrating over X a function which is not defined on all of X. The following terminology overcomes these technical problems and assists us in formulating the theorem precisely.

Let $(X, \mathfrak{R}, \mu)$ be a positive measure space. Let f be a function whose domain of definition D_f is a subset of X with $X \setminus D_f \in \overline{\mathfrak{N}}(\mu)$. Suppose that there is a $g \in \overline{\mathcal{L}}(\mu)$ with $\{g|_{D_f} \neq f\} \in \overline{\mathfrak{N}}(\mu)$. Then each function $h \in \overline{\mathbb{R}}^X$ for which $\{h|_{D_f} \neq f\} \in \overline{\mathfrak{N}}(\mu)$ is also contained in $\overline{\mathcal{L}}(\mu)$. Moreover, $\int h \, d\mu = \int g \, d\mu$. We define the integral of such functions f by

$$\int f \, d\mu := \int g \, d\mu,$$

where g is an arbitrary element of $\overline{\mathcal{L}}(\mu)$ with $\{g|_{D_f} \neq f\} \in \overline{\mathfrak{N}}(\mu)$. We say that f is **essentially contained in** $\overline{\mathcal{L}}(\mu)$.

We are now in a position to formulate Fubini's theorem.

Theorem 4.50 (Fubini) *Take $f \in \overline{\mathcal{L}}(\mu \otimes \nu)$ and define*

$$X_f := \{x \in X \mid f(x, \cdot) \notin \overline{\mathcal{L}}(\nu)\}$$
$$Y_f := \{y \in Y \mid f(\cdot, y) \notin \overline{\mathcal{L}}(\mu)\}.$$

Then

(a) $X_f \in \overline{\mathfrak{N}}(\mu)$, $Y_f \in \overline{\mathfrak{N}}(\nu)$;

(b) the map

$$X \setminus X_f \longrightarrow \mathbb{R}, \quad x \longmapsto \int f(x, y) \, d\nu(y)$$

is essentially contained in $\overline{\mathcal{L}}(\mu)$ and

$$Y \setminus Y_f \longrightarrow \mathbb{R}, \quad y \longmapsto \int f(x, y) \, d\mu(x)$$

is essentially contained in $\overline{\mathcal{L}}(\nu)$;

(c) $\displaystyle\int f\, d(\mu \otimes \nu)$

$$= \int \Big(\int f(x,y)\, d\nu(y)\Big) d\mu(x) = \int \Big(\int f(x,y)\, d\mu(x)\Big) d\nu(y).$$

Proof. We use the induction principle (Theorem 3.38). Let $\mathcal{F}$ be the set of all $f \in \overline{\mathcal{L}}(\mu \otimes \nu)$ for which (a)–(c) are true.

Take $f \in \mathcal{L}(\mathfrak{R} \otimes \mathfrak{S})$. Then f is of the form given by

$$f(x,y) = \sum_{\iota \in I} \alpha_\iota e_{A_\iota}(x)\, e_{B_\iota}(y), \qquad ((x,y) \in X \times Y)$$

where $(A_\iota \times B_\iota)_{\iota \in I}$ is a finite family in $\mathfrak{R} \Box \mathfrak{S}$ and $(\alpha_\iota)_{\iota \in I}$ a corresponding family in $\mathbb{R}$. Then $f(x,\cdot) \in \mathcal{L}(\mathfrak{S})$ for each $x \in X$ and $f(\cdot, y) \in \mathcal{L}(\mathfrak{R})$ for each $y \in Y$. Hence $X_f = Y_f = \emptyset$. Furthermore,

$$\int f(x,y)\, d\nu(y) = \sum_{\iota \in I} \alpha_\iota \nu(B_\iota) e_{A_\iota}(x) \qquad \text{for each } x \in X,$$

$$\int f(x,y)\, d\mu(x) = \sum_{\iota \in I} \alpha_\iota \mu(A_\iota) e_{B_\iota}(y) \qquad \text{for each } y \in Y.$$

Hence $\big(x \mapsto \int f(x,y)\, d\nu(y)\big) \in \mathcal{L}(\mathfrak{R})$ and $\big(y \mapsto \int f(x,y)\, d\mu(x)\big) \in \mathcal{L}(\mathfrak{S})$. It follows that

$$\int f\, d(\mu \otimes \nu) = \sum_{\iota \in I} \alpha_\iota \mu(A_\iota) \nu(B_\iota) = \int \Big(\int f(x,y)\, d\nu(y)\Big) d\mu(x)$$

$$= \int \Big(\int f(x,y)\, d\mu(x)\Big) d\nu(y).$$

Thus $\mathcal{L}(\mathfrak{R} \otimes \mathfrak{S}) \subset \mathcal{F}$.

Now let $(f_n)_{n \in \mathbb{N}}$ be an increasing sequence in $\mathcal{F}$ with $\uparrow \int f_n d(\mu \otimes \nu) < \infty$. Put $f := \uparrow f_n$. Define

$$A' := \bigcup_{n \in \mathbb{N}} X_{f_n}, \quad B' := \bigcup_{n \in \mathbb{N}} Y_{f_n}.$$

Then $A' \in \overline{\mathfrak{N}}(\mu)$ and $B' \in \overline{\mathfrak{N}}(\nu)$. If $x \in X \setminus A'$, then $f_n(x, \cdot) \in \overline{\mathcal{L}}(\nu)$ for every $n \in \mathbb{N}$. Similarly, if $y \in Y \setminus B'$, then $f_n(\cdot, y) \in \overline{\mathcal{L}}(\mu)$ for every $n \in \mathbb{N}$. Given $n \in \mathbb{N}$, define

$$g_n : X \longrightarrow \mathbb{R}, \quad x \longmapsto \begin{cases} \int f_n(x,y)\, d\nu(y) & \text{for } x \in X \setminus A' \\ 0 & \text{for } x \in A', \end{cases}$$

$$h_n : Y \longrightarrow \mathbb{R}, \quad y \longmapsto \begin{cases} \int f_n(x,y)\, d\mu(x) & \text{for } y \in Y \setminus B' \\ 0 & \text{for } y \in B'. \end{cases}$$

$(g_n)_{n\in\mathbb{N}}$ and $(h_n)_{n\in\mathbb{N}}$ are increasing sequences in $\overline{\mathcal{L}}(\mu)$ and $\overline{\mathcal{L}}(\nu)$ respectively. Moreover,

$$\int g_n \, d\mu = \int \Big(\int f_n(x,y) \, d\nu(y) \Big) d\mu(x) = \int f_n d(\mu \otimes \nu) \le \int f \, d(\mu \otimes \nu)$$

and

$$\int h_n \, d\nu = \int \Big(\int f_n(x,y) \, d\mu(x) \Big) d\nu(y) = \int f_n d(\mu \otimes \nu) \le \int f \, d(\mu \otimes \nu).$$

By the monotone convergence theorem $g := \uparrow g_n \in \overline{\mathcal{L}}(\mu)$ and $h := \uparrow h_n \in \overline{\mathcal{L}}(\nu)$. Thus

$$A'' := \{x \in X \mid |g|(x) = \infty\} \in \overline{\mathfrak{N}}(\mu),$$
$$B'' := \{y \in Y \mid |h|(y) = \infty\} \in \overline{\mathfrak{N}}(\nu).$$

Take $x \in X \setminus (A' \cup A'')$. Then $\big(f_n(x,\cdot)\big)_{n\in\mathbb{N}}$ is an increasing sequence in $\overline{\mathcal{L}}(\nu)$ and

$$\sup_{n\in\mathbb{N}} \int f_n(x,y) \, d\nu(y) = g(x) \in \mathbb{R}.$$

Consequently, $f(x,\cdot) \in \overline{\mathcal{L}}(\nu)$ and

$$\int f(x,y) \, d\nu(y) = \uparrow \int f_n(x,y) \, d\nu(y) = g(x).$$

Dually, $f(\cdot,y) \in \overline{\mathcal{L}}(\mu)$ and

$$\int f(x,y) \, d\mu(x) = \uparrow \int f_n(x,y) \, d\mu(x) = h(x)$$

for every $y \in Y \setminus (B' \cup B'')$. It follows that $X_f \subset A' \cup A''$ and hence $X_f \in \overline{\mathfrak{N}}(\mu)$. Dually, $Y_f \in \overline{\mathfrak{N}}(\nu)$. For each $x \in X \setminus X_f$, $\int f(x,y) \, d\nu(y)$ is well defined and, taking

$$\tilde{g} : X \setminus X_f \longrightarrow \mathbb{R}, \quad x \longmapsto \int f(x,y) \, d\nu(y),$$

we have that $\{\tilde{g} \ne g|_{X\setminus X_f}\} \in \overline{\mathfrak{N}}(\mu)$. $\tilde{g}$ is therefore essentially contained in $\overline{\mathcal{L}}(\mu)$ and dually

$$\tilde{h} : Y \setminus Y_f \longrightarrow \mathbb{R}, \quad y \longmapsto \int f(x,y) \, d\mu(x)$$

is essentially contained in $\overline{\mathcal{L}}(\nu)$. Finally, by the monotone convergence theorem,

$$\int f \, d(\mu \otimes \nu) = \uparrow \int f_n d(\mu \otimes \nu) = \uparrow \int \Big(\int f_n(x,y) \, d\nu(y) \Big) d\mu(x)$$
$$= \int g(x) \, d\mu(x) = \int \Big(\int f(x,y) \, d\nu(y) \Big) d\mu(x)$$

and dually

$$\int f \, d(\mu \otimes \nu) = \int \left(\int f(x,y) \, d\mu(x) \right) d\nu(y).$$

Hence $f \in \mathcal{F}$. Similarly, it can be shown that $\downarrow f_n \in \mathcal{F}$ for decreasing sequences $(f_n)_{n \in \mathbb{N}}$ in $\mathcal{F}$, where $\downarrow \int f_n d(\mu \otimes \nu) > -\infty$. Condition (ii) of the induction principle is therefore satisfied.

Finally, take $f, g \in \mathcal{F}$ with $f \le g$ such that

$$\int f \, d(\mu \otimes \nu) = \int g \, d(\mu \otimes \nu).$$

Take $h \in \overline{\mathcal{L}}(\mu \otimes \nu)$ such that $f \le h \le g$. Then

$$\int \left(\int f(x,y) \, d\nu(y) \right) d\mu(x) = \int \left(\int g(x,y) \, d\nu(y) \right) d\mu(x).$$

From this we conclude that the set of those $x \in X$ for which either $\int f(x,y) \, d\nu(y)$ or $\int g(x,y) \, d\nu(y)$ is not defined or the integrals are defined but unequal is contained in $\overline{\mathfrak{N}}(\mu)$ (Theorem 3.22). For any other x,

$$\int f(x,y) \, d\nu(y) = \int g(x,y) \, d\nu(y),$$

implying that $h(x, \cdot) \in \overline{\mathcal{L}}(\nu)$ and

$$\int h(x,y) \, d\nu(y) = \int f(x,y) \, d\nu(y) = \int g(x,y) \, d\nu(y).$$

Thus $X_h \in \overline{\mathfrak{N}}(\mu)$. Dually, $Y_h \in \overline{\mathfrak{N}}(\nu)$ and

$$\int h(x,y) \, d\mu(x) = \int f(x,y) \, d\mu(x) = \int g(x,y) \, d\mu(x)$$

whenever y is not an element of a certain set of $\overline{\mathfrak{N}}(\nu)$. Property (b) follows, as do

$$\int h \, d(\mu \otimes \nu) = \int f \, d(\mu \otimes \nu) = \int \left(\int f(x,y) \, d\nu(y) \right) d\mu(x)$$

$$= \int \left(\int h(x,y) \, d\nu(y) \right) d\mu(x)$$

and

$$\int h \, d(\mu \otimes \nu) = \int \left(\int h(x,y) \, d\mu(x) \right) d\nu(y).$$

Hence $h \in \mathcal{F}$, completing the proof of the theorem. $\square$

Fubini's theorem was stated for functions in $\overline{\mathcal{L}}(\mu \otimes \nu)$. The obvious question is whether it holds for every $\mu \otimes \nu$-integrable function. This is unfortunately not the case. The exercises contain the construction of a counterexample.

While we do not have Fubini's theorem in the most general case, it does hold in the most common case, namely that of σ-finite measure spaces.

Theorem 4.51 *If $(X, \mathfrak{R}, \mu)$ and $(Y, \mathfrak{S}, \nu)$ are σ-finite, then*

(a) *$(X \times Y, \mathfrak{R} \otimes \mathfrak{S}, \mu \otimes \nu)$ is σ-finite and*

(b) *Theorem 4.50 holds for every $f \in \mathcal{L}^1(\mu \otimes \nu)$.*

Proof. If $(A_n)_{n \in \mathbb{N}}$ and $(B_n)_{n \in \mathbb{N}}$ are sequences in $\mathfrak{R}$ and $\mathfrak{S}$ respectively such that $X = \bigcup_{n \in \mathbb{N}} A_n$ and $Y = \bigcup_{n \in \mathbb{N}} B_n$, then

$$X \times Y = \bigcup_{(n,m) \in \mathbb{N} \times \mathbb{N}} (A_n \times B_m).$$

(a) follows. Part (b) follows from (a) by Theorem 4.18.   □

Another obvious question is the following. Suppose that for some $f \in \overline{\mathbb{R}}^{X \times Y}$ parts (a) and (b) of Fubini's theorem hold and that

$$\int \left(\int f(x,y) \, d\nu(y) \right) d\mu(x) = \int \left(\int f(x,y) \, d\mu(x) \right) d\nu(y).$$

Is it true that $f \in \mathcal{L}^1(\mu \otimes \nu)$? The answer is negative even in the case of σ-finite measure spaces. An example is provided in the exercises. However, if f is positive and $\mu \otimes \nu$-measurable and $\{f \neq 0\}$ is contained in the union of some sequence from $\mathfrak{R} \otimes \mathfrak{S}$, then the answer is affirmative. This is the substance of Tonelli's theorem which is presented in the exercises.

The starting point for our definition of the product measure $\mu \otimes \nu$ was the requirement that

$$(\mu \otimes \nu)(A \times B) = \mu(A)\nu(B) \qquad \text{for all } A \in \mathfrak{R} \text{ and } B \in \mathfrak{S}.$$

It is natural to expect this relation to extend to sets in $\mathfrak{L}(\mu)$ and $\mathfrak{L}(\nu)$, i.e. that

$$(\mu \otimes \nu)^{X \times Y}(A \times B) = \mu^X(A)\nu^Y(B) \qquad \text{for all } A \in \mathfrak{L}(\mu) \text{ and all } B \in \mathfrak{L}(\nu).$$

We pose the problem more generally. Define the 'product' $f \otimes g$ of the functions $f \in \overline{\mathbb{R}}^X$ and $g \in \overline{\mathbb{R}}^Y$ by

$$f \otimes g : X \times Y \longrightarrow \overline{\mathbb{R}}, \quad (x,y) \longmapsto f(x)g(y).$$

(Thus, for example, $e_A \otimes e_B = e_{A \times B}$ for $A \subset X$, $B \subset Y$.) Is it then true that for $f \in \mathcal{L}^1(\mu)$ and $g \in \mathcal{L}^1(\nu)$ the function $f \otimes g$ is $\mu \otimes \nu$-integrable and its integral is $(\int f \, d\mu)(\int g \, d\nu)$? Our next theorem confirms our expectation.

Theorem 4.52

(a) *If $A \in \mathfrak{N}(\mu)$ and $B \in \mathfrak{N}(\nu)$, then $A \times Y \in \mathfrak{N}(\mu \otimes \nu)$ and $X \times B \in \mathfrak{N}(\mu \otimes \nu)$.*

(b) *Take $A \in \mathfrak{L}(\mu)$ and $B \in \mathfrak{L}(\nu)$. Then $A \times B \in \mathfrak{L}(\mu \otimes \nu)$ and*

$$(\mu \otimes \nu)^{X \times Y}(A \times B) = \mu^X(A)\,\nu^Y(B).$$

(c) *Take* $f \in \mathcal{L}^1(\mu)$ *and* $g \in \mathcal{L}^1(\nu)$. *Then* $f \otimes g \in \mathcal{L}^1(\mu \otimes \nu)$ *and*

$$\int (f \otimes g) \, d(\mu \otimes \nu) = \left(\int f \, d\mu \right) \left(\int g \, d\nu \right).$$

Proof. As ever, we use the induction principle. Take $g \in \mathcal{L}(\mathfrak{S})_+$, $g = \sum_{\lambda \in L} \beta_\lambda e_{B_\lambda}$. Define

$$\mathcal{F} := \left\{ f \in \overline{\mathcal{L}}(\mu) \, \middle| \, f \otimes g \in \mathcal{L}^1(\mu \otimes \nu), \int (f \otimes g) \, d(\mu \otimes \nu) = \left(\int f \, d\mu \right) \left(\int g \, d\nu \right) \right\}.$$

If $f \in \mathcal{L}(\mathfrak{R})$, $f = \sum_{\iota \in I} \alpha_\iota e_{A_\iota}$, then

$$f \otimes g = \sum_{(\iota, \lambda) \in I \times L} \alpha_\iota \beta_\lambda e_{A_\iota \times B_\lambda} \in \mathcal{L}(\mathfrak{R} \otimes \mathfrak{S})$$

and

$$\int (f \otimes g) \, d(\mu \otimes \nu) = \sum_{(\iota, \lambda) \in I \times L} \alpha_\iota \beta_\lambda \mu(A_\iota) \nu(B_\lambda)$$

$$= \left(\sum_{\iota \in I} \alpha_\iota \mu(A_\iota) \right) \left(\sum_{\lambda \in L} \beta_\lambda \nu(B_\lambda) \right) = \left(\int f \, d\mu \right) \left(\int g \, d\nu \right).$$

Thus $\mathcal{L}(\mathfrak{R}) \subset \mathcal{F}$.

Let $(f_n)_{n \in \mathbb{N}}$ be a monotone sequence in $\mathcal{F}$ such that $\lim_{n \to \infty} \int f_n d\mu$ is finite. Then $(f_n \otimes g)_{n \in \mathbb{N}}$ is a monotone sequence in $\mathcal{L}^1(\mu \otimes \nu)$. Moreover, for each $n \in \mathbb{N}$

$$\int (f_n \otimes g) \, d(\mu \otimes \nu) = \left(\int f_n d\mu \right) \left(\int g \, d\nu \right).$$

From this we conclude that $\lim_{n \to \infty} \int (f_n \otimes g) \, d(\mu \otimes \nu)$ is finite. Furthermore, by the monotone convergence theorem

$$\left(\lim_{n \to \infty} f_n \right) \otimes g = \lim_{n \to \infty} f_n \otimes g \in \mathcal{L}^1(\mu \otimes \nu)$$

and

$$\int \left(\left(\lim_{n \to \infty} f_n \right) \otimes g \right) d(\mu \otimes \nu) = \lim_{n \to \infty} \int (f_n \otimes g) \, d(\mu \otimes \nu)$$

$$= \lim_{n \to \infty} \left(\int f_n d\mu \right) \left(\int g \, d\nu \right) = \left(\int \left(\lim_{n \to \infty} f_n \right) d\mu \right) \left(\int g \, d\nu \right).$$

Hence, $\lim_{n \to \infty} f_n \in \mathcal{F}$.

Given $f_1, f_2 \in \mathcal{F}$ with $f_1 \leq f_2$ and such that $\int f_1 d\mu = \int f_2 d\mu$,

$$\int (f_1 \otimes g) \, d(\mu \otimes \nu) = \int (f_2 \otimes g) \, d(\mu \otimes \nu).$$

If $h \in \overline{\mathcal{L}}(\mu)$ satisfies $f_1 \leq h \leq f_2$, then $f_1 \otimes g \leq h \otimes g \leq f_2 \otimes g$. Hence

$h \otimes g \in \mathcal{L}^1(\mu \otimes \nu)$ and

$$\int (h \otimes g) \, d(\mu \otimes \nu) = \int (f_1 \otimes g) \, d(\mu \otimes \nu)$$
$$= \left(\int f_1 d\mu \right) \left(\int g \, d\nu \right) = \left(\int h \, d\mu \right) \left(\int g \, d\nu \right).$$

We conclude that $h \in \mathcal{F}$. By the induction principle (Theorem 3.38), $\overline{\mathcal{L}}(\mu) \subset \mathcal{F}$.

Choose a fixed $f \in \overline{\mathcal{L}}(\mu)_+ \cap \mathbb{R}^X$. A completely analogous argument shows that $f \otimes g \in \mathcal{L}^1(\mu \otimes \nu)$ for every $g \in \overline{\mathcal{L}}(\nu)$ and that

$$\int (f \otimes g) \, d(\mu \otimes \nu) = \left(\int f \, d\mu \right) \left(\int g \, d\nu \right).$$

Use the decomposition $f = f^+ - f^-$ to extend these results to arbitrary $f \in \overline{\mathcal{L}}(\mu) \cap \mathbb{R}^X$.

Take $A \in \mathfrak{N}(\mu)$. Then for any $B \times C \in \mathfrak{R} \square \mathfrak{S}$,

$$(A \times Y) \cap (B \times C) = (A \cap B) \times C.$$

But $e_{A \cap B} \in \overline{\mathcal{L}}(\mu)$ and $e_C \in \overline{\mathcal{L}}(\nu)$, so that

$$e_{(A \cap B) \times C} = e_{A \cap B} \otimes e_C \in \mathcal{L}^1(\mu \otimes \nu)$$

and

$$\int e_{(A \cap B) \times C} d(\mu \otimes \nu) = \left(\int e_{A \cap B} d\mu \right) \left(\int e_C d\nu \right) = 0.$$

Thus $(A \times Y) \cap (B \times C) \in \overline{\mathfrak{N}}(\mu \otimes \nu)$ and we see that $(A \times Y) \cap D \in \overline{\mathfrak{N}}(\mu \otimes \nu)$ for every $D \in \mathfrak{R} \otimes \mathfrak{S}$. By Proposition 4.6, this means that $A \times Y \in \mathfrak{N}(\mu \otimes \nu)$. A similar argument proves that $X \times B \in \mathfrak{N}(\mu \otimes \nu)$ for every $B \in \mathfrak{N}(\nu)$. This completes the proof of (a).

Take $f \in \mathcal{L}^1(\mu)$ and $g \in \mathcal{L}^1(\nu)$. Then there are functions $f' \in \overline{\mathcal{L}}(\mu) \cap \mathbb{R}^X$ and $g' \in \overline{\mathcal{L}}(\nu)$ with

$$\{f \neq f'\} \in \mathfrak{N}(\mu) \quad \text{and} \quad \{g \neq g'\} \in \mathfrak{N}(\nu).$$

But $f' \otimes g' \in \mathcal{L}^1(\mu \otimes \nu)$ and

$$\{f \otimes g \neq f' \otimes g'\} \subset (\{f \neq f'\} \times Y) \cup (X \times \{g \neq g'\}).$$

Thus $\{f \otimes g \neq f' \otimes g'\} \in \mathfrak{N}(\mu \otimes \nu)$ and it follows that $f \otimes g \in \mathcal{L}^1(\mu \otimes \nu)$ and

$$\int (f \otimes g) \, d(\mu \otimes \nu) = \int (f' \otimes g') \, d(\mu \otimes \nu)$$
$$= \left(\int f' d\mu \right) \left(\int g' d\nu \right) = \left(\int f \, d\mu \right) \left(\int g \, d\nu \right).$$

This completes the proof of (c).

(b) follows from (c). $\square$

Our observations about products can readily be extended to any finite number of 'factors' $(X_k, \mathfrak{R}_k, \mu_k)$ $(k \leq n)$. We may proceed inductively by defining

$$\bigotimes_{k=1}^{1} \mathfrak{R}_k := \mathfrak{R}_1, \quad \bigotimes_{k=1}^{1} \mu_k := \mu_1,$$

and

$$\bigotimes_{k=1}^{m} \mathfrak{R}_k := \left(\bigotimes_{k=1}^{m-1} \mathfrak{R}_k \right) \otimes \mathfrak{R}_m, \quad \bigotimes_{k=1}^{m} \mu_k := \left(\bigotimes_{k=1}^{m-1} \mu_k \right) \otimes \mu_m$$

whenever $m > 1$. This yields a positive measure space

$$\left(X, \bigotimes_{k=1}^{n} \mathfrak{R}_k, \bigotimes_{k=1}^{n} \mu_k \right)$$

on the Cartesian product $X = \prod_{k=1}^{n} X_k$. Recall that, for example, the set $(X_1 \times X_2) \times X_3$ may be identified with $X_1 \times X_2 \times X_3$ – the set of all ordered triples (x_1, x_2, x_3) with $x_k \in X_k$ – and also with $\prod_{k \in \{1,2,3\}} X_k$ – the Cartesian product of the sets X_1, X_2, X_3. See Chapter 1.

An important question arises at this point, namely that of the dependence of the product on the ordering of the factors. We now pursue this question.

Let $((X_\iota, \mathfrak{R}_\iota, \mu_\iota))_{\iota \in I}$ be a finite, non-empty family of positive measure spaces. We cannot speak of the order of the factors until an ordering on I is introduced. For each such ordering we can use the above construction. Do the products so defined vary with the ordering of I?

The commutativity of multiplication of real numbers justifies the following definition of $\square_{\iota \in I} \mu_\iota$.

$$\square_{\iota \in I} \mathfrak{R}_\iota := \left\{ \prod_{\iota \in I} A_\iota \,\middle|\, A_\iota \in \mathfrak{R}_\iota \text{ for all } \iota \in I \right\}$$

and

$$\square_{\iota \in I} \mu_\iota : \square_{\iota \in I} \mathfrak{R}_\iota \longrightarrow \mathbb{R}, \quad \prod_{\iota \in I} A_\iota \longmapsto \prod_{\iota \in I} \mu_\iota(A_\iota).$$

These are natural generalizations from the case with two factors. We call the elements of $\square_{\iota \in I} \mathfrak{R}_\iota$ **rectangles**.

Proposition 4.53

(a) $\square_{\iota \in I} \mathfrak{R}_\iota$ *is a semi-ring of sets.*

(b) *Given a finite disjoint family* $(A_\lambda)_{\lambda \in L}$ *in* $\square_{\iota \in I} \mathfrak{R}_\iota$ *for which* $\bigcup_{\lambda \in L} A_\lambda$ *belongs to* $\square_{\iota \in I} \mathfrak{R}_\iota$,

$$\left(\square_{\iota \in I} \mu_\iota \right) \left(\bigcup_{\lambda \in L} A_\lambda \right) = \sum_{\lambda \in L} \left(\square_{\iota \in I} \mu_\iota \right) (A_\lambda).$$

Proof. (a) We proceed by induction on the number n of elements of I. For $n = 1$ the assertion is trivial. We assume that the assertion is true whenever I contains n elements and consider the family $((X_\iota, \mathfrak{R}_\iota, \mu_\iota))_{\iota \in I}$ whose indexing set I contains $n + 1$ elements. Choose a fixed index $\kappa \in I$. Let $(A_\iota)_{\iota \in I}$ and $(B_\iota)_{\iota \in I}$ be elements of $\prod_{\iota \in I} \mathfrak{R}_\iota$. It is immediate that

$$\left(\prod_{\iota \in I} A_\iota \right) \cap \left(\prod_{\iota \in I} B_\iota \right) = \prod_{\iota \in I} (A_\iota \cap B_\iota).$$

To form the difference, note that

$$\left(\prod_{\iota \in I} A_\iota \right) \setminus \left(\prod_{\iota \in I} B_\iota \right)$$

$$= \left(\left(\prod_{\iota \in I \setminus \{\kappa\}} A_\iota \right) \times A_\kappa \right) \setminus \left(\left(\prod_{\iota \in I \setminus \{\kappa\}} B_\iota \right) \times B_\kappa \right)$$

$$= \left(\left(\prod_{\iota \in I \setminus \{\kappa\}} A_\iota \right) \setminus \left(\prod_{\iota \in I \setminus \{\kappa\}} B_\iota \right) \right) \times A_\kappa$$

$$\cup \left(\left(\prod_{\iota \in I \setminus \{\kappa\}} A_\iota \right) \cap \left(\prod_{\iota \in I \setminus \{\kappa\}} B_\iota \right) \right) \times (A_\kappa \setminus B_\kappa).$$

By hypothesis, there is a finite disjoint family $(C_\lambda)_{\lambda \in L}$ of rectangles in $\square_{\iota \in I \setminus \{\kappa\}} \mathfrak{R}_\iota$ such that

$$\left(\left(\prod_{\iota \in I \setminus \{\kappa\}} A_\iota \right) \setminus \left(\prod_{\iota \in I \setminus \{\kappa\}} B_\iota \right) \right) = \bigcup_{\lambda \in L} C_\lambda.$$

$(C_\lambda \times A_\kappa)_{\lambda \in L}$ is then a finite disjoint family in $\square_{\iota \in I} \mathfrak{R}_\iota$ and

$$\left(\left(\prod_{\iota \in I \setminus \{\kappa\}} A_\iota \right) \setminus \left(\prod_{\iota \in I \setminus \{\kappa\}} B_\iota \right) \right) \times A_\kappa = \bigcup_{\lambda \in L} (C_\lambda \times A_\kappa).$$

This set is disjoint from

$$\left(\left(\prod_{\iota \in I \setminus \{\kappa\}} A_\iota \right) \cap \left(\prod_{\iota \in I \setminus \{\kappa\}} B_\iota \right) \right) \times (A_\kappa \setminus B_\kappa)$$

$$= \left(\prod_{\iota \in I \setminus \{\kappa\}} (A_\iota \cap B_\iota) \right) \times (A_\kappa \setminus B_\kappa) \in \square_{\iota \in I} \mathfrak{R}_\iota.$$

But then $\left(\prod_{\iota \in I} A_\iota \right) \setminus \left(\prod_{\iota \in I} B_\iota \right)$ is a disjoint union of finitely many sets in $\square_{\iota \in I} \mathfrak{R}_\iota$, proving (a).

(b) is proven by the method used in Proposition 4.48 for two factors. We leave the details to the reader. $\qquad \square$

Part (b) of the next theorem states what we were hoping for. The product is, in fact, independent of the order of the factors. Moreover, both routes for defining a product measure, namely the inductive one we presented first and the alternative one starting with the definition of $\square_{\iota \in I} \mu_\iota$, yield the same object.

Theorem 4.54

(a) *There is a unique positive measure $\bigotimes_{\iota \in I} \mu_\iota$ on the ring of sets $\bigotimes_{\iota \in I} \mathfrak{R}_\iota$ generated by $\square_{\iota \in I} \mathfrak{R}_\iota$ having the property that*

$$\left(\bigotimes_{\iota \in I} \mu_\iota \right)\bigg|_{\square_{\iota \in I} \mathfrak{R}_\iota} = \square_{\iota \in I} \mu_\iota.$$

(b) *Let n be the number of elements in I. Let $\iota_1, \iota_2, \ldots, \iota_n$ be an arbitrary numbering of the elements of I. Then, identifying $\prod_{\iota \in I} X_\iota$ and $\prod_{k=1}^n X_{\iota_k}$,*

$$\left(\prod_{\iota \in I} X_\iota, \bigotimes_{\iota \in I} \mathfrak{R}_\iota, \bigotimes_{\iota \in I} \mu_\iota \right) = \left(\prod_{k=1}^n X_{\iota_k}, \bigotimes_{k=1}^n \mathfrak{R}_{\iota_k}, \bigotimes_{k=1}^n \mu_{\iota_k} \right).$$

Proof. (a) The existence and uniqueness of a positive content $\bigotimes_{\iota \in I} \mu_\iota$ on $\bigotimes_{\iota \in I} \mathfrak{R}_\iota$ with the required property are results of Theorem 2.27. That $\bigotimes_{\iota \in I} \mu_\iota$ is a positive measure is proven by the method used in Theorem 4.49.

(b) We first prove that

$$\bigotimes_{\iota \in I} \mathfrak{R}_\iota = \bigotimes_{k=1}^n \mathfrak{R}_{\iota_k}.$$

Note that there really is something to prove, since $\bigotimes_{\iota \in I} \mathfrak{R}_\iota$ was defined as the ring of sets generated by the rectangles, while $\bigotimes_{k=1}^n \mathfrak{R}_{\iota_k}$ was defined inductively.

Clearly $\square_{\iota \in I} \mathfrak{R}_\iota \subset \bigotimes_{k=1}^n \mathfrak{R}_{\iota_k}$, and since the object on the right is a ring of sets, $\bigotimes_{\iota \in I} \mathfrak{R}_\iota \subset \bigotimes_{k=1}^n \mathfrak{R}_{\iota_k}$.

Take $m \in \mathbb{N}$ with $m < n$. We assume that $B \times A_{m+1} \times \cdots \times A_n \in \bigotimes_{\iota \in I} \mathfrak{R}_\iota$ for all $B \in \bigotimes_{k=1}^m \mathfrak{R}_{\iota_k}$ and $A_i \in \mathfrak{R}_{\iota_i}$, where $i \in \{m+1, \ldots, n\}$. Let $\mathfrak{S}$ be the set of those $A \in \bigotimes_{k=1}^{m+1} \mathfrak{R}_{\iota_k}$ for which $A \times A_{m+2} \times \cdots \times A_n \in \bigotimes_{\iota \in I} \mathfrak{R}_\iota$. Since $\bigotimes_{\iota \in I} \mathfrak{R}_\iota$ is a ring of sets, it is easy to see that $\mathfrak{S}$ is also a ring of sets. By hypothesis $\left(\bigotimes_{k=1}^m \mathfrak{R}_{\iota_k} \right) \square \mathfrak{R}_{\iota_{m+1}} \subset \mathfrak{S}$. Then $\bigotimes_{k=1}^{m+1} \mathfrak{R}_{\iota_k} \subset \mathfrak{S}$. It follows that $B \times A_{m+2} \times \cdots \times A_n \in \bigotimes_{\iota \in I} \mathfrak{R}_\iota$ for all $B \in \bigotimes_{k=1}^{m+1} \mathfrak{R}_{\iota_k}$ and all $A_i \in \mathfrak{R}_{\iota_i}$, where $i \in \{m+2, \ldots, n\}$. If $m = 1$, then trivially $B \times A_{m+1} \times \cdots \times A_n \in \bigotimes_{\iota \in I} \mathfrak{R}_\iota$ for all $B \in \mathfrak{R}_{\iota_1}$ and all $A_i \in \mathfrak{R}_{\iota_i}$, where $i \in \{m+1, \ldots, n\}$. It follows by induction from the considerations above that $\bigotimes_{k=1}^n \mathfrak{R}_{\iota_k} \subset \bigotimes_{\iota \in I} \mathfrak{R}_\iota$. Hence

$$\bigotimes_{\iota \in I} \mathfrak{R}_\iota = \bigotimes_{k=1}^n \mathfrak{R}_{\iota_k}.$$

It follows by the commutativity of multiplication of numbers that for arbitrary rectangles $\prod_{\iota \in I} A_\iota$

$$\left(\bigotimes_{\iota \in I} \mu_\iota \right)\left(\prod_{\iota \in I} A_\iota \right) = \prod_{\iota \in I} \mu_\iota(A_\iota) = \prod_{k=1}^n \mu_{\iota_k}(A_{\iota_k}) = \left(\bigotimes_{k=1}^n \mu_{\iota_k} \right)\left(\prod_{k=1}^n A_{\iota_k} \right).$$

The uniqueness of the extension to the ring of sets generated implies that

$$\bigotimes_{\iota \in I} \mu_\iota = \bigotimes_{k=1}^{n} \mu_{\iota_k}.$$

$\square$

$\left(\prod_{\iota \in I} X_\iota, \bigotimes_{\iota \in I} \mathfrak{R}_\iota, \bigotimes_{\iota \in I} \mu_\iota \right)$ is called the **product** of $\left((X_\iota, \mathfrak{R}_\iota, \mu_\iota) \right)_{\iota \in I}$.

We need the following proposition for the statement of Fubini's theorem for finite products.

Proposition 4.55 *Let J, K be non-empty disjoint subsets of I such that $I = J \cup K$. Identifying $\left(\prod_{\iota \in J} X_\iota \right) \times \left(\prod_{\iota \in K} X_\iota \right)$ with $\prod_{\iota \in I} X_\iota$, we have*

(a) $\left(\bigotimes_{\iota \in J} \mathfrak{R}_\iota \right) \otimes \left(\bigotimes_{\iota \in K} \mathfrak{R}_\iota \right) = \bigotimes_{\iota \in I} \mathfrak{R}_\iota.$

(b) $\left(\bigotimes_{\iota \in J} \mu_\iota \right) \otimes \left(\bigotimes_{\iota \in K} \mu_\iota \right) = \bigotimes_{\iota \in I} \mu_\iota.$

Proof. (a) Since

$$\square_{\iota \in I} \mathfrak{R}_\iota \subset \left(\bigotimes_{\iota \in J} \mathfrak{R}_\iota \right) \otimes \left(\bigotimes_{\iota \in K} \mathfrak{R}_\iota \right),$$

it follows that

$$\bigotimes_{\iota \in I} \mathfrak{R}_\iota \subset \left(\bigotimes_{\iota \in J} \mathfrak{R}_\iota \right) \otimes \left(\bigotimes_{\iota \in K} \mathfrak{R}_\iota \right).$$

Take $A \in \bigotimes_{\iota \in J} \mathfrak{R}_\iota$ and $B \in \bigotimes_{\iota \in K} \mathfrak{R}_\iota$. There are finite disjoint families $(A_\gamma)_{\gamma \in C}$ in $\square_{\iota \in J} \mathfrak{R}_\iota$ and $(B_\delta)_{\delta \in D}$ in $\square_{\iota \in K} \mathfrak{R}_\iota$ such that $A = \bigcup_{\gamma \in C} A_\gamma$ and $B = \bigcup_{\delta \in D} B_\delta$. It follows that

$$A \times B = \bigcup_{\substack{\gamma \in C \\ \delta \in D}} (A_\gamma \times B_\delta) \in \bigotimes_{\iota \in I} \mathfrak{R}_\iota.$$

Since A and B were arbitrary,

$$\left(\bigotimes_{\iota \in J} \mathfrak{R}_\iota \right) \square \left(\bigotimes_{\iota \in K} \mathfrak{R}_\iota \right) \subset \bigotimes_{\iota \in I} \mathfrak{R}_\iota$$

and hence

$$\left(\bigotimes_{\iota \in J} \mathfrak{R}_\iota \right) \otimes \left(\bigotimes_{\iota \in K} \mathfrak{R}_\iota \right) = \bigotimes_{\iota \in I} \mathfrak{R}_\iota.$$

(b) Since the measures agree on $\square_{\iota \in I} \mathfrak{R}_\iota$ they also agree on $\bigotimes_{\iota \in I} \mathfrak{R}_\iota$.

$\square$

We are now ready to formulate and prove Fubini's theorem for a finite number of factors. The statement and comments are analogous to those in the case of two factors. The theorem justifies the calculation of $\int f \, d(\bigotimes_{\iota \in I} \mu_\iota)$, for $f \in \overline{\mathcal{L}}(\bigotimes_{\iota \in I} \mu_\iota)$, by evaluating the iterated integrals with respect to the individual measures μ_ι. Moreover the order is irrelevant: any will do.

Theorem 4.56 (Fubini) *Take $f \in \overline{\mathcal{L}}(\bigotimes_{\iota \in I} \mu_\iota)$ and let J and K be non-empty disjoint subsets of I such that $I = J \cup K$.*

Denote the elements of $\prod_{\iota \in J} X_\iota$ and $\prod_{\iota \in K} X_\iota$ by $x := (x_\iota)_{\iota \in J}$ and $y := (y_\iota)_{\iota \in K}$ respectively. Identify $\left(\prod_{\iota \in J} X_\iota\right) \times \left(\prod_{\iota \in K} X_\iota\right)$ with $\prod_{\iota \in I} X_\iota$ and define

$$A := \left\{ x \in \prod_{\iota \in J} X_\iota \,\Big|\, f(x, \cdot) \notin \overline{\mathcal{L}}\left(\bigotimes_{\iota \in K} \mu_\iota \right) \right\}$$

$$B := \left\{ y \in \prod_{\iota \in K} X_\iota \,\Big|\, f(\cdot, y) \notin \overline{\mathcal{L}}\left(\bigotimes_{\iota \in J} \mu_\iota \right) \right\}.$$

Then:

(a) $A \in \overline{\mathfrak{N}}(\bigotimes_{\iota \in J} \mu_\iota)$, $B \in \overline{\mathfrak{N}}(\bigotimes_{\iota \in K} \mu_\iota)$.

(b) The map

$$\left(\prod_{\iota \in J} X_\iota \right) \setminus A \longrightarrow \mathbb{R}, \quad x \longmapsto \int f(x, y) \, d\left(\bigotimes_{\iota \in K} \mu_\iota \right)(y)$$

is essentially contained in $\overline{\mathcal{L}}(\bigotimes_{\iota \in J} \mu_\iota)$ and the map

$$\left(\prod_{\iota \in K} X_\iota \right) \setminus B \longrightarrow \mathbb{R}, \quad y \longmapsto \int f(x, y) \, d\left(\bigotimes_{\iota \in J} \mu_\iota \right)(x)$$

is essentially contained in $\overline{\mathcal{L}}(\bigotimes_{\iota \in K} \mu_\iota)$.

(c)
$$\int f \, d\left(\bigotimes_{\iota \in I} \mu_\iota \right) = \int \left(\int f(x, y) \, d\left(\bigotimes_{\iota \in J} \mu_\iota \right)(x) \right) d\left(\bigotimes_{\iota \in K} \mu_\iota \right)(y)$$
$$= \int \left(\int f(x, y) \, d\left(\bigotimes_{\iota \in K} \mu_\iota \right)(y) \right) d\left(\bigotimes_{\iota \in J} \mu_\iota \right)(x).$$

(d) Let $\iota_1, \ldots, \iota_n$ be a numbering of the elements of I. Then

$$\int f \, d\left(\bigotimes_{\iota \in I} \mu_\iota \right) =$$
$$\int \int \cdots \int f(z_{\iota_1}, \ldots, z_{\iota_n}) \, d\mu_{\iota_1}(z_{\iota_1}) \ldots d\mu_{\iota_{n-1}}(z_{\iota_{n-1}}) d\mu_{\iota_n}(z_{\iota_n}).$$

Proof. (a)–(c) follow from Fubini's theorem for two factors together with the above proposition.

Given (a) and (b), (d) follows from (c) by induction. $\qquad \square$

As examples, consider products of Stieltjes measures. These lead to important measures on the spaces $\mathbb{R}^n$. We emphasize in particular the n-fold product of the Lebesgue measure spaces $(\mathbb{R}, \mathfrak{J}, \lambda)$. We denote this by $(\mathbb{R}^n, \mathfrak{J}^n, \lambda^n)$. We show in the next chapter that λ^n provides a satisfactory solution of the measure problem for $\mathbb{R}^n$ mentioned in the Introduction.

Exercises

1. Show that in general $\mathfrak{R}\square\mathfrak{S}$ need not be a ring of sets.

2. Verify that $\mathfrak{R}_\delta \otimes \mathfrak{S}_\delta \subset (\mathfrak{R}\otimes\mathfrak{S})_\delta$ for any rings of sets $\mathfrak{R}$ and $\mathfrak{S}$.

3. Let X,Y be sets, g (or h) be a positive real function on X (or Y) and $\mathfrak{R}$ (or $\mathfrak{S}$) the set of finite subsets of X (or Y).

   (a) Describe $\mathfrak{R}\otimes\mathfrak{S}$ and $\mu^g \otimes \mu^h$.

   (b) Show that the following are equivalent.

   (b1) Given any positive measure λ on $\mathfrak{R}\otimes\mathfrak{S}$, there are g and h such that $\lambda = \mu^g \otimes \mu^h$.
   (b2) The cardinality of X or Y is at most 1.

4. Take $X := Y := \mathbb{R}$. Let λ be Lebesgue measure on X and ν the 0-measure on $(Y, \mathfrak{P}(Y))$. Define $f := e_{\mathbb{R}\times\mathbb{R}}$. Verify that $f \in \overline{\mathcal{L}}(\lambda\otimes\nu)$, but $Y_f = \mathbb{R}$.

5. For $(m,n) \in \mathbb{Z}^2$ define

$$\alpha_{mn} := \begin{cases} 1 & \text{if } m-n=0 \\ -1 & \text{if } m-n=1 \\ 0 & \text{otherwise.} \end{cases}$$

   Prove the following.

   (a) $\sum_{n\in\mathbb{Z}}\alpha_{mn}=0$ for every $m\in\mathbb{Z}$ and $\sum_{m\in\mathbb{Z}}\alpha_{mn}=0$ for every $n\in\mathbb{Z}$.
   (b) $\sum_{m\in\mathbb{Z}}\left(\sum_{n\in\mathbb{Z}}\alpha_{mn}\right) = \sum_{n\in\mathbb{Z}}\left(\sum_{m\in\mathbb{Z}}\alpha_{mn}\right) = 0.$
   (c) $\sum_{(m,n)\in\mathbb{Z}^2}\alpha_{mn}$ does not exist.
   (d) $\sum_{m\in\mathbb{N}}\alpha_{mn}=0$ for every $n\in\mathbb{N}$, $\sum_{n\in\mathbb{N}}\alpha_{mn}=0$ for every $m\in\mathbb{N}\setminus\{1\}$ and $\sum_{n\in\mathbb{N}}\alpha_{1n}=1.$
   (e) $\sum_{m\in\mathbb{N}}\left(\sum_{n\in\mathbb{N}}\alpha_{mn}\right)=1$ and $\sum_{n\in\mathbb{N}}\left(\sum_{m\in\mathbb{N}}\alpha_{mn}\right)=0.$

   What does this exercise have to do with product measures and Fubini's theorem?

6. Let λ denote Lebesgue measure on $\mathfrak{J}$. Put $\mathfrak{F}:=\{A\subset\mathbb{R}\,|\,A \text{ is finite}\}$ and for each $A\in\mathfrak{F}$ let $\mu(A)$ be the number of elements in A.

   (a) Describe $\mathcal{L}(\mathfrak{J}\otimes\mathfrak{F})$, $\mathcal{L}(\mathfrak{J}\otimes\mathfrak{F})^\uparrow$, $\overline{\mathcal{L}}(\lambda\otimes\mu)$ and $\mathfrak{S}:=\{A\subset\mathbb{R}\times\mathbb{R}\,|\,e_A\in\overline{\mathcal{L}}(\lambda\otimes\mu)\}$ in terms of $\mathcal{L}(\mathfrak{J})^\uparrow$, $\overline{\mathcal{L}}(\lambda)$ and $\{C\subset\mathbb{R}\,|\,e_C\in\overline{\mathcal{L}}(\lambda)\}$. In particular, prove that for each $A\in\mathfrak{S}$ there is a countable $B\subset\mathbb{R}$ with $A\subset\mathbb{R}\times B$.

   (b) Let $D:=\{(x,x)\,|\,x\in\mathbb{R}\}$ be the diagonal in $\mathbb{R}\times\mathbb{R}$. Prove that $D\in\mathfrak{N}(\lambda\otimes\mu)$, but $D\notin\overline{\mathfrak{N}}(\lambda\otimes\mu)$.

(c) Show that

$$\int e_D d(\lambda \otimes \mu) = \int \left(\int e_D(x,y) \, d\lambda(x) \right) d\mu(y) = 0,$$

but

$$x \longmapsto \int e_D(x,y) \, d\mu(y)$$

is not λ-integrable. Thus $\int (\int e_D(x,y) \, d\mu(y)) \, d\lambda(x)$ does not exist.

(d) Why does Fubini's theorem not hold in this case?

7. Let λ be Lebesgue measure on $\mathbb{R}$ and $(X, \mathfrak{R}, \mu)$ be a σ-finite positive measure space. Take $f \in \overline{\mathbb{R}}_+^X$ and define

$$A := \{(x,y) \in X \times \mathbb{R} \mid 0 \le y < f(x)\}.$$

Prove that $f \in \mathcal{L}^1(\mu)$ if and only if $e_A \in \mathcal{L}^1(\mu \otimes \lambda)$, and that in this case $\int f d\mu = \int e_A d(\mu \otimes \lambda)$.

8. Given $f \in \mathbb{R}^X$, let

$$\mathcal{G}(f) := \{(x, f(x)) \mid x \in X\}$$

be the graph of f. Prove the following.

(a) If λ is Lebesgue measure on $\mathbb{R}$, then $\mathcal{G}(f) \in \mathfrak{N}(\lambda \otimes \lambda)$ for every $f \in \mathbb{R}^{\mathbb{R}}$. (First consider the sets $\mathcal{G}(f) \cap ([n, n+1[\times [m, m+1[)$, for $n, m \in \mathbb{Z}$.)

(b) Put $\mathfrak{R} := \{\emptyset, \mathbb{R}\}$ and $\mu(\emptyset) := 0$, $\mu(\mathbb{R}) := 1$. Then there is an $f \in \mathbb{R}^{\mathbb{R}}$ such that $\mathcal{G}(f) \notin \mathfrak{N}(\mu \otimes \lambda)$.

9. Let $(X, \mathfrak{R}, \mu)$ and $(Y, \mathfrak{S}, \nu)$ be positive measure spaces. Given $A \subset X \times Y$ and $x \in X$, let $A_x := \{y \in Y \mid (x,y) \in A\}$ denote the 'cut of A at level x'. Assume that A and B are subsets of $X \times Y$ with $e_A, e_B \in \overline{\mathcal{L}}(\mu \otimes \nu)$ and $\nu^Y(A_x) = \nu^Y(B_x)$ for μ-almost every $x \in X$. Show that

$$(\mu \otimes \nu)^{X \times Y}(A) = (\mu \otimes \nu)^{X \times Y}(B).$$

Now consider the special case $X := \mathbb{R}$, $Y := \mathbb{R}^2$, $\mu := \lambda$, $\nu := \lambda \otimes \lambda$, where as usual λ denotes Lebesgue measure. Interpret the result just obtained (**Cavalieri's principle**).

10. Prove the following.

(a) Let φ be a congruence map of $\mathbb{R}^2$ (i.e. a bijective length-preserving map from $\mathbb{R}^2$ to $\mathbb{R}^2$). Let λ be Lebesgue measure on $\mathbb{R}$ and take $f \in \mathcal{L}^1(\lambda \otimes \lambda)$. Then

$$f \circ \varphi \in \mathcal{L}^1(\lambda \otimes \lambda) \quad \text{and} \quad \int f \circ \varphi \, d(\lambda \otimes \lambda) = \int f \, d(\lambda \otimes \lambda).$$

(This is the **congruence invariance of the two-dimensional Lebesgue measure**.)

(b) Let A and B be congruent subsets of $\mathbb{R}^2$ with $A \in \mathcal{L}(\lambda \otimes \lambda)$. Then

$$B \in \mathcal{L}(\lambda \otimes \lambda) \quad \text{and} \quad (\lambda \otimes \lambda)^{\mathbb{R} \times \mathbb{R}}(B) = (\lambda \otimes \lambda)^{\mathbb{R} \times \mathbb{R}}(A).$$

Hint: The result can be obtained along the following lines:

(i) Each straight segment belongs to $\overline{\mathfrak{N}}(\lambda \otimes \lambda)$.

(ii) If D is a triangle with two sides parallel to coordinate axes and of length a and b, then $D \in \mathcal{L}(\lambda \otimes \lambda)$ and $\overline{\lambda \otimes \lambda}(D) = \frac{ab}{2}$.

(iii) If R is a rectangle with sides of length a and b, then $R \in \mathcal{L}(\lambda \otimes \lambda)$ and $\overline{\lambda \otimes \lambda}(R) = ab$.

Now define

$$\mathcal{F} := \left\{ g \in \overline{\mathcal{L}}(\lambda \otimes \lambda) \,\middle|\, g \circ \varphi \in \overline{\mathcal{L}}(\lambda \otimes \lambda), \int g \circ \varphi \, d(\lambda \otimes \lambda) = \int g \, d(\lambda \otimes \lambda) \right\}$$

and use the induction principle.

11. For $A \in \overline{\mathfrak{N}}(\mu \otimes \nu)$ prove the following.

$$\{x \in X \mid \{y \in Y \mid (x,y) \in A\} \notin \overline{\mathfrak{N}}(\nu)\} \in \overline{\mathfrak{N}}(\mu),$$
$$\{y \in Y \mid \{x \in X \mid (x,y) \in A\} \notin \overline{\mathfrak{N}}(\mu)\} \in \overline{\mathfrak{N}}(\nu).$$

12. Generalize Fubini's theorem to functions essentially contained in $\overline{\mathcal{L}}(\mu \otimes \nu)$.

13. Let f be a positive $\mu \otimes \nu$-measurable function on $X \times Y$ such that $\{f \neq 0\} \subset \bigcup_{n \in \mathbb{N}} A_n$ for some increasing sequence $(A_n)_{n \in \mathbb{N}}$ in $\mathfrak{R} \otimes \mathfrak{S}$. Suppose that $X_f \in \overline{\mathfrak{N}}(\mu)$ and that

$$X \setminus X_f \longrightarrow \mathbb{R}, \quad x \longmapsto \int f(x,y) \, d\nu(y)$$

is essentially contained in $\overline{\mathcal{L}}(\mu)$. Prove that $f \in \overline{\mathcal{L}}(\mu \otimes \nu)$ and

$$\int f \, d(\mu \otimes \nu) = \int \left(\int f(x,y) \, d\nu(y) \right) d\mu(x).$$

(Of course, the dual assertion is also true.) This is **Tonelli's theorem**. Hint: Note that $f = \bigvee_{n \in \mathbb{N}} f \wedge ne_{A_n}$. Apply Exercise 4 of Section 4.4 to see that $f \wedge ne_{A_n} \in \overline{\mathcal{L}}(\mu \otimes \nu)$ for every $n \in \mathbb{N}$ and verify that

$$\int f \wedge ne_{A_n} \, d(\mu \otimes \nu) = \int \left(\int f \wedge ne_{A_n}(x,y) \, d\nu(y) \right) d\mu(x).$$

Show that

$$\sup_{n \in \mathbb{N}} \int \left(\int f \wedge ne_{A_n}(x,y) \, d\nu(y) \right) d\mu(x) = \int \left(\int f(x,y) \, d\nu(y) \right) d\mu(x).$$

Conclude that $\bigvee_{n \in \mathbb{N}} f \wedge n e_{A_n} \in \overline{\mathcal{L}}(\mu \otimes \nu)$ and

$$\int \left(\bigvee_{n \in \mathbb{N}} f \wedge n e_{A_n} \right) d(\mu \otimes \nu) = \int \left(\int f(x, y) \, d\nu(y) \right) d\mu(x).$$

14. Let $(X, \mathfrak{R}_j, \mu_j)$ and $(Y, \mathfrak{S}_j, \nu_j)$ be positive measure spaces $(j = 1, 2)$ with

$$
\begin{aligned}
(X, \mathfrak{R}_1, \mu_1) &\preccurlyeq \left(X, (\mathfrak{R}_2)_\delta, \mu_2^X \big|_{(\mathfrak{R}_2)_\delta} \right), \\
(X, \mathfrak{R}_2, \mu_2) &\preccurlyeq \left(X, (\mathfrak{R}_1)_\delta, \mu_1^X \big|_{(\mathfrak{R}_1)_\delta} \right), \\
(Y, \mathfrak{S}_1, \nu_1) &\preccurlyeq \left(Y, (\mathfrak{S}_2)_\delta, \nu_2^Y \big|_{(\mathfrak{S}_2)_\delta} \right), \\
(Y, \mathfrak{S}_2, \nu_2) &\preccurlyeq \left(Y, (\mathfrak{S}_1)_\delta, \nu_1^Y \big|_{(\mathfrak{S}_1)_\delta} \right).
\end{aligned}
$$

Prove the following.

$$\left(X \times Y, \, \mathcal{L}^1(\mu_1 \otimes \nu_1), \, \int \cdot \, d(\mu_1 \otimes \nu_1) \right)$$
$$= \left(X \times Y, \, \mathcal{L}^1(\mu_2 \otimes \nu_2), \, \int \cdot \, d(\mu_2 \otimes \nu_2) \right),$$

$$\left(X \times Y, \, \mathfrak{L}(\mu_1 \otimes \nu_1), \, (\mu_1 \otimes \nu_1)^{X \times Y} \right)$$
$$= \left(X \times Y, \, \mathfrak{L}(\mu_2 \otimes \nu_2), \, (\mu_2 \otimes \nu_2)^{X \times Y} \right).$$

Hint: Use Theorem 4.16 and Exercise 2.

15. Given a non-empty finite family $\left((X_\iota, \mathfrak{R}_\iota, \mu_\iota) \right)_{\iota \in I}$ of positive measure spaces, prove the following.

   (a) $\bigotimes_{\iota \in I} \mu_\iota = 0$ if and only if there is some $\iota \in I$ with $\mu_\iota = 0$.
   (b) If each μ_ι is bounded, then $\bigotimes_{\iota \in I} \mu_\iota$ is bounded.
   (c) If $\bigotimes_{\iota \in I} \mu_\iota$ is bounded and $\mu_\iota \neq 0$ for every $\iota \in I$, then every μ_ι is bounded.

16. Let $\left((X_\iota, \mathfrak{R}_\iota, \mu_\iota) \right)_{\iota \in I}$ and $\left((X_\iota, \mathfrak{R}_\iota, \nu_\iota) \right)_{\iota \in I}$ be two non-empty finite families of positive measure spaces with $\bigotimes_{\iota \in I} \mu_\iota = \bigotimes_{\iota \in I} \nu_\iota \neq 0$. Show that there is a unique family $(\alpha_\iota)_{\iota \in I}$ in $\mathbb{R}_+$ such that $\prod_{\iota \in I} \alpha_\iota = 1$ and $\nu_\iota = \alpha_\iota \mu_\iota$ for every $\iota \in I$.

17. Let $\left((X_\iota, \mathfrak{R}_\iota, \mu_\iota) \right)_{\iota \in I}$ be a non-empty finite family of positive measure spaces and for each $\iota \in I$ take $f_\iota \in \mathcal{L}^1(\mu_\iota)$. Define

$$\bigotimes_{\iota \in I} f_\iota : \prod_{\iota \in I} X_\iota \longrightarrow \overline{\mathbb{R}}, \quad (x_\iota)_{\iota \in I} \longmapsto \prod_{\iota \in I} f_\iota(x_\iota).$$

Prove that

$$\bigotimes_{\iota \in I} f_\iota \in \mathcal{L}^1 \left(\bigotimes_{\iota \in I} \mu_\iota \right) \quad \text{and} \quad \int \left(\bigotimes_{\iota \in I} f_\iota \right) d \left(\bigotimes_{\iota \in I} \mu_\iota \right) = \prod_{\iota \in I} \int f_\iota \, d\mu_\iota.$$

5

Measures on Hausdorff spaces

Two branches of measure theory developed separately for a long time. One was abstract measure theory, dealing with measures on rings of subsets of a given set X, without restrictions on X. This theory was influenced by probability theory. The other branch, spurred on by problems in analysis, is the theory of measures on topological spaces, especially on locally compact spaces.

Interesting results arise when the measure and the topology are 'compatible' in some sense. Measure theory on topological spaces developed its own constructions using such compatibility conditions and these constructions differ to some extent from those of abstract measure theory, as for example in the classical Daniell construction.

It turns out that in the context of Hausdorff spaces, the more general integral developed here leads to the integral used in the literature. However, our development is completely abstract and independent of the particular structures on the base sets X. This notion of integration therefore synthesizes the two perspectives. While we do not enter into all the details in this book, we do highlight a few important aspects.

5.1 Regular measures

We attend to some necessary preliminaries concerning the extension of mappings defined on lattices of sets.

A set $\mathfrak{S}$ of subsets of the set X is called a **lattice of sets** if

(i) $A \cup B \in \mathfrak{S}$ and $A \cap B \in \mathfrak{S}$ whenever $A, B \in \mathfrak{S}$;

(ii) $\emptyset \in \mathfrak{S}$.

The most important example of a lattice of sets for this chapter is the set $\mathfrak{K}(X)$ of compact subsets of a Hausdorff space X.

It is not difficult to describe the ring of sets $\mathfrak{S}_r$ generated by a lattice of sets $\mathfrak{S}$.

Proposition 5.1 *Let $\mathfrak{S}$ be a lattice of sets. Then*

$$\mathfrak{S}' := \{A \setminus B \mid A, B \in \mathfrak{S}, \, B \subset A\}$$

is a semi-ring of sets, and $\mathfrak{S} \subset \mathfrak{S}' \subset \mathfrak{S}_r$.

Proof. Take $A, B, C, D \in \mathfrak{S}$, with $B \subset A$, $D \subset C$. Then

$$(A \setminus B) \cap (C \setminus D) = (A \cap C) \setminus (B \cup D) = (A \cap C) \setminus ((B \cup D) \cap (A \cap C)) \in \mathfrak{S}'.$$

Furthermore,

$$\begin{aligned}
(A \setminus B) \setminus (C \setminus D) &= (A \setminus (B \cup C)) \cup ((A \cap D) \setminus B) \\
&= (A \setminus ((B \cup C) \cap A)) \cup ((A \cap D) \setminus (B \cap (A \cap D))).
\end{aligned}$$

We have

$$A \setminus ((B \cup C) \cap A) \in \mathfrak{S}', \quad (A \cap D) \setminus (B \cap (A \cap D)) \in \mathfrak{S}',$$

and

$$(A \setminus ((B \cup C) \cap A)) \cap ((A \cap D) \setminus (B \cap (A \cap D))) = \emptyset.$$

Thus $\mathfrak{S}'$ is a semi-ring of sets. If $A \in \mathfrak{S}$, then $A = A \setminus \emptyset \in \mathfrak{S}'$. Therefore $\mathfrak{S} \subset \mathfrak{S}'$. Clearly $\mathfrak{S}' \subset \mathfrak{S}_r$. $\square$

Corollary 5.2 *Let $\mathfrak{S}$ be a lattice of sets. Then $A \in \mathfrak{S}_r$ if and only if there are finite, non-empty families $(A_\iota)_{\iota \in I}$ and $(B_\iota)_{\iota \in I}$ in $\mathfrak{S}$ such that for each $\iota \in I$ $B_\iota \subset A_\iota$, that $(A_\iota \setminus B_\iota) \cap (A_{\iota'} \setminus B_{\iota'}) = \emptyset$ for distinct $\iota, \iota' \in I$ and that*

$$A = \bigcup_{\iota \in I} (A_\iota \setminus B_\iota).$$

This is an immediate consequence of Propositions 5.1 and 2.26.

Let $\mathfrak{S}$ be a lattice of sets as above. We now consider an additive map μ on $\mathfrak{S}_r$. Given $A, B \in \mathfrak{S}_r$,

$$\mu(A \cup B) + \mu(A \cap B) = \mu(A) + \mu(B).$$

This relation is called **modularity** and also holds if μ acts only on $\mathfrak{S}$, since $A \cap B \in \mathfrak{S}$ and $A \cup B \in \mathfrak{S}$ for all $A, B \in \mathfrak{S}$. Thus the restriction to $\mathfrak{S}$ of an additive map μ defined on $\mathfrak{S}_r$ is modular. It is interesting to note that modularity is also essentially sufficient to guarantee that a map ν defined on $\mathfrak{S}$ can be extended to an additive map μ on $\mathfrak{S}_r$. These facts are formulated more precisely in the next theorem.

Theorem 5.3 *Let $\nu : \mathfrak{S} \to \mathbb{R}$ be a modular map defined on the lattice of sets $\mathfrak{S}$. Suppose that $\nu(\emptyset) = 0$.*

(a) There is a unique additive map μ defined on $\mathfrak{S}_r$ such that $\mu|_{\mathfrak{S}} = \nu$.

(b) If ν is increasing, then μ is a positive content.

Proof. (a) Take $A, B, C, D \in \mathfrak{S}$ with $B \subset A$, $D \subset C$ such that $A \setminus B = C \setminus D$. Then $A \cup D = B \cup C$ and $A \cap D = B \cap C$. Thus

$$\nu(A) + \nu(D) - \nu(A \cap D) = \nu(A \cup D) = \nu(B \cup C) = \nu(B) + \nu(C) - \nu(B \cap C)$$

and

$$\nu(A \cap D) = \nu(B \cap C).$$

It follows that

$$\nu(A) - \nu(B) = \nu(C) - \nu(D).$$

This enables us to define an extension ν' of ν on

$$\mathfrak{S}' := \{A \setminus B \mid A, B \in \mathfrak{S}, \, B \subset A\} :$$

for each $A, B \in \mathfrak{S}$, $B \subset A$, let

$$\nu'(A \setminus B) := \nu(A) - \nu(B).$$

We already know that $\mathfrak{S}'$ is a semi-ring of sets. We next show that ν' has the additive property given in Theorem 2.27. We use complete induction on n to prove that

$$\nu(A) - \nu(B) = \sum_{k=1}^{n} \big(\nu(A_k) - \nu(B_k)\big)$$

for any pair of families in $\mathfrak{S}$, $(A_k)_{k \le n}$ and $(B_k)_{k \le n}$, with $B_k \subset A_k$ for every $k \le n$, $(A_k \setminus B_k) \cap (A_{k'} \setminus B_{k'}) = \emptyset$ whenever $k, k' \le n$ are distinct, and for $A, B \in \mathfrak{S}$ such that $B \subset A$ and

$$A \setminus B = \bigcup_{k \le n} (A_k \setminus B_k).$$

Consider the case $n = 1$. Then $A \setminus B = A_1 \setminus B_1$, $B \subset A$ and $B_1 \subset A_1$. Hence

$$\nu(A) - \nu(B) = \nu(A_1) - \nu(B_1),$$

as we have already shown.

For the inductive step, first consider

$$A = \bigcup_{k \le n+1} (A_k \setminus B_k)$$

where we assume that the above conditions are fulfilled and that $B = \emptyset$. Then

$$A \setminus (A \cap A_{n+1}) = \bigcup_{k \le n} \big(A_k \setminus ((B_k \cup A_{n+1}) \cap A_k)\big).$$

It therefore follows from our hypothesis that

$$\nu(A) - \nu(A \cap A_{n+1}) = \sum_{k=1}^{n} \Big(\nu(A_k) - \nu((B_k \cup A_{n+1}) \cap A_k)\Big).$$

By modularity, if $k \leq n$, then

$$\nu\big((B_k \cup A_{n+1}) \cap A_k\big) = \nu(B_k \cup A_{n+1}) + \nu(A_k) - \nu(A_k \cup A_{n+1})$$
$$= \nu(B_k) - \nu(B_k \cap A_{n+1}) + \nu(A_k \cap A_{n+1}).$$

Substitution yields

$$\nu(A) - \nu(A \cap A_{n+1})$$
$$= \sum_{k=1}^{n} \big(\nu(A_k) - \nu(B_k)\big) - \sum_{k=1}^{n} \big(\nu(A_k \cap A_{n+1}) - \nu(B_k \cap A_{n+1})\big).$$

Note that

$$A \cap B_{n+1} = \bigcup_{k \leq n} \big((A_k \cap A_{n+1}) \setminus (B_k \cap A_{n+1})\big).$$

Hence our assumption and the fact that $\nu(\emptyset) = 0$ imply that

$$\nu(A \cap B_{n+1}) = \sum_{k=1}^{n} \big(\nu(A_k \cap A_{n+1}) - \nu(B_k \cap A_{n+1})\big).$$

Finally, $A_{n+1} \setminus B_{n+1} = (A \cap A_{n+1}) \setminus (A \cap B_{n+1})$. Therefore

$$\nu(A_{n+1}) - \nu(B_{n+1}) = \nu(A \cap A_{n+1}) - \nu(A \cap B_{n+1}).$$

Substituting, we see that

$$\nu(A) = \sum_{k=1}^{n+1} \big(\nu(A_k) - \nu(B_k)\big).$$

We now consider the general case

$$A \setminus B = \bigcup_{k \leq n+1} (A_k \setminus B_k).$$

It is immediate that

$$A \cup B_{n+1} = (A_{n+1} \cup B) \cup \left(\bigcup_{k \leq n} \big(A_k \setminus ((B_k \cup A_{n+1}) \cap A_k)\big) \right).$$

As we have already shown,

$$\nu(A \cup B_{n+1}) = \nu(A_{n+1} \cup B) + \sum_{k=1}^{n} \big(\nu(A_k) - \nu((B_k \cup A_{n+1}) \cap A_k)\big).$$

By modularity,

$$\nu(A \cup B_{n+1}) = \nu(A) + \nu(B_{n+1}) - \nu(A \cap B_{n+1}),$$
$$\nu(A_{n+1} \cup B) = \nu(A_{n+1}) + \nu(B) - \nu(A_{n+1} \cap B),$$

and if $k \leq n$, then

$$\nu\big((B_k \cup A_{n+1}) \cap A_k\big) = \nu(B_k \cup A_{n+1}) + \nu(A_k) - \nu(A_k \cup A_{n+1})$$
$$= \nu(B_k) - \nu(B_k \cap A_{n+1}) + \nu(A_k \cap A_{n+1}).$$

Substitution yields

$$\nu(A) - \nu(B) = \sum_{k=1}^{n+1} \big(\nu(A_k) - \nu(B_k)\big) + \nu(A \cap B_{n+1}) - \nu(A_{n+1} \cap B)$$

$$- \sum_{k=1}^{n} \big(\nu(A_k \cap A_{n+1}) - \nu(B_k \cap A_{n+1})\big).$$

But

$$A \cap B_{n+1} = (A_{n+1} \cap B) \cup \Big(\bigcup_{k \leq n} \big((A_k \setminus B_k) \cap A_{n+1}\big) \Big).$$

Thus

$$\nu(A \cap B_{n+1}) = \nu(A_{n+1} \cap B) + \sum_{k=1}^{n} \big(\nu(A_k \cap A_{n+1}) - \nu(B_k \cap A_{n+1})\big),$$

and hence

$$\nu(A) - \nu(B) = \sum_{k=1}^{n+1} \big(\nu(A_k) - \nu(B_k)\big).$$

This verifies the additive property of Theorem 2.27 and (a) follows.

(b) is obvious, since if ν is increasing, then $\nu' \geq 0$. By the definition of μ, $\mu \geq 0$. $\qquad\square$

We are now ready to investigate measures on Hausdorff spaces.

Let X be a Hausdorff space. We write $\mathfrak{K}(X)$ for the set of all compact subsets of X and $\mathfrak{R}(X)$ for the ring of sets generated by $\mathfrak{K}(X)$. If $U \subset X$ is relatively compact (i.e. if $\overline{U} \in \mathfrak{K}(X)$) and open, then $U \in \mathfrak{R}(X)$ because $U = \overline{U} \setminus (\overline{U} \setminus U)$.

When seeking to unite measure and topology, positive measures on $\mathfrak{R}(X)$ – or on rings of sets containing $\mathfrak{R}(X)$ – are of particular importance. In such cases the compatibility condition mentioned in the introduction to this chapter is 'regularity'. Let $\mathfrak{R}$ be a ring of subsets of X containing $\mathfrak{K}(X)$. Then the positive content μ on $\mathfrak{R}$ is called **regular** if

$$\mu(A) = \sup\{\mu(K) \,|\, K \in \mathfrak{K}(X), \, K \subset A\}$$

for every $A \in \mathfrak{R}$. Regularity thus means that the sets of $\mathfrak{R}$ can be approximated in measure from below by compact sets. If μ is a regular measure and $A \in \mathfrak{R}$, then obviously there is an increasing sequence $(K_n)_{n \in \mathbb{N}}$ of compact subsets of A such that $A \setminus \bigcup_{n \in \mathbb{N}} K_n \in \mathfrak{N}(\mu)$. Note that there is also a disjoint sequence $(L_n)_{n \in \mathbb{N}}$ of compact subsets of A with $A \setminus \bigcup_{n \in \mathbb{N}} L_n \in \mathfrak{N}(\mu)$. (Use a recursive construction.) For example, every Stieltjes measure $\mu_g{}^{\mathbb{R}}$ is regular, as we shall see.

There are of course other important compatibility conditions as well. Nevertheless, we limit our investigation to regular positive contents, since these are of particular importance for numerous applications.

We begin with a readily verified and yet remarkable result.

Theorem 5.4 *Let μ be a regular positive content on the ring of sets $\mathfrak{R} \supset \mathfrak{R}(X)$, $\mathfrak{R} \subset \mathfrak{P}(X)$. Then μ is a positive measure.*

Proof. To show that μ is null-continuous, let $(A_n)_{n\in\mathbb{N}}$ be a decreasing sequence in $\mathfrak{R}$ with $\downarrow A_n = \emptyset$. Take $\varepsilon > 0$. For each $n \in \mathbb{N}$ there is a $K_n \in \mathfrak{R}(X)$, $K_n \subset A_n$, with

$$\mu(A_n) \le \mu(K_n) + \varepsilon/2^n.$$

Then $\bigcap_{n\in\mathbb{N}} K_n = \emptyset$ and therefore there is an $m \in \mathbb{N}$ for which $\bigcap_{n\le m} K_n = \emptyset$. Thus

$$A_m \subset \bigcup_{n\le m} (A_n \setminus K_n),$$

and hence

$$\downarrow \mu(A_n) \le \mu(A_m) \le \sum_{n\le m} \big(\mu(A_n) - \mu(K_n)\big) < \varepsilon.$$

Since ε is arbitrary, it follows that $\downarrow \mu(A_n) = 0$. $\qquad\square$

According to the next theorem regularity of a positive content μ on $\mathfrak{R}(X)$ is equivalent to a property which might be described as 'generalized null-continuity on compact sets'.

Theorem 5.5 *Let μ be a positive content on $\mathfrak{R}(X)$. Then the following are equivalent.*

(a) μ is regular.

(b) If $(K_\iota)_{\iota\in I}$ is a non-empty downward directed family in $\mathfrak{R}(X)$, then

$$\mu\Big(\bigcap_{\iota\in I} K_\iota\Big) = \inf_{\iota\in I} \mu(K_\iota).$$

Proof. (a)$\Rightarrow$(b). Put $K := \bigcap_{\iota\in I} K_\iota$. Take $\varepsilon > 0$ and $\iota_0 \in I$. Then there is an $L \in \mathfrak{R}(X)$ with $L \subset K_{\iota_0} \setminus K$ and

$$\mu(K_{\iota_0}) - \mu(K) = \mu(K_{\iota_0} \setminus K) \le \mu(L) + \varepsilon.$$

There is a $\lambda \in I$ such that $K_\lambda \cap L = \emptyset$ and $K_\lambda \subset K_{\iota_0}$. Then

$$\mu(K_\lambda) + \mu(L) \le \mu(K_{\iota_0}) \le \mu(K) + \mu(L) + \varepsilon$$

and therefore

$$\mu(K_\lambda) \le \mu(K) + \varepsilon.$$

Since ε is arbitrary,

$$\inf_{\iota \in I} \mu(K_\iota) \le \mu(K).$$

The reverse inequality is trivial, since the positive content μ is monotone.

(b)$\Rightarrow$(a). Take $K, L \in \mathfrak{K}(X)$ with $K \subset L$. Define

$$\mathfrak{C} := \{C \in \mathfrak{K}(X) \,|\, K \subset C \subset L, \ K \cap \overline{L \setminus C} = \emptyset\}.$$

Then $\mathfrak{C}$ is directed down and the reader can easily verify that $\bigcap_{C \in \mathfrak{C}} C = K$. Hence

$$\inf_{C \in \mathfrak{C}} \mu(C) = \mu(K).$$

But for each $C \in \mathfrak{C}$, $L \subset C \cup \overline{L \setminus C}$ and

$$\mu(L) \le \mu(C) + \mu(\overline{L \setminus C}) \le \mu(C) + \sup\{\mu(D) \,|\, D \in \mathfrak{K}(X), \ D \subset L \setminus K\}.$$

It follows that

$$\mu(L) - \mu(K) \le \sup\{\mu(D) \,|\, D \in \mathfrak{K}(X), \ D \subset L \setminus K\}.$$

Take $A \in \mathfrak{R}(X)$. Then by Corollary 5.2, there are families $(K_k)_{k \le n}$ and $(L_k)_{k \le n}$ in $\mathfrak{K}(X)$ with $K_k \subset L_k$ for every $k \le n$, $(L_k \setminus K_k) \cap (L_{k'} \setminus K_{k'}) = \emptyset$ for distinct $k, k' \le n$ and

$$A = \bigcup_{k \le n} (L_k \setminus K_k).$$

If $\varepsilon > 0$, then for each $k \le n$ we can find a $D_k \in \mathfrak{K}(X)$ such that

$$D_k \subset L_k \setminus K_k \quad \text{and} \quad \mu(L_k) - \mu(K_k) \le \mu(D_k) + \varepsilon/n.$$

Put $D := \bigcup_{k \le n} D_k$. Then $D \subset A$ and $\mu(A) - \mu(D) < \varepsilon$. Since ε is arbitrary,

$$\mu(A) \le \sup\{\mu(D) \,|\, D \in \mathfrak{K}(X), \ D \subset A\}.$$

The reverse inequality is clear since μ is monotone. Thus μ is regular. $\square$

The property presented in Theorem 5.5(b) is important, because it enables us to construct regular measures on $\mathfrak{R}(X)$ by extending set-mappings which are defined only on $\mathfrak{K}(X)$.

Theorem 5.6 (Existence Theorem) *Let $\nu : \mathfrak{K}(X) \to \mathbb{R}_+$ be a mapping with the following properties.*

(i) $\nu(K \cup L) \le \nu(K) + \nu(L)$ for all $K, L \in \mathfrak{K}(X)$.

(ii) $\nu(K \cup L) = \nu(K) + \nu(L)$ for all disjoint $K, L \in \mathfrak{K}(X)$.

(iii) $\nu(\bigcap_{\iota \in I} K_\iota) = \inf_{\iota \in I} \nu(K_\iota)$ for every non-empty downward directed family $(K_\iota)_{\iota \in I}$ in $\mathfrak{K}(X)$.

Then there is a unique regular positive measure μ on $\mathfrak{R}(X)$ with $\mu(K) = \nu(K)$ for every $K \in \mathfrak{K}(X)$.

Proof. By Theorems 5.3–5.5, it suffices to prove that $\nu(\emptyset) = 0$, that ν is modular and that ν is increasing. The last of these is trivial, since (iii) implies that

$$\nu(K) = \nu(K \cap L) \leq \nu(L)$$

whenever $K, L \in \mathfrak{K}(X)$, $K \subset L$. Moreover, it follows from (ii) that $\nu(\emptyset) = 0$.

The proof of Theorem 5.5 (b)$\Rightarrow$(a) can be adapted to prove that for arbitrary $K, L \in \mathfrak{K}(X)$, $K \subset L$,

$$\nu(L) - \nu(K) \leq \sup\{\nu(D) \,|\, D \in \mathfrak{K}(X), \ D \subset L \setminus K\}.$$

The proof uses hypotheses (i) and (iii). Conversely, since ν is increasing and (ii) holds, it follows that

$$\nu(L) \geq \nu(D \cup K) = \nu(D) + \nu(K)$$

for every $D \in \mathfrak{K}(X)$, $D \subset L \setminus K$. We conclude that

$$\nu(L) - \nu(K) \geq \sup\{\nu(D) \,|\, D \in \mathfrak{K}(X), \ D \subset L \setminus K\}.$$

Thus these quantities coincide. Hence, for $K, L \in \mathfrak{K}(X)$,

$$\begin{aligned}
\nu(K \cup L) - \nu(K) &= \sup\{\nu(D) \,|\, D \in \mathfrak{K}(X), \ D \subset (K \cup L) \setminus K\} \\
&= \sup\{\nu(D) \,|\, D \in \mathfrak{K}(X), \ D \subset L \setminus (K \cap L)\} \\
&= \nu(L) - \nu(K \cap L),
\end{aligned}$$

and therefore

$$\nu(K \cup L) + \nu(K \cap L) = \nu(K) + \nu(L).$$

This shows that ν is modular. ◻

Since μ-integrable sets can be approximated by sets from the original domain $\mathfrak{R}$ of the measure μ (see Theorem 4.25) and since regularity is also an approximation property, it is reasonable to anticipate that regularity for $\mathfrak{R}$ carries over to μ-integrable sets. The next theorem confirms our expectation.

Theorem 5.7

(a) *Let μ be a positive measure on the ring of sets $\mathfrak{R}$ such that for every $A \in \mathfrak{R}$*

$$\mu(A) = \sup\{\mu^X(K) \,|\, K \in \mathfrak{K}(X) \cap \mathfrak{L}(\mu), \ K \subset A\}.$$

Then for every $A \in \mathfrak{L}(\mu)$

$$\mu^X(A) = \sup\{\mu^X(K) \,|\, K \in \mathfrak{K}(X) \cap \mathfrak{L}(\mu), \ K \subset A\}.$$

(b) *Let μ be a regular positive measure on the ring of sets $\mathfrak{R} \supset \mathfrak{R}(X)$, $\mathfrak{R} \subset \mathfrak{P}(X)$. Then μ^X is also regular.*

Proof. (a) Take $A \in \mathfrak{L}(\mu)$ and $\varepsilon > 0$. By Theorem 4.25(a) there is a decreasing sequence $(B_n)_{n \in \mathbb{N}}$ in $\mathfrak{R}$ with $\bigcap_{n \in \mathbb{N}} B_n \subset A$ and

$$\mu^X(A) - \mu^X\left(\bigcap_{n \in \mathbb{N}} B_n\right) < \frac{\varepsilon}{2}.$$

Take $n \in \mathbb{N}$ and choose $K_n \in \mathfrak{K}(X) \cap \mathfrak{L}(\mu)$ with $K_n \subset B_n$ such that

$$\mu(B_n) - \mu^X(K_n) < \frac{\varepsilon}{2^{n+1}}.$$

Then $K := \bigcap_{n \in \mathbb{N}} K_n$ is compact and μ-integrable. Moreover, $K \subset A$ and

$$\mu^X(A) - \mu^X(K) = \mu^X(A) - \mu^X\left(\bigcap_{n \in \mathbb{N}} B_n\right) + \mu^X\left(\bigcap_{n \in \mathbb{N}} B_n\right) - \mu^X(K)$$

$$< \frac{\varepsilon}{2} + \sum_{n \in \mathbb{N}} \left(\mu(B_n) - \mu^X(K_n)\right) < \varepsilon.$$

(b) follows from (a). $\qquad\qquad\qquad\qquad\qquad\qquad\qquad\qquad\qquad\qquad\quad\square$

For a set A to be a μ-null set, it is sufficient that it be locally a μ-null set, i.e. that $A \cap B \in \mathfrak{N}(\mu)$ for every $B \in \mathfrak{R}$ (see Proposition 4.6). It should therefore come as no surprise to learn that when μ is a regular measure it is already sufficient that $A \cap K \in \mathfrak{N}(\mu)$ for every compact set K, as we prove in the next theorem. This type of principle is often met with regular measures: if a property can be described in terms of local behaviour, then it is enough to study it on (the intersections with) the compact sets. This is quite natural, since a regular measure is determined by its values on compact sets. We shall meet other instances of this principle later.

Theorem 5.8 *Let μ be a regular positive measure on the ring of sets $\mathfrak{R} \supset \mathfrak{R}(X)$, $\mathfrak{R} \subset \mathfrak{P}(X)$. Given $A \subset X$, the following are equivalent.*

(a) A is a μ-null set.

(b) Given $K \in \mathfrak{K}(X)$ and $\varepsilon > 0$, there is an $L \in \mathfrak{K}(X)$ such that $L \subset K \backslash A$ and $\mu(K) - \mu(L) < \varepsilon$.

(c) Given $K \in \mathfrak{K}(X)$, $A \cap K \in \mathfrak{N}(\mu)$.

Proof. (a)$\Rightarrow$(b). Let A be a μ-null set. Then $A \cap K \in \mathfrak{N}(\mu)$ for every $K \in \mathfrak{K}(X)$. Thus $\mu(K) = \mu^X(K \backslash A)$ and the conclusion now follows immediately from Theorem 5.7.

(b)$\Rightarrow$(c). Assume (b) and take $K \in \mathfrak{K}(X)$. Then there is an increasing sequence $(L_n)_{n \in \mathbb{N}}$ in $\mathfrak{K}(X)$ with $L_n \subset K \backslash A$ and

$$\mu(K) \leq \mu(L_n) + 1/n$$

for every $n \in \mathbb{N}$. $L := \bigcup_{n \in \mathbb{N}} L_n$ is μ-integrable, $L \subset K \backslash A$ and $\mu(K) = \mu^X(L)$. Thus $K \backslash L \in \mathfrak{N}(\mu)$, and $A \cap K \subset K \backslash L$ implies that $A \cap K \in \mathfrak{N}(\mu)$.

(c)$\Rightarrow$(a). Take $B \in \mathfrak{R}$. Since μ is regular, there is a sequence $(K_n)_{n \in \mathbb{N}}$ in $\mathfrak{K}(X)$ with $K_n \subset B$ for every $n \in \mathbb{N}$ and

$$B \setminus \bigcup_{n \in \mathbb{N}} K_n \in \mathfrak{N}(\mu).$$

We see that

$$B \cap A = \left(\left(B \setminus \bigcup_{n \in \mathbb{N}} K_n \right) \cap A \right) \cup \left(\bigcup_{n \in \mathbb{N}} K_n \cap A \right) \in \mathfrak{N}(\mu).$$

(a) now follows by Proposition 4.6.   □

We note two remarkable consequences of this theorem.

Corollary 5.9 *Let μ be a regular positive measure on the ring of sets $\mathfrak{R} \supset \mathfrak{R}(X)$, $\mathfrak{R} \subset \mathfrak{P}(X)$, and Φ the set of all regular positive measure spaces $(X, \mathfrak{S}, \nu)$ for which*

$$(X, \mathfrak{R}, \mu) \preccurlyeq (X, \mathfrak{S}, \nu).$$

Then $(X, \mathfrak{L}(\mu), \mu^X)$ is the largest element in Φ.

Proof. Take $(X, \mathfrak{S}, \nu) \in \Phi$ and consider a set $A \in \mathfrak{S}$. There is a sequence $(K_n)_{n \in \mathbb{N}}$ in $\mathfrak{K}(X)$ with $K_n \subset A$ for each $n \in \mathbb{N}$ and $A \setminus \bigcup_{n \in \mathbb{N}} K_n \in \mathfrak{N}(\nu)$. $\bigcup_{n \in \mathbb{N}} K_n$ is also in $\mathfrak{L}(\mu)$ and we show that $A \setminus \bigcup_{n \in \mathbb{N}} K_n \in \mathfrak{N}(\mu)$.

Take $K \in \mathfrak{K}(X)$. Then there is a sequence $(L_n)_{n \in \mathbb{N}}$ in $\mathfrak{K}(X)$ for which

$$L_n \subset K \setminus \left(A \setminus \bigcup_{n \in \mathbb{N}} K_n \right)$$

whenever $n \in \mathbb{N}$ and

$$K \setminus \bigcup_{n \in \mathbb{N}} L_n \in \mathfrak{N}(\nu).$$

We may assume that $(L_n)_{n \in \mathbb{N}}$ is increasing. Then

$$\mu(K) - \mu^X \left(\bigcup_{n \in \mathbb{N}} L_n \right) = \mu(K) - \sup_{n \in \mathbb{N}} \mu(L_n)$$

$$= \nu(K) - \sup_{n \in \mathbb{N}} \nu(L_n)$$

$$= \nu(K) - \nu^X \left(\bigcup_{n \in \mathbb{N}} L_n \right) = 0.$$

Thus $K \setminus \bigcup_{n \in \mathbb{N}} L_n \in \mathfrak{N}(\mu)$. But

$$K \cap \left(A \setminus \bigcup_{n \in \mathbb{N}} K_n \right) \subset K \setminus \bigcup_{n \in \mathbb{N}} L_n$$

and so by Theorem 5.8 (c)$\Rightarrow$(a), $A \setminus \bigcup_{n \in \mathbb{N}} K_n \in \mathfrak{N}(\mu)$. Hence

$$A = \bigcup_{n \in \mathbb{N}} K_n \cup \left(A \setminus \bigcup_{n \in \mathbb{N}} K_n \right) \in \mathfrak{L}(\mu)$$

and

$$\mu^X(A) = \mu^X\left(\bigcup_{n\in\mathbb{N}} K_n\right) = \nu^X\left(\bigcup_{n\in\mathbb{N}} K_n\right) = \nu(A).$$

Thus $(X, \mathfrak{S}, \nu) \preccurlyeq (X, \mathfrak{L}(\mu), \mu^X)$. The regularity of μ^X was established in Theorem 5.7. $\qquad\square$

Corollary 5.9 is particularly important: it is one of the justifications for our definition of the integral. It shows that even in the case of regular measures on Hausdorff spaces the abstractly defined integral provides the proper extension of $(X, \mathcal{L}(\mathfrak{R}), \ell_\mu)$. No additional constructions are required.

We stress that $(X, \mathfrak{L}(\mu), \mu^X)$ is not only a maximal, but even the largest element in the set of all regular extensions of the regular positive measure space $(X, \mathfrak{R}, \mu)$. (Recall that an element of an ordered set is maximal if there is no strictly larger element in this set. There may well be many maximal elements, which then, of course, are not related to one another by the order relation: they are incommensurable. A largest element, however, is always uniquely determined and, in fact, larger than every other element of the given set.)

A characterization of the integral in the sense of Corollary 5.9 can also be given in the abstract case. We leave this as an exercise for the reader (Exercise 10).

The next corollary justifies restricting our investigation of regular positive measures to $\mathfrak{R}(X)$.

Corollary 5.10 *Let μ be a regular positive measure on the ring of sets $\mathfrak{R} \supset \mathfrak{R}(X)$, $\mathfrak{R} \subset \mathfrak{P}(X)$, and ν the restriction of μ to $\mathfrak{R}(X)$. Then*

$$(X, \mathfrak{L}(\nu), \nu^X) = (X, \mathfrak{L}(\mu), \mu^X)$$

and

$$\left(X, \mathcal{L}^1(\nu), \int \cdot\, d\nu\right) = \left(X, \mathcal{L}^1(\mu), \int \cdot\, d\mu\right).$$

Proof. Since $(X, \mathfrak{L}(\mu), \mu^X)$ is a regular extension of $(X, \mathfrak{R}(X), \nu)$, it follows from Corollary 5.9 that

$$(X, \mathfrak{L}(\mu), \mu^X) \preccurlyeq (X, \mathfrak{L}(\nu), \nu^X).$$

Thus in particular $(X, \mathfrak{L}(\nu), \nu^X)$ is a regular extension of $(X, \mathfrak{R}, \mu)$ and it follows by Corollary 5.9 again that

$$(X, \mathfrak{L}(\nu), \nu^X) \preccurlyeq (X, \mathfrak{L}(\mu), \mu^X).$$

Hence

$$(X, \mathfrak{L}(\nu), \nu^X) = (X, \mathfrak{L}(\mu), \mu^X).$$

Now appeal to Corollary 4.11 and the monotone convergence theorem to draw the second conclusion. $\qquad\square$

We proceed with a criterion similar to Theorem 5.8 for μ-measurability of a set.

Proposition 5.11 *Let μ be a regular positive measure on the ring of sets $\mathfrak{R} \supset \mathfrak{R}(X)$, $\mathfrak{R} \subset \mathfrak{P}(X)$, and take $A \subset X$. Then the following are equivalent.*

(a) $A \in \mathfrak{M}(\mu)$.

(b) *Given $K \in \mathfrak{K}(X)$ and $\varepsilon > 0$, there are $L', L'' \in \mathfrak{K}(X)$ with*

$$L' \subset A \cap K, \quad L'' \subset K \setminus A \quad \text{and} \quad \mu(K \setminus (L' \cup L'')) < \varepsilon.$$

(c) $A \cap K \in \mathfrak{L}(\mu)$ *for every $K \in \mathfrak{K}(X)$.*

Proof. (a)$\Rightarrow$(b). Take $A \in \mathfrak{M}(\mu)$, $K \in \mathfrak{K}(X)$ and $\varepsilon > 0$. Then $A \cap K \in \mathfrak{L}(\mu)$ and $K \setminus A \in \mathfrak{L}(\mu)$. By Theorem 5.7(b) there are $L', L'' \in \mathfrak{K}(X)$ with

$$L' \subset A \cap K \quad \text{and} \quad \mu(L') > \mu^X(A \cap K) - \frac{\varepsilon}{2},$$

$$L'' \subset K \setminus A \quad \text{and} \quad \mu(L'') > \mu^X(K \setminus A) - \frac{\varepsilon}{2}.$$

It follows that

$$\mu(K \setminus (L' \cup L'')) = \mu^X((A \cap K) \setminus L') + \mu^X((K \setminus A) \setminus L'') < \varepsilon.$$

(b)$\Rightarrow$(c). Given $K \in \mathfrak{K}(X)$ and $n \in \mathbb{N}$, there are $L'_n, L''_n \in \mathfrak{K}(X)$ such that

$$L'_n \subset A \cap K, \quad L''_n \subset K \setminus A \quad \text{and} \quad \mu(K \setminus (L'_n \cup L''_n)) < \frac{1}{n}.$$

Putting $L' := \bigcup_{n \in \mathbb{N}} L'_n$ and $L'' := \bigcup_{n \in \mathbb{N}} L''_n$, we see that $L', L'' \in \mathfrak{L}(\mu)$ and $K \setminus (L' \cup L'') \in \mathfrak{N}(\mu)$. Thus $A \cap K \setminus L' \in \mathfrak{N}(\mu)$ and hence $A \cap K \in \mathfrak{L}(\mu)$.

(c)$\Rightarrow$(a). Take $B \in \mathfrak{L}(\mu)$. There is a sequence $(K_n)_{n \in \mathbb{N}}$ of compact subsets of X contained in B with $B \setminus \bigcup_{n \in \mathbb{N}} K_n \in \mathfrak{N}(\mu)$. Then

$$A \cap B \setminus \left(\bigcup_{n \in \mathbb{N}} (A \cap K_n) \right) \in \mathfrak{N}(\mu).$$

By (c) $A \cap K_n \in \mathfrak{L}(\mu)$ for every $n \in \mathbb{N}$ and thus $\bigcup_{n \in \mathbb{N}} (A \cap K_n) \in \mathfrak{L}(\mu)$. Hence $A \cap B \in \mathfrak{L}(\mu)$. $\square$

Recall that $\mathfrak{B}(X)$, the set of all Borel sets of X, is the σ-algebra on X generated by the open sets of X, or, equivalently, the σ-algebra on X generated by the closed sets of X.

Corollary 5.12 *Take a regular positive measure μ on the ring of sets $\mathfrak{R} \supset \mathfrak{R}(X)$, $\mathfrak{R} \subset \mathfrak{P}(X)$. Then:*

(a) $\mathfrak{B}(X) \subset \mathfrak{M}(\mu)$. *In particular, every open and every closed subset of X is μ-measurable.*

(b) *Every relatively compact Borel set of X belongs to $\mathfrak{L}(\mu)$.*

Proof. (a) For every closed subset A of X and every $K \in \mathfrak{K}(X)$, we have $A \cap K \in \mathfrak{K}(X) \subset \mathfrak{L}(\mu)$. Thus the set of all closed sets of X is contained in $\mathfrak{M}(\mu)$. Since $\mathfrak{M}(\mu)$ is a σ-algebra, it also contains $\mathfrak{B}(X)$.

(b) follows from (a). $\qquad\qquad\qquad\qquad\qquad\qquad\qquad\qquad\qquad\qquad\qquad\square$

The 'natural' functions on topological spaces are the continuous ones. How are these related to regular measures? It is of course unreasonable to expect *every* continuous function to be integrable in general since not even constant functions need be integrable. Before formulating Lusin's theorem – which describes the relationship between continuity and measure, characterizing in particular μ-measurable functions whenever μ is regular – we first explain what we mean by a 'continuous extended real-valued function'. This presupposes a topology on $\overline{\mathbb{R}}$. We always consider $\overline{\mathbb{R}}$ with its natural topology, i.e. $U \subset \overline{\mathbb{R}}$ is open if and only if

(i) $U \cap \mathbb{R}$ is open in $\mathbb{R}$;

(ii) if $\infty \in U$, then there is some $a \in \mathbb{R}$ such that $]a, \infty] \subset U$;

(iii) if $-\infty \in U$, then there is some $a \in \mathbb{R}$ such that $[-\infty, a[\subset U$.

Of course the topology of $\mathbb{R}$ is just the relative topology induced by the topology on $\overline{\mathbb{R}}$.

We now state and prove the fundamental theorem due to Lusin.

Theorem 5.13 (Lusin) *Let μ be a regular positive measure on the ring of sets $\mathfrak{R} \supset \mathfrak{R}(X)$, $\mathfrak{R} \subset \mathfrak{P}(X)$, and take $f \in \overline{\mathbb{R}}^X$. Then the following are equivalent.*

(a) f is μ-measurable.

(b) Given $K \in \mathfrak{K}(X)$ and $\varepsilon > 0$, there is an $L \in \mathfrak{K}(X)$ such that

$$L \subset K, \quad \mu(K \setminus L) < \varepsilon \quad \text{and} \quad f|_L \text{ is continuous.}$$

(c) Given $K \in \mathfrak{K}(X)$, there is a disjoint sequence $(L_n)_{n \in \mathbb{N}}$ of compact sets such that

$$\bigcup_{n \in \mathbb{N}} L_n \subset K, \quad K \setminus \bigcup_{n \in \mathbb{N}} L_n \in \mathfrak{N}(\mu) \quad \text{and} \quad f|_{L_n} \text{ is continuous } (n \in \mathbb{N}).$$

Proof. (a)$\Rightarrow$(b). First suppose that $f = e_A$ for some $A \in \mathfrak{M}(\mu)$, and take L', L'' as in Proposition 5.11(b). Then $L := L' \cup L''$ has the required properties.

Now assume that $f = \sum_{\iota \in I} e_{A_\iota}$ for some finite family $(A_\iota)_{\iota \in I}$ in $\mathfrak{M}(\mu)$ with n elements. Take $\iota \in I$. By the above considerations there is an $L_\iota \in \mathfrak{K}(X)$ such that

$$L_\iota \subset K, \quad \mu(K \setminus L_\iota) < \frac{\varepsilon}{n+1} \quad \text{and} \quad e_{A_\iota}|_{L_\iota} \text{ is continuous.}$$

Then $L := \bigcap_{\iota \in I} L_\iota$ has the required properties.

For the third step suppose that $0 \leq f \leq 1$. Let $(f_n)_{n \in \mathbb{N}}$ be the sequence defined in Proposition 4.10. Take $n \in \mathbb{N}$. By the second step there is an $L_n \in \mathfrak{K}(X)$ such that

$$L_n \subset K, \quad \mu(K \setminus L_n) < \frac{\varepsilon}{2^n} \text{ and } f_n|_{L_n} \text{ is continuous.}$$

Then $L := \bigcap_{n \in \mathbb{N}} L_n$ is compact and contained in K. Moreover,

$$\mu(K \setminus L) \leq \sum_{n \in \mathbb{N}} \mu(K \setminus L_n) < \varepsilon.$$

Since f is bounded, it is easy to see that $(f_n)_{n \in \mathbb{N}}$ converges uniformly to f (cf. the proof of Proposition 4.10). Thus $f|_L$ is continuous.

Finally, let f be an arbitrary μ-measurable function. Put

$$g(t) := \frac{1}{\pi} \left(\arctan t + \frac{\pi}{2} \right).$$

Then g is an increasing homeomorphism from $\overline{\mathbb{R}}$ onto $[0, 1]$. By Proposition 4.36, if $\alpha \in \mathbb{R}$ then

$$\{g \circ f < \alpha\} = f^{-1}\left(g^{-1}([-\infty, \alpha[) \right) \in \mathfrak{M}(\mu),$$

i.e. $g \circ f$ is μ-measurable. By the third step there is an $L \in \mathfrak{K}(X)$ such that

$$L \subset K, \quad \mu(K \setminus L) < \varepsilon \quad \text{and} \quad g \circ f|_L \text{ is continuous.}$$

Then $f|_L$ is also continuous.

(b)$\Rightarrow$(c). We use recursion to construct a sequence $(L_n)_{n \in \mathbb{N}}$ in $\mathfrak{K}(X)$ such that for each $n \in \mathbb{N}$

$$L_n \subset K \setminus \bigcup_{m=1}^{n-1} L_m, \quad \mu\left(K \setminus \bigcup_{m=1}^{n} L_m \right) < \frac{1}{n} \quad \text{and} \quad f|_{L_n} \text{ is continuous.}$$

Suppose that for some $n \in \mathbb{N}$ the sets $L_1, \ldots, L_{n-1}$ have been constructed. Regularity implies that there is some $L \in \mathfrak{K}(X)$ satisfying

$$L \subset K \setminus \bigcup_{m=1}^{n-1} L_m \text{ and } \mu(L) > \mu\left(K \setminus \bigcup_{m=1}^{n-1} L_m \right) - \frac{1}{2n}.$$

By (b) there is an $L_n \in \mathfrak{K}(X)$ such that

$$L_n \subset L, \quad \mu(L \setminus L_n) < \frac{1}{2n} \quad \text{and} \quad f|_{L_n} \text{ is continuous.}$$

Then $L_n \subset K \setminus \bigcup_{m=1}^{n-1} L_m$ and

$$\mu\left(K \setminus \bigcup_{m=1}^{n} L_m \right) \leq \mu\left(\left(K \setminus \bigcup_{m=1}^{n-1} L_m \right) \setminus L \right) + \mu(L \setminus L_n) < \frac{1}{n}.$$

This completes the recursive construction. The sequence $(L_n)_{n \in \mathbb{N}}$ clearly has the required properties.

(c)$\Rightarrow$(a). Take $\alpha \in \mathbb{R}$ and put $A := \{f < \alpha\}$. Given $B \in \mathfrak{L}(\mu)$, use Theorem 5.7(b) and hypothesis (c) to find a sequence $(K_n)_{n \in \mathbb{N}}$ in $\mathfrak{K}(X)$ with

$$\bigcup_{n \in \mathbb{N}} K_n \subset B, \quad B \setminus \bigcup_{n \in \mathbb{N}} K_n \in \mathfrak{N}(\mu) \quad \text{and} \quad f|_{K_n} \text{ continuous for every } n \in \mathbb{N}.$$

Take $n \in \mathbb{N}$. Since $A' := A \cap K_n = \{f|_{K_n} < \alpha\}$, A' is an open subset of the topological space K_n. Hence there is an open subset U of X such that $A' = U \cap K_n$. Thus $A' \in \mathfrak{B}(X)$. By Corollary 5.12(b), $A \cap K_n \in \mathfrak{L}(\mu)$. Since $A \cap B \setminus \bigcup_{n \in \mathbb{N}} K_n \in \mathfrak{N}(\mu)$,

$$A \cap B = \left(\bigcup_{n \in \mathbb{N}} (A \cap K_n) \right) \cup \left((A \cap B) \setminus \bigcup_{n \in \mathbb{N}} K_n \right) \in \mathfrak{L}(\mu).$$

Thus $A \in \mathfrak{M}(\mu)$. □

Proposition 4.37 and Theorem 4.38 now immediately imply the following result on integrable functions.

Corollary 5.14 *Let μ be a regular positive measure on the ring of sets $\mathfrak{R} \supset \mathfrak{R}(X)$, $\mathfrak{R} \subset \mathfrak{P}(X)$. Take $f \in \overline{\mathbb{R}}^X$. Then the following are equivalent.*

(a) $f \in \mathcal{L}^1(\mu)$.

(b) There is a $g \in \mathcal{L}^1(\mu)$ with $|f| \le g$ μ-a.e. Given $K \in \mathfrak{R}(X)$ and $\varepsilon > 0$, there is an $L \in \mathfrak{R}(X)$ such that

$$L \subset K, \quad \mu(K \setminus L) < \varepsilon \quad \text{and} \quad f|_L \text{ is continuous.}$$

(c) There is a $g \in \mathcal{L}^1(\mu)$ with $|f| \le g$ μ-a.e. and for each $K \in \mathfrak{R}(X)$ there is a disjoint sequence $(L_n)_{n \in \mathbb{N}}$ of compact sets such that

$$\bigcup_{n \in \mathbb{N}} L_n \subset K, \quad K \setminus \bigcup_{n \in \mathbb{N}} L_n \in \mathfrak{N}(\mu) \quad \text{and} \quad f|_{L_n} \text{ is continuous } (n \in \mathbb{N}).$$

The preceding results might create the impression that if μ is a regular measure, then μ-integrable functions are 'almost' continuous. But this would be a misconception. Note that Lusin's theorem does *not* state that f (as a function on X) is continuous at every point in L or L_n. It only states that the restriction $f|_L$, or $f|_{L_n}$, is continuous. But this is a completely different matter. Consider, for example, Lebesgue measure $\lambda^{\mathbb{R}}$ and the function $f := e_{\mathbb{Q} \cap [0,1]}$. (We show below that $\lambda^{\mathbb{R}}$ is regular.) f is of course $\lambda^{\mathbb{R}}$-integrable but it is discontinuous at every point in $[0,1]$!

However, we have the following natural result.

Corollary 5.15 *Let μ be a regular positive measure on the ring of sets $\mathfrak{R} \supset \mathfrak{R}(X)$, $\mathfrak{R} \subset \mathfrak{P}(X)$. Take $f \in \overline{\mathbb{R}}^X$ with*

$$\{x \in X \mid f \text{ is discontinuous at } x\} \in \mathfrak{N}(\mu).$$

Then f is μ-measurable. If in addition $|f| \leq g$ μ-a.e. for some $g \in \mathcal{L}^1(\mu)$, then $f \in \mathcal{L}^1(\mu)$.

These assertions apply in particular to continuous extended real-valued functions f on X.

Proof. Put $A := \{x \in X \mid f$ is discontinuous at $x\}$. Take $\varepsilon > 0$ and $K \in \mathfrak{K}(X)$. Since μ is regular, there is an $L \in \mathfrak{K}(X)$ with $L \subset K \setminus A$ and $\mu^X((K \setminus A) \setminus L) < \varepsilon$. Then $f|_L$ is continuous and $\mu(K \setminus L) < \varepsilon$. Hence, by Lusin's theorem, f is μ-measurable. The second assertion follows from Theorem 4.38. $\square$

Exercises

1. Find examples of a modular map ν for which $\nu(\emptyset) \neq 0$.

2. Let X be a Hausdorff space and μ a positive measure on the ring of sets $\mathfrak{R} \supset \mathfrak{R}(X)$, $\mathfrak{R} \subset \mathfrak{P}(X)$. Prove that if f is an upper (lower) semicontinuous positive (negative) function on X with compact support, then $f \in \overline{\mathcal{L}}(\mu)$.

3. Let Y be a subset of the Hausdorff space X and $\mathfrak{R} \supset \mathfrak{R}(X)$, $\mathfrak{R} \subset \mathfrak{P}(X)$, a ring of sets. Define $\mathfrak{S} := \{A \in \mathfrak{R} \mid A \subset Y\}$. Prove the following.

   (a) $\mathfrak{R}(Y) \subset \mathfrak{S}$.

   (b) If μ is a regular positive measure on $\mathfrak{R}$, then $\mu|_{\mathfrak{S}}$ is a regular positive measure on $\mathfrak{S}$.

4. Let X be a Hausdorff space and $(\mu_n)_{n \in \mathbb{N}}$ a sequence of regular positive measures on the ring of sets $\mathfrak{R} \supset \mathfrak{R}(X)$, $\mathfrak{R} \subset \mathfrak{P}(X)$. Prove the following.

   (a) If $(\mu_n(A))_{n \in \mathbb{N}}$ is an increasing convergent sequence for every $A \in \mathfrak{R}$, then
   $$\mathfrak{R} \longrightarrow \mathbb{R}, \quad A \longmapsto \lim_{n \to \infty} \mu_n(A)$$
   defines a regular positive measure on $\mathfrak{R}$.

   (b) If for each $A \in \mathfrak{R}$ $\sum_{n \in \mathbb{N}} \mu_n(A)$ is finite, then
   $$\mathfrak{R} \longrightarrow \mathbb{R}, \quad A \longmapsto \sum_{n \in \mathbb{N}} \mu_n(A)$$

   is a regular positive measure on $\mathfrak{R}$.

5. Let X, Y be Hausdorff spaces, μ a regular positive measure on the ring of sets $\mathfrak{R} \supset \mathfrak{R}(X)$, $\mathfrak{R} \subset \mathfrak{P}(X)$, and ν a regular positive measure on the ring of sets $\mathfrak{S} \supset \mathfrak{R}(Y)$, $\mathfrak{S} \subset \mathfrak{P}(Y)$. Take $X \times Y$ with the product topology. Show that
   $$(\mu \otimes \nu)(A) = \sup\{(\mu \otimes \nu)(K) \mid K \in \mathfrak{K}(X \times Y), \ K \subset A\}$$
   whenever $A \in \mathfrak{R} \otimes \mathfrak{S}$. Find examples to show that $\mathfrak{K}(X \times Y) \subset \mathfrak{R} \otimes \mathfrak{S}$ need not hold.

6. Take the set X with the discrete topology.

   (a) Determine which positive measures on $\mathfrak{R}(X)$ are regular.
   (b) Let Y be a Hausdorff space. Take a ring of sets $\mathfrak{S} \supset \mathfrak{R}(Y)$, $\mathfrak{S} \subset \mathfrak{P}(Y)$. Show that $\mathfrak{R}(X \times Y) \subset \mathfrak{R}(X) \otimes \mathfrak{S}$ when $X \times Y$ has the product topology.
   (c) Let μ be a positive measure on $\mathfrak{R}(X)$ and ν a regular positive measure on $\mathfrak{S}$. Prove that $\mu \otimes \nu$ is regular.

7. Let X be a Hausdorff space and μ a regular positive measure on the ring of sets $\mathfrak{R} \supset \mathfrak{R}(X)$, $\mathfrak{R} \subset \mathfrak{P}(X)$. Take $f \in \mathcal{L}^1(\mu)$. Prove the following.

   (a) If $f \geq 0$, then
   $$\int f d\mu = \sup_{K \in \mathfrak{K}(X)} \int f e_K d\mu = \sup_{\substack{K \in \mathfrak{K}(X) \\ K \subset \{f>0\}}} \int f e_K d\mu.$$

   (b)
   $$\int f d\mu = \sup_{K \in \mathfrak{K}(X)} \int f e_K d\mu + \inf_{K \in \mathfrak{K}(X)} \int f e_K d\mu$$
   $$= \sup_{\substack{K \in \mathfrak{K}(X) \\ K \subset \{f>0\}}} \int f e_K d\mu + \inf_{\substack{K \in \mathfrak{K}(X) \\ K \subset \{f<0\}}} \int f e_K d\mu.$$

   Hint for (a): Use Corollary 4.28 and Exercise 10 of Section 4.1, and observe that f is μ-measurable.

8. Let X be a Hausdorff space and μ a regular positive measure on the ring of sets $\mathfrak{R} \supset \mathfrak{R}(X)$, $\mathfrak{R} \subset \mathfrak{P}(X)$. Define $U_0 := \bigcup\{U \mid U \in \mathfrak{N}(\mu), U \text{ open}\}$. Prove the following.

   (a) $U_0 \in \mathfrak{N}(\mu)$. Hence there exists a greatest open μ-null set. The set $\operatorname{supp}\mu := X \setminus U_0$ is called the **support** of μ.
   (b) If U is open and $U \cap \operatorname{supp}\mu \neq \emptyset$, then $U \notin \mathfrak{N}(\mu)$.

   Hint for (a): Take $K \in \mathfrak{K}(X)$. Show that $\mu(L) = 0$ for every $L \in \mathfrak{K}(X)$ with $L \subset U_0 \cap K$.

9. Let $\mathfrak{R}$ and $\mathfrak{S}$ be lattices of sets with $\mathfrak{S} \subset \mathfrak{R}$. Let $\nu : \mathfrak{R} \to \mathbb{R}_+$ be a modular map. Call ν **$\mathfrak{S}$-regular** if
   $$\nu(A) - \nu(B) = \sup\{\nu(C) \mid C \in \mathfrak{S}, C \subset A \setminus B\}$$
   for all $A, B \in \mathfrak{R}$ with $B \subset A$. Prove the following.

   (a) If $\mathfrak{R}$ is a ring of sets and ν a positive content on $\mathfrak{R}$, then ν is $\mathfrak{S}$-regular if and only if
   $$\nu(A) = \sup\{\nu(C) \mid C \in \mathfrak{S}, C \subset A\}$$
   for every $A \in \mathfrak{R}$.

(b) Suppose that $\nu : \mathfrak{S} \to \mathbb{R}_+$ is increasing, $\mathfrak{S}$-regular and null-continuous (i.e. $\nu(A_n) \downarrow 0$ for every sequence in $\mathfrak{S}$ with $A_n \downarrow \emptyset$). Then the positive content μ described in Theorem 5.3 is an $\mathfrak{S}$-regular positive measure on $\mathfrak{S}_r$.

Hint for (b): The $\mathfrak{S}$-regularity of μ is easy to see. Break the proof of the null-continuity into several steps. First, let $(C_n)_{n \in \mathbb{N}}$ be a sequence in $\mathfrak{S}'$ with $C := \bigcup_{n \in \mathbb{N}} C_n \in \mathfrak{S}$. To show that $\mu(C) = \sup_{n \in \mathbb{N}} \mu\big(\bigcup_{m \leq n} C_m\big)$, note that

$$C \setminus \bigcup_{m \leq n} C_m = \bigcap_{m \leq n} \big(B_m \cup (C \setminus A_m)\big)$$

where $A_m, B_m \in \mathfrak{S}$ with $C_m = A_m \setminus B_m$ and $B_m \subset A_m \subset C$. Given $\varepsilon > 0$ and $n \in \mathbb{N}$, there is a $D_n \in \mathfrak{S}$ with $D_n \subset B_n \cup (C \setminus A_n)$ and

$$\mu\big(B_n \cup (C \setminus A_n)\big) - \nu(D_n) < \varepsilon/2^n.$$

Since ν is null-continuous, $\inf_{n \in \mathbb{N}} \nu\big(\bigcap_{m \leq n} D_m\big) = 0$. Conclude that

$$\inf_{n \in \mathbb{N}} \mu\bigg(\bigcap_{m \leq n} \big(B_m \cup (C \setminus A_m)\big)\bigg) < \varepsilon.$$

Hence this infimum is 0, which completes the first step.

As a second step, prove that the same is true for $C \in \mathfrak{S}'$. To do so, note that if $C = A \setminus B$ with $A, B \in \mathfrak{S}$, $B \subset A$, then $A = \bigcup_{n \in \mathbb{N}} (B \cup C_n)$.

As a third step, show that $\mu(C) = \sup_{n \in \mathbb{N}} \mu(C_n)$ for every increasing sequence $(C_n)_{n \in \mathbb{N}}$ in $\mathfrak{S}_r$ for which $C := \bigcup_{n \in \mathbb{N}} C_n \in \mathfrak{S}'$.

Finally prove the same is true for $C \in \mathfrak{S}_r$.

10. Let $(X, \mathfrak{R}, \mu)$ be a positive measure space. Prove the following.

(a) μ^X is $\mathfrak{R}_\delta$-regular.
(b) Let Φ denote the set of positive measure spaces $(X, \mathfrak{S}, \nu) \succcurlyeq (X, \mathfrak{R}, \mu)$ such that $\mathfrak{S} \supset \mathfrak{R}_\delta$ and ν is $\mathfrak{R}_\delta$-regular. Then $(X, \mathcal{L}(\mu), \mu^X)$ is the largest element of Φ.

5.2 Measures on metric and locally compact spaces

In this section we specialize to Hausdorff spaces with additional topological properties. Their richer structure makes for results which are not available in arbitrary Hausdorff spaces.

We restrict our attention to measures on metrizable spaces for the moment. The following topological property of these spaces is the linchpin for the elegant result that every positive measure on $\mathfrak{R}(X)$ is necessarily regular when X is metrizable. Thus to find examples of a non-regular measure on $\mathfrak{R}(X)$, we must look beyond the realm of metrizable spaces. Such an example is presented in the exercises. Note also that the result above cannot be generalized to measures on arbitrary rings of sets $\mathfrak{R} \supset \mathfrak{R}(X), \mathfrak{R} \subset \mathfrak{P}(X)$. The exercises contain an example of this as well.

Proposition 5.16 *Let $(K_\iota)_{\iota \in I}$ be a non-empty downward directed family of compact sets in the metrizable space X. Then there is a sequence $(\iota_n)_{n \in \mathbb{N}}$ in I for which $(K_{\iota_n})_{n \in \mathbb{N}}$ is decreasing and*

$$\bigcap_{\iota \in I} K_\iota = \bigcap_{n \in \mathbb{N}} K_{\iota_n}.$$

Proof. Put $K := \bigcap_{\iota \in I} K_\iota$. Given $n \in \mathbb{N}$, define

$$B_n := \{x \in X \mid d(x,y) < 1/n \text{ for some } y \in K\},$$

where d denotes a metric generating the topology on X. Take $n \in \mathbb{N}$. Then B_n is an open neighbourhood of K and

$$\bigcap_{\iota \in I} K_\iota \cap (X \setminus B_n) = \emptyset.$$

Then $K_{\lambda_n} \cap (X \setminus B_n) = \emptyset$ for some $\lambda_n \in I$. Hence $K_{\lambda_n} \subset B_n$. Choose $\iota_1 := \lambda_1$ and choose $(\iota_n)_{n \in \mathbb{N}}$ recursively from I so that

$$K_{\iota_n} \subset K_{\iota_{n-1}} \cap K_{\lambda_n}$$

for every $n \geq 2$. Then

$$K \subset \bigcap_{n \in \mathbb{N}} K_{\iota_n} \subset \bigcap_{n \in \mathbb{N}} B_n = K,$$

which implies the conclusion. $\qquad\qquad\square$

Corollary 5.17 *If X is metrizable, then every positive measure on $\mathfrak{R}(X)$ is regular.*

Proof. Let $(K_\iota)_{\iota \in I}$ be a non-empty downward directed family in $\mathfrak{K}(X)$. By Proposition 5.16, there is a sequence $(\iota_n)_{n \in \mathbb{N}}$ for which $(K_{\iota_n})_{n \in \mathbb{N}}$ is decreasing and $\bigcap_{\iota \in I} K_\iota = \bigcap_{n \in \mathbb{N}} K_{\iota_n}$. Then

$$\mu\left(\bigcap_{\iota \in I} K_\iota\right) \leq \inf_{\iota \in I} \mu(K_\iota) \leq \inf_{n \in \mathbb{N}} \mu(K_{\iota_n}) = \mu\left(\bigcap_{n \in \mathbb{N}} K_{\iota_n}\right) = \mu\left(\bigcap_{\iota \in I} K_\iota\right),$$

and hence $\mu\left(\bigcap_{\iota \in I} K_\iota\right) = \inf_{\iota \in I} \mu(K_\iota)$. An appeal to Theorem 5.5 completes the proof. $\qquad\qquad\square$

The case $X = \mathbb{R}^n$ ($n \in \mathbb{N}$) is of particular interest. For $n = 1$, we have the following corollary.

Corollary 5.18 *If μ_g is a positive Stieltjes measure, then $\mu_g{}^{\mathbb{R}}$ is regular.*

Proof. Put $\nu := \mu_g{}^{\mathbb{R}}|_{\mathfrak{R}(\mathbb{R})}$. Then ν is regular by Corollary 5.17. So, by Theorem 5.7, is $\nu^{\mathbb{R}}$. Note that $\mathfrak{I} \subset (\mathfrak{R}(\mathbb{R}))_\delta$ and $\mathfrak{R}(\mathbb{R}) \subset \mathfrak{I}_\delta$. It follows from Theorem 4.16 that $\nu^{\mathbb{R}} = \mu_g{}^{\mathbb{R}}$. Thus $\mu_g{}^{\mathbb{R}}$ is regular. $\qquad\qquad\square$

Another special case of particular importance is that of locally compact spaces. We treat this somewhat more thoroughly.

Given a locally compact space X, let $\mathcal{K}(X)$ denote the vector lattice of continuous real-valued functions on X with compact support. We first show that every function in $\mathcal{K}(X)$ is μ-integrable, for any positive measure μ on a ring of sets containing $\mathfrak{R}(X)$. (Regularity is therefore not required here.)

Proposition 5.19 *Let X be a locally compact space and μ a positive measure on the ring of sets $\mathfrak{R} \supset \mathfrak{R}(X)$, $\mathfrak{R} \subset \mathfrak{P}(X)$. Then $\mathcal{K}(X) \subset \overline{\mathcal{L}}(\mu)$ and*

$$\mathcal{K}(X) \longrightarrow \mathbb{R}, \quad f \longmapsto \int f \, d\mu$$

defines a positive linear functional on $\mathcal{K}(X)$.

Proof. Take $f \in \mathcal{K}(X)_+$ and consider the sequence $(f_n)_{n \in \mathbb{N}}$ defined in Proposition 4.10. Take $n \in \mathbb{N}$. Then $f_n \in \mathcal{L}(\mathfrak{R})$, since each of the sets $\{f \geq \alpha\}$ $(\alpha > 0)$ is compact. Since

$$\int f_n d\mu \leq \big(\sup_{x \in X} f(x) \big) \mu(\overline{\{f > 0\}}),$$

the monotone convergence theorem implies that $f \in \overline{\mathcal{L}}(\mu)$. For an arbitrary $f \in \mathcal{K}(X)$, use the decomposition $f = f^+ - f^-$.

The rest is obvious.   □

The great significance of the continuous functions with compact support on a locally compact space rests on the fact that they may be used to approximate closely the μ-integrable functions when μ is a regular measure. We postpone the proof of this fact until the next chapter, where we shall prove a more general statement (Theorem 6.15). This should nevertheless motivate a closer examination of $\mathcal{K}(X)$.

Our next theorem is one of the best-known ones. It is due to F. Riesz and states that *every* positive linear functional on $\mathcal{K}(X)$ is an integral of the form described in the preceding proposition and, as such, enjoys all the strong properties of integrals. Moreover, the measure μ can always be chosen to be a regular one. Thus, in particular, if we start with a positive measure ν on $\mathfrak{R}(X)$ which is not regular, then we can still find a regular positive measure μ on $\mathfrak{R}(X)$ for which $\int f \, d\nu = \int f \, d\mu$ whenever $f \in \mathcal{K}(X)$.

A further extension of the Riesz theorem is found in Chapter 8.

Theorem 5.20 (Riesz Representation Theorem) *Let X be a locally compact space and ℓ a positive linear functional on $\mathcal{K}(X)$. Given $K \in \mathfrak{K}(X)$, put*

$$\nu(K) := \inf\{\ell(f) \mid f \in \mathcal{K}(X), \ f \geq e_K\}.$$

Then

(a) ν meets conditions (i)–(iii) of Theorem 5.6;

(b) there is a unique regular positive measure μ_ℓ on $\mathfrak{R}(X)$ for which $\mu_\ell(K) = \nu(K)$ whenever $K \in \mathfrak{K}(X)$;

(c) $\mathcal{K}(X) \subset \overline{\mathcal{L}}(\mu_\ell)$ and for each $f \in \mathcal{K}(X)$

$$\int f \, d\mu_\ell = \ell(f).$$

Proof. (a) First note that for each $K \in \mathfrak{K}(X)$, there is by Urysohn's theorem an $f \in \mathcal{K}(X)$ with $e_X \geq f \geq e_K$. Hence $\nu(K) \in \mathbb{R}_+$.

Take $K, L \in \mathfrak{K}(X)$ and $f, g \in \mathcal{K}(X)$ with $f \geq e_K$ and $g \geq e_L$. Then $f + g \geq e_{K \cup L}$ and so

$$\nu(K \cup L) \leq \ell(f + g) = \ell(f) + \ell(g),$$

whence

$$\nu(K \cup L) \leq \nu(K) + \nu(L).$$

Suppose that $K \cap L = \emptyset$. Take $h \in \mathcal{K}(X)$ with $e_{K \cup L} = e_K + e_L \leq h$. Then by Urysohn's theorem there are functions $f, g \in \mathcal{K}(X)$ such that $e_K \leq f$, $e_L \leq g$ and $f + g \leq h$. It follows that

$$\nu(K) + \nu(L) \leq \ell(f) + \ell(g) \leq \ell(h).$$

Since h was arbitrary,

$$\nu(K) + \nu(L) \leq \nu(K \cup L).$$

Since the converse inequality has already been proved,

$$\nu(K \cup L) = \nu(K) + \nu(L).$$

Finally, let $(K_\iota)_{\iota \in I}$ be a non-empty downward directed family in $\mathfrak{K}(X)$. Put $K := \bigcap_{\iota \in I} K_\iota$. Take $f \in \mathcal{K}(X)$ with $f \geq e_K$. If $\alpha > 0$, then $(1 + \alpha)f \in \mathcal{K}(X)$. Moreover, $\{(1 + \alpha)f > 1\}$ is open and $K \subset \{(1 + \alpha)f > 1\}$. Thus there is an $\iota \in I$ with $K_\iota \subset \{(1 + \alpha)f > 1\}$. It follows that

$$\nu(K_\iota) \leq (1 + \alpha) \, \ell(f).$$

Since f is arbitrary, this implies that

$$\inf_{\iota \in I} \nu(K_\iota) \leq (1 + \alpha) \, \nu(K),$$

and since α is also arbitrary, it follows that

$$\inf_{\iota \in I} \nu(K_\iota) \leq \nu(K).$$

The converse inequality being obvious, this completes the proof of (i)–(iii).

(b) follows immediately from Theorem 5.6.

(c) Take $f \in \mathcal{K}(X)_+$. Given any $n \in \mathbb{N}$ and any $k \in \mathbb{N}$ with $k \leq n2^n$, define

$$A_{n,k} := \{f \geq k/2^n\} \quad \text{and} \quad A_{n,0} := \overline{\{f > 0\}}.$$

The sets $A_{n,0}$ and $A_{n,k}$ are all compact. Given $n \in \mathbb{N}$, put

$$f_n := \frac{1}{2^n} \sum_{k=1}^{n2^n} e_{A_{n,k}} \quad \text{and} \quad g_n := \frac{1}{2^n} \sum_{k=1}^{n2^n} e_{A_{n,k-1}}.$$

Then $(f_n)_{n \in \mathbb{N}}$ and $(g_n)_{n \in \mathbb{N}}$ are sequences of $\mathfrak{R}(X)$-step functions. By Proposition 4.10, $f_n \uparrow f$. Choose $m \in \mathbb{N}$ with $m \geq \sup_{x \in X} f(x)$. Then $g_n \geq f$ whenever $n \geq m$. If $n \geq m$ and $k \leq n2^n$, then

$$\frac{1}{2^n} e_{A_{n,k}} \leq f \wedge \frac{k}{2^n} - f \wedge \frac{k-1}{2^n} \leq \frac{1}{2^n} e_{A_{n,k-1}},$$

and

$$f \wedge \frac{k}{2^n} - f \wedge \frac{k-1}{2^n} \in \mathcal{K}(X),$$

which implies that

$$\frac{1}{2^n} \mu_\ell(A_{n,k}) \leq \ell\left(f \wedge \frac{k}{2^n} - f \wedge \frac{k-1}{2^n}\right) \leq \frac{1}{2^n} \mu_\ell(A_{n,k-1}).$$

Note that

$$\sum_{k=1}^{n2^n} \left(f \wedge \frac{k}{2^n} - f \wedge \frac{k-1}{2^n}\right) = f.$$

Thus

$$\int f_n d\mu_\ell \leq \ell(f) \leq \int g_n d\mu_\ell.$$

Note also that

$$g_n - f_n = \frac{1}{2^n} e_{A_{n,0}},$$

so that

$$\int g_n d\mu_\ell - \int f_n d\mu_\ell = \frac{1}{2^n} \mu_\ell(A_{n,0}).$$

Thus, f is actually in $\overline{\mathcal{L}}(\mu_\ell)$ and

$$\ell(f) = \uparrow \int f_n d\mu_\ell = \int f \, d\mu_\ell.$$

For arbitrary $f \in \mathcal{K}(X)$, decompose f in the usual manner. $\square$

As a simple example, consider the positive linear functional on $\mathcal{C}([a,b])$ which assigns to f its Riemann integral on $A := [a,b]$. The Riesz theorem asserts that there is a regular positive measure μ on A such that

$$\int f \, d\mu = \int_a^b f(x) \, dx \qquad \text{for every } f \in \mathcal{C}(A).$$

This is, of course, no surprise – μ is simply Lebesgue measure on A!

Note also that the Riesz theorem may be used for another derivation of the fact proved in Theorem 2.20, namely that every positive linear functional on $\mathcal{K}(X)$ is null-continuous.

The following is in a certain sense a converse to the Riesz representation theorem.

Theorem 5.21 *Let X be a locally compact space and μ a regular positive measure on $\mathfrak{R}(X)$.*

(a) Define

$$\ell : \mathcal{K}(X) \longrightarrow \mathbb{R}, \quad f \longmapsto \int f \, d\mu.$$

Then $\mu_\ell = \mu$, where μ_ℓ is the measure described in Theorem 5.20.

(b) Given $K \in \mathfrak{K}(X)$,

$$\mu(K) = \inf \left\{ \int f \, d\mu \,\middle|\, f \in \mathcal{K}(X), \ f \geq e_K \right\}.$$

(c) If X is metrizable, then regularity of μ need not be assumed for (a) and (b).

Proof. (a) Take $K \in \mathfrak{K}(X)$ and $f \in \mathcal{K}(X)$ such that $f(X) \subset [0,1]$ and $e_K \leq f$. Then $C := \{f > 0\} \setminus K$ belongs to $\mathfrak{R}(X)$. Take $\varepsilon > 0$. By the regularity of μ there is an $L \in \mathfrak{K}(X)$, $L \subset C$, with

$$\mu(C) - \mu(L) < \varepsilon.$$

Now take $g \in \mathcal{K}(X)_+$ with $g \leq f$, $g|_K = 1$ and $g|_L = 0$. Then $e_K \leq g \leq e_{K \cup (C \setminus L)}$. It follows that

$$\mu_\ell(K) \leq \ell(g) \leq \mu(K \cup (C \setminus L)) = \mu(K) + \mu(C \setminus L) \leq \mu(K) + \varepsilon.$$

Since $\varepsilon > 0$ is arbitrary, it follows that $\mu_\ell(K) \leq \mu(K)$.

Conversely, given $f \in \mathcal{K}(X)$ with $f \geq e_K$,

$$\mu(K) = \int e_K d\mu \leq \int f \, d\mu = \ell(f).$$

It follows that

$$\mu(K) \leq \inf\{\ell(f) \mid f \in \mathcal{K}(X), \ f \geq e_K\} = \mu_\ell(K).$$

Hence $\mu(K) = \mu_\ell(K)$ for every $K \in \mathfrak{K}(X)$. Corollary 5.2 now implies that $\mu = \mu_\ell$.

(b) By Theorem 5.20(b),

$$\mu_\ell(K) = \inf \left\{ \int f \, d\mu \,\middle|\, f \in \mathcal{K}(X), \ f \geq e_K \right\},$$

and the result follows from (a).

(c) follows from Corollary 5.17. $\qquad\square$

The following is an immediate consequence of (b). It shows, in particular, that the regular positive measure μ_ℓ described in the Riesz representation theorem is uniquely determined.

Corollary 5.22 *Let X be a locally compact space. For $i = 1, 2$ let μ_i be a regular positive measure on the ring of sets $\mathfrak{R}_i \supset \mathfrak{R}(X)$, $\mathfrak{R}_i \subset \mathfrak{P}(X)$, such that*

$$\int f \, d\mu_1 = \int f \, d\mu_2 \qquad \text{for every } f \in \mathcal{K}(X).$$

Then

$$\mu_1|_{\mathfrak{R}_1 \cap \mathfrak{R}_2} = \mu_2|_{\mathfrak{R}_1 \cap \mathfrak{R}_2}.$$

We made a detailed study in the earlier chapters of when it is possible to interchange taking limits with integration. All our results, such as the monotone convergence theorem and the Lebesgue convergence theorem, could only be proved, however, for countable families of integrable functions. With locally compact spaces the situation is different. Here we can often omit the assumption of countability, due to the fact that the topology and the measure are compatible. We establish one theorem in this vein. This theorem may be regarded an application of Lusin's theorem.

Theorem 5.23 *Let X be a locally compact space and μ a regular positive measure on the ring of sets $\mathfrak{R} \supset \mathfrak{R}(X)$, $\mathfrak{R} \subset \mathfrak{P}(X)$. Let $(f_\iota)_{\iota \in I}$ be a nonempty downward directed family in $\mathcal{K}(X)_+$. Then*

$$\bigwedge_{\iota \in I} f_\iota \in \mathcal{L}^1(\mu) \quad \text{and} \quad \int \left(\bigwedge_{\iota \in I} f_\iota \right) d\mu = \inf_{\iota \in I} \int f_\iota d\mu.$$

Proof. Put $f := \bigwedge_{\iota \in I} f_\iota$. To show that f is μ-measurable, take $\alpha \in \mathbb{R}$. Since $\emptyset$ is clearly in $\mathfrak{M}(\mu)$, we may suppose that $\{f < \alpha\} \neq \emptyset$. So take $x \in \{f < \alpha\}$. Then there is some $\iota \in I$ with $f_\iota(x) < \alpha$. Since f_ι is continuous, the set $\{f_\iota < \alpha\}$ is an open neighbourhood of x contained in $\{f < \alpha\}$. Hence $\{f < \alpha\}$ is open and consequently μ-measurable (Corollary 5.12). Having thus established that f is μ-measurable, we use Theorem 4.38(a) to conclude that $f \in \mathcal{L}^1(\mu)$.

We choose a fixed $\iota_0 \in I$. Let K denote the support of f_{ι_0} and put $\alpha := \sup_{x \in X} f_{\iota_0}(x)$. Take $\varepsilon > 0$. By Lusin's theorem there is an $L \in \mathfrak{R}(X)$ such that

$$L \subset K, \quad \mu(K \setminus L) < \frac{\varepsilon}{\alpha + 1} \quad \text{and} \quad f|_L \text{ is continuous.}$$

Apply Theorem 2.13 to the family $(f_\iota|_L - f|_L)_{\iota \in I}$ to obtain an $\iota \in I$ with $f_\iota|_L - f|_L < \frac{\varepsilon}{1 + \mu(L)}$. Since $(f_\iota)_{\iota \in I}$ is directed down, we may suppose that $f_\iota \leq f_{\iota_0}$. Thus

$$\int f_\iota d\mu = \int_L f_\iota d\mu + \int_{K \setminus L} f_\iota d\mu$$

$$\leq \int_L \left(f + \frac{\varepsilon}{1 + \mu(L)} \right) d\mu + \int_{K \setminus L} f_{\iota_0} d\mu$$

$$\leq \int f\, d\mu + \varepsilon \frac{\mu(L)}{1 + \mu(L)} + \alpha\mu(K \setminus L)$$

$$< \int f\, d\mu + 2\varepsilon.$$

Hence $\inf_{\iota \in I} \int f_\iota d\mu < \int f\, d\mu + 2\varepsilon$, and since ε was arbitrary,

$$\inf_{\iota \in I} \int f_\iota d\mu \leq \int f\, d\mu.$$

The reverse inequality being trivial, the proof is complete. $\qquad\square$

We now consider a special class of locally compact spaces.

Theorem 5.24 *Let X be a locally compact space with a countable base and μ a positive measure on $\mathfrak{R}(X)$.*

(a) X is σ-compact, i.e. there is a sequence $(K_n)_{n \in \mathbb{N}}$ of compact subsets of X whose union is X.

(b) Define

$$\ell : \mathcal{K}(X) \longrightarrow \mathbb{R}, \quad f \longmapsto \int f\, d\mu.$$

Then

$$\left(X, \mathcal{L}^1(\mu), \int \cdot\, d\mu\right) = (X, \overline{\mathcal{L}}(\mu), \overline{\ell_\mu}) = (X, \overline{\mathcal{L}}(\ell), \overline{\ell}).$$

Proof. (a) Let $\mathfrak{B}$ be a countable base of X. Since X is locally compact, it follows easily that $\mathfrak{C} := \{B \in \mathfrak{B} \mid B \text{ is relatively compact}\}$ is also a countable base of X. But then $X = \bigcup_{B \in \mathfrak{C}} \overline{B}$ expresses X as the union of countably many compact sets.

(b) Using (a) and Theorem 4.18,

$$\left(X, \mathcal{L}^1(\mu), \int \cdot\, d\mu\right) = (X, \overline{\mathcal{L}}(\mu), \overline{\ell_\mu}),$$

and so it only remains to show that

$$(X, \overline{\mathcal{L}}(\mu), \overline{\ell_\mu}) = (X, \overline{\mathcal{L}}(\ell), \overline{\ell}).$$

But by Proposition 5.19

$$(X, \mathcal{K}(X), \ell) \preccurlyeq (X, \overline{\mathcal{L}}(\mu), \overline{\ell_\mu}).$$

Since X has a countable base, there is a metric d generating the topology on X. Take $K \in \mathfrak{R}(X)$. Given $n \in \mathbb{N}$, put

$$B_n := \{x \in X \mid d(x, y) < 1/n \text{ for some } y \in K\}.$$

$(B_n)_{n \in \mathbb{N}}$ is a decreasing sequence of open neighbourhoods of K and $K = \downarrow B_n$. For each $n \in \mathbb{N}$ there is a function $f_n \in \mathcal{K}(X)$ with $f_n(X) \subset [0, 1]$,

$f_n|_K = 1$ and $f_n|_{X \setminus B_n} = 0$. Moreover, f_n may be so chosen that $(f_n)_{n \in \mathbb{N}}$ is decreasing. Then $e_K = \downarrow f_n$. It follows that $e_K \in \overline{\mathcal{L}}(\ell)$ and

$$\overline{\ell}(e_K) = \downarrow \ell(f_n) = \downarrow \int f_n d\mu = \int e_K d\mu = \ell_\mu(e_K).$$

Thus

$$(X, \mathcal{L}(\mathfrak{R}(X)), \ell_\mu) \preccurlyeq (X, \overline{\mathcal{L}}(\ell), \overline{\ell}).$$

An appeal to Theorem 3.41 completes the proof. □

The preceding theorem applies in particular to positive measures on $\mathfrak{R}(\mathbb{R}^n)$ $(n \in \mathbb{N})$.

Next is a brief description of a form of 'outer' regularity defined in terms of open sets, in contrast to the 'inner' regularity defined in terms of compact sets.

Theorem 5.25 *Let X be a locally compact space and μ a regular positive measure on the ring of sets $\mathfrak{R} \supset \mathfrak{R}(X)$, $\mathfrak{R} \subset \mathfrak{P}(X)$.*

(a) *Let $A \in \mathfrak{L}(\mu)$ be contained in a countable union of compact subsets of X and take $\varepsilon > 0$. Then there is an open set $B \in \mathfrak{L}(\mu)$ with $A \subset B$ and $\mu^X(B) - \mu^X(A) < \varepsilon$.*

(b) *If X is σ-compact (in particular if X has a countable base), then (a) holds for every $A \in \mathfrak{L}(\mu)$.*

Proof. (a) First suppose that A is relatively compact (i.e. $\overline{A} \in \mathfrak{K}(X)$). Given $x \in \overline{A}$, let U_x be a relatively compact open neighbourhood of x. Being compact, $\overline{A}$ (and hence also A) is covered by finitely many of the sets U_x. Then their union U is open and relatively compact. Hence it is in $\mathfrak{R}(X)$. By regularity, there is a $K \in \mathfrak{K}(X)$ with $K \subset U \setminus A$ and $\mu^X(U \setminus A) - \mu(K) < \varepsilon$. Then $B := U \setminus K$ meets the requirements.

Now take a sequence $(K_n)_{n \in \mathbb{N}}$ in $\mathfrak{K}(X)$ whose union contains A. Take $n \in \mathbb{N}$. In view of the first part of the proof, there is an open set $B_n \in \mathfrak{L}(\mu)$ with

$$B_n \supset A \cap K_n \quad \text{and} \quad \mu^X(B_n) - \mu^X(A \cap K_n) < \frac{\varepsilon}{2^n}.$$

Put $B := \bigcup_{n \in \mathbb{N}} B_n$. Then $B \supset A$ and $B \setminus A \subset \bigcup_{n \in \mathbb{N}}(B_n \setminus (A \cap K_n))$. It follows by Theorem 4.4 and the μ-measurability of $B \setminus A$ that $B \setminus A \in \mathfrak{L}(\mu)$ (and hence $B \in \mathfrak{L}(\mu)$) and that $\mu^X(B \setminus A) < \varepsilon$.

(b) follows from (a) using Theorem 5.24(a). □

The exercises contain an example showing that the hypothesis in (a) is essential.

The preceding chapter considered products of positive measure spaces. We now ask how products of regular measures behave in relation to the product topology. A comprehensive treatment of this topic is too extensive for the scope of our book. However, it is not difficult to prove the following theorem, which, though not definitive, is frequently adequate for practical purposes.

Theorem 5.26 *Let $(X_\iota)_{\iota \in I}$ be a finite, non-empty family of Hausdorff spaces. For each $\iota \in I$ let $\mathfrak{R}_\iota$ be a ring of sets, $\mathfrak{R}_\iota \subset \mathfrak{P}(X_\iota)$, and μ_ι a positive measure on $\mathfrak{R}_\iota$. Define $X := \prod_{\iota \in I} X_\iota$ and consider X with the product topology.*

(a) If for every $\iota \in I$ and every $A \in \mathfrak{R}_\iota$

$$\mu_\iota(A) = \sup \{\mu_\iota^{X_\iota}(K) \mid K \in \mathfrak{K}(X_\iota) \cap \mathfrak{L}(\mu_\iota),\ K \subset A\},$$

then

$$\Big(\bigotimes_{\iota \in I} \mu_\iota\Big)^X(A) = \sup \Big\{\Big(\bigotimes_{\iota \in I} \mu_\iota\Big)^X(K) \,\Big|\, K \in \mathfrak{K}(X) \cap \mathfrak{L}\Big(\bigotimes_{\iota \in I} \mu_\iota\Big),\ K \subset A\Big\}$$

for every $A \in \mathfrak{L}\big(\bigotimes_{\iota \in I} \mu_\iota\big)$.

(b) Let each X_ι be locally compact and metrizable. If $\mathfrak{K}(X_\iota) \subset \mathfrak{R}_\iota$ and if μ_ι is regular for each $\iota \in I$, then $\mathfrak{R}(X) \subset \mathfrak{L}\big(\bigotimes_{\iota \in I} \mu_\iota\big)$ and $\big(\bigotimes_{\iota \in I} \mu_\iota\big)^X$ is regular.

Proof. (a) First take $\prod_{\iota \in I} A_\iota \in \square_{\iota \in I} \mathfrak{R}_\iota$ and $\varepsilon > 0$. Put $\alpha_\iota := \mu_\iota(A_\iota)$ for $\iota \in I$. Since

$$\mathbb{R}^I \longrightarrow \mathbb{R}, \quad (\beta_\iota)_{\iota \in I} \longmapsto \prod_{\iota \in I} \beta_\iota$$

is continuous, there is a $\delta > 0$ such that

$$\Big| \prod_{\iota \in I} \alpha_\iota - \prod_{\iota \in I} \beta_\iota \Big| < \varepsilon$$

whenever $|\alpha_\iota - \beta_\iota| < \delta$ for every $\iota \in I$. By hypothesis there is a $K_\iota \in \mathfrak{K}(X_\iota) \cap \mathfrak{L}(\mu_\iota)$ with $K_\iota \subset A_\iota$ such that $\alpha_\iota - \mu_\iota^{X_\iota}(K_\iota) < \delta$ for every $\iota \in I$. Then

$$\prod_{\iota \in I} K_\iota \in \mathfrak{K}(X) \cap \mathfrak{L}\Big(\bigotimes_{\iota \in I} \mu_\iota\Big)$$

and so, by Theorem 4.52(b),

$$\Big(\bigotimes_{\iota \in I} \mu_\iota\Big)^X \Big(\prod_{\iota \in I} K_\iota\Big) = \prod_{\iota \in I} \mu_\iota^{X_\iota}(K_\iota).$$

Thus

$$\Big(\bigotimes_{\iota \in I} \mu_\iota\Big)^X \Big(\prod_{\iota \in I} A_\iota\Big) - \Big(\bigotimes_{\iota \in I} \mu_\iota\Big)^X \Big(\prod_{\iota \in I} K_\iota\Big) < \varepsilon.$$

For $A \in \bigotimes_{\iota \in I} \mathfrak{R}_\iota$, the conclusion follows by Proposition 2.26. For $A \in \mathfrak{L}\big(\bigotimes_{\iota \in I} \mu_\iota\big)$ it follows by Theorem 5.7(a).

(b) It suffices to show $\mathfrak{K}(X) \subset \mathfrak{L}\big(\bigotimes_{\iota \in I} \mu_\iota\big)$, for then $\mathfrak{R}(X) \subset \mathfrak{L}\big(\bigotimes_{\iota \in I} \mu_\iota\big)$.

Take $K \in \mathfrak{K}(X)$. For $\iota \in I$, let d_ι be a metric generating the topology on X_ι. For each $n \in \mathbb{N}$, let $\mathfrak{R}^n$ denote the set of all sets of the form $\prod_{\iota \in I} C_\iota$, where, for each $\iota \in I$, $C_\iota \subset X_\iota$ is compact with non-empty interior and

$$d_\iota(C_\iota) := \sup\{d_\iota(x, y) \mid x, y \in C_\iota\} < 1/n.$$

Clearly $\mathfrak{R}^n \subset \bigsqcup_{\iota \in I} \mathfrak{R}_\iota$ and so $(\mathfrak{R}^n)_r \subset \mathfrak{L}(\bigotimes_{\iota \in I} \mu_\iota)$. Given $n \in \mathbb{N}$, there is a finite family $(A_\lambda^n)_{\lambda \in L_n}$ in $\mathfrak{R}^n$ such that $A_\lambda^n \cap K \neq \emptyset$ for every $\lambda \in L_n$ and

$$K \subset \bigcup_{\lambda \in L_n} A_\lambda^n.$$

We have $\bigcup_{\lambda \in L_n} A_\lambda^n \in (\mathfrak{R}^n)_r$ for all $n \in \mathbb{N}$, and

$$\bigcap_{n \in \mathbb{N}} \left(\bigcup_{\lambda \in L_n} A_\lambda^n \right) = K.$$

Applying Theorem 4.2(a), it follows that $K \in \mathfrak{L}(\bigotimes_{\iota \in I} \mu_\iota)$. $\square$

Exercises

1. Let X be a locally compact space, ℓ a positive linear functional on $\mathcal{K}(X)$ and $\mu := \mu_\ell$ the measure on $\mathfrak{R}(X)$ associated to ℓ by the Riesz theorem. Prove the following propositions.

   (a) $(X, \overline{\mathcal{L}}(\ell), \overline{\ell}) \preccurlyeq (X, \overline{\mathcal{L}}(\mu), \overline{\ell_\mu})$.

   (b) Take $K \in \mathfrak{K}(X)$ and suppose that there is a decreasing sequence $(U_n)_{n \in \mathbb{N}}$ of open sets with $K = \bigcap_{n \in \mathbb{N}} U_n$. Then $K \in \{C \mid e_C \in \overline{\mathcal{L}}(\mu)\}$.

   (c) Suppose that for each $K \in \mathfrak{K}(X)$ there is a sequence $(U_n)_{n \in \mathbb{N}}$ of open sets with $K = \bigcap_{n \in \mathbb{N}} U_n$. Then

   $$(X, \overline{\mathcal{L}}(\mu), \overline{\ell_\mu}) \preccurlyeq (X, \overline{\mathcal{L}}(\ell), \overline{\ell}).$$

2. Let X be a Hausdorff space, μ a regular positive measure on the ring of sets $\mathfrak{R} \supset \mathfrak{R}(X)$, $\mathfrak{R} \subset \mathfrak{P}(X)$, and $A \in \mathfrak{L}(\mu)$ contained in some open $U \in \mathfrak{L}(\mu)$. Show that for each $\varepsilon > 0$ there is an open $B \in \mathfrak{L}(\mu)$ which contains A and for which $\mu^X(B) - \mu^X(A) < \varepsilon$.

3. Let X be an infinite discrete topological space. Define

   $$\mathfrak{R} := \{A \subset X \mid A \text{ is finite or } X \setminus A \text{ is finite}\},$$

   $$\mu : \mathfrak{R} \longrightarrow \mathbb{R}, \quad A \longmapsto \begin{cases} 0 & \text{if } A \text{ is finite} \\ 1 & \text{if } X \setminus A \text{ is finite}. \end{cases}$$

   Show that μ is outer regular but not regular.

4. Show that in general

   $$\lambda^{\mathbb{R}}(A) \neq \sup\{\lambda^{\mathbb{R}}(U) \mid U \subset A, \ U \text{ open}\}$$

   for a Lebesgue integrable set A.

5. Let X be a locally compact space and μ a regular positive measure on the ring of sets $\mathfrak{R} \supset \mathfrak{R}(X)$, $\mathfrak{R} \subset \mathfrak{P}(X)$. Suppose that $f \in \overline{\mathbb{R}}^X$ has compact support. Prove that the following are equivalent.

   (a) f is μ-measurable.

(b) Given $\varepsilon > 0$, there is a continuous function $g : X \to \overline{\mathbb{R}}$ with compact support such that $\{f \neq g\} \in \mathcal{L}(\mu)$ and $\mu^X(\{f \neq g\}) < \varepsilon$.

Hint: Use Lusin's theorem and the Tietze extension theorem.

6. Let $\mathfrak{F}$ be a non-trivial δ-stable ultrafilter on the set X. Define

$$\mu : \mathfrak{P}(X) \longrightarrow \mathbb{R}, \quad A \longmapsto \begin{cases} 0 & \text{if } A \notin \mathfrak{F}, \\ 1 & \text{if } A \in \mathfrak{F}. \end{cases}$$

(a) Show that μ is a positive measure.

Consider X with the discrete topology. Choose ω not in X, put $X^* := X \cup \{\omega\}$ and define

$$\mathfrak{T} := \mathfrak{P}(X) \cup \{X^* \setminus A \mid A \subset X, \ A \text{ finite}\},$$
$$\nu : \mathfrak{P}(X^*) \longrightarrow \mathbb{R}, \quad B \longmapsto \mu(B \setminus \{\omega\}).$$

Prove the following.

(b) $\mathfrak{T}$ is a topology on X^* which renders X^* compact. (It is called the 'Alexandroff' or 'one-point compactification of X').

(c) $\mathfrak{R}(X^*) = \mathfrak{P}(X^*)$.

(c) ν is a positive measure on $\mathfrak{R}(X^*)$ which is not regular.

7. Let X be an uncountable discrete topological space and ν counting measure on X. Write $\mathfrak{S}$ for the set of all finite subsets of X. Take Lebesgue measure λ on $\mathfrak{J}$. Put $\mathfrak{R} := \mathcal{L}(\lambda)$ and $\mu := \lambda^{\mathbb{R}}$. Take $\mathbb{R} \times X$ with the product topology. Prove the following.

(a) $\mathbb{R} \times X$ is a locally compact space with $\mathfrak{K}(\mathbb{R} \times X) \subset \mathfrak{R} \otimes \mathfrak{S}$.

(b) $\mu \otimes \nu$ is regular.

(c) $\{0\} \times X \in \mathfrak{N}(\mu \otimes \nu)$.

(d) No open set $U \in \mathcal{L}(\mu \otimes \nu)$ contains $\{0\} \times X$. Hence $\mu \otimes \nu$ is not outer regular.

Hint for (d): Suppose there is some open $U \in \mathcal{L}(\mu\otimes\nu)$ containing $\{0\}\times X$. We may assume that $U \subset \,]{-}1,1[\,\times X$. Let $\pi : \mathbb{R} \times X \to \mathbb{R}$ be the canonical projection and put $U_x := \pi(U \cap (\mathbb{R} \times \{x\}))$ for every $x \in X$. Show that $U_x \in \mathfrak{R}$ and that $\mu(U_x) > 0$. Writing α_x for $\mu(U_x)$, find a sequence $(x_n)_{n\in\mathbb{N}}$ of distinct elements of X such that $\sum_{n\in\mathbb{N}} \alpha_{x_n} = \infty$. Then $(U_{x_n} \times \{x_n\})_{n\in\mathbb{N}}$ is a disjoint sequence in $\mathfrak{R} \otimes \mathfrak{S}$, whose union V is contained in U. Use this to conclude that $V \in \mathcal{L}(\mu \otimes \nu)$ and obtain a contradiction to $\sum_{n\in\mathbb{N}} \alpha_{x_n} = \infty$.

5.3 The congruence invariance of the n-dimensional Lebesgue measure

We now examine products of Lebesgue measures somewhat more closely, as mentioned at the end of the last chapter. Note that the relationship between n-dimensional Lebesgue measure

$$\lambda^n := \left(\bigotimes_{k=1}^{n} \lambda^{\mathbb{R}} \right)^{\mathbb{R}^n}$$

and the topology of $\mathbb{R}^n$ is clarified by the general theorems in this chapter. But the invariance of λ^n under congruence mappings has not yet been established. For the following choose a fixed $n \in \mathbb{N}$.

We begin with several observations before proving the invariance of λ^n under congruence mappings.

We define the distance $d_n(x,y)$ between the two points $x = (x_1, \ldots, x_n)$ and $y = (y_1, \ldots, y_n)$ of $\mathbb{R}^n$ by

$$d_n(x,y) := \left(\sum_{k=1}^{n} (y_k - x_k)^2 \right)^{1/2}.$$

Thus d_n is simply the usual Euclidean distance in $\mathbb{R}^n$. $\mathbb{R}^n$ is a metric space with respect to d_n. For $z \in \mathbb{R}^n$ and $r \in \mathbb{R}$, $r > 0$, we denote by

$$B(z,r) := \{y \in \mathbb{R}^n \mid d_n(z,y) < r\}$$

the sphere with centre z and radius r. Let $\overline{B(z,r)}$ be the closure of the sphere. Then

$$\overline{B(z,r)} = \{y \in \mathbb{R}^n \mid d_n(z,y) \leq r\}.$$

A sphere already has a volume given by Riemann integration. The volume of $B(z,r)$ is, as the reader is aware,

$$\frac{\pi^{n/2}}{\Gamma(\frac{n}{2}+1)} r^n,$$

where Γ denotes the gamma function (see also Exercise 1). Thus in particular the volume is independent of the centre z of the sphere. We conclude from the Riemann integrability of $e_{B(z,r)}$ that $B(z,r) \in \mathcal{L}(\lambda^n)$ and

$$\lambda^n \big(B(z,r) \big) = \frac{\pi^{n/2}}{\Gamma(\frac{n}{2}+1)} r^n$$

for an arbitrary sphere $B(z,r)$.

Our proof of the invariance of λ^n makes use of certain properties of the volume of a sphere. While these may be derived readily from the above formula, we shall not make use of it. We choose a different path.

We begin by noting that each sphere $B(z,r)$ can be approximated using n-dimensional interval forms, that is, by finite unions of pairwise disjoint rectangular prisms of the form

$$\prod_{k=1}^{n} [\alpha_k, \beta_k[.$$

For each sphere $B(z,r)$ there is an increasing sequence $(A_k)_{k\in\mathbb{N}}$ of such interval forms with

$$B(z,r) = \bigcup_{k\in\mathbb{N}} A_k.$$

By the monotone convergence theorem, $B(z,r) \in \mathcal{L}(\lambda^n)$ and

$$\lambda^n\big(B(z,r)\big) = \sup_{k\in\mathbb{N}} \lambda^n(A_k).$$

We consider two spheres $B(z_1, r)$ and $B(z_2, r)$ with the same radius r. Let $(A_k)_{k\in\mathbb{N}}$ be an increasing sequence of n-dimensional interval forms such that

$$B(z_1, r) = \bigcup_{k\in\mathbb{N}} A_k.$$

Given $k \in \mathbb{N}$, let the set A_k' be defined as the translation of A_k by the vector $z_2 - z_1$. The sets A_k' are interval forms and since the volume of an n-dimensional rectangular prism is invariant under translation, $\lambda^n(A_k) = \lambda^n(A_k')$ for every $k \in \mathbb{N}$. The sequence $(A_k')_{k\in\mathbb{N}}$ is increasing and

$$B(z_2, r) = \bigcup_{k\in\mathbb{N}} A_k'.$$

Thus

$$\lambda^n\big(B(z_2, r)\big) = \lambda^n\big(B(z_1, r)\big),$$

and so the volume of $B(z, r)$ depends only upon r.

Now consider the sphere $B(0, r)$. Let σ_r denote a radial dilation by a factor of r whose centre is 0. Then

$$B(0, r) = \sigma_r\big(B(0, 1)\big).$$

Given an n-dimensional interval form A, $\sigma_r(A)$ is also an n-dimensional interval form. If

$$A = \bigcup_{\iota\in I} \left(\prod_{k=1}^{n} [\alpha_{\iota_k}, \beta_{\iota_k}[\right),$$

then

$$\sigma_r(A) = \bigcup_{\iota\in I} \left(\prod_{k=1}^{n} [r\alpha_{\iota_k}, r\beta_{\iota_k}[\right).$$

Hence

$$\lambda^n\big(\sigma_r(A)\big) = r^n \lambda^n(A).$$

Thus, if $(A_k)_{k \in \mathbb{N}}$ is an increasing sequence of n-dimensional interval forms such that

$$B(0,1) = \bigcup_{k \in \mathbb{N}} A_k,$$

then

$$B(0,r) = \bigcup_{k \in \mathbb{N}} \sigma_r(A_k),$$

and the sequence $(\sigma_r(A_k))_{k \in \mathbb{N}}$ is also increasing. Hence

$$\lambda^n\big(B(0,r)\big) = r^n \lambda^n\big(B(0,1)\big).$$

Note that

$$\overline{B(z,r)} = \bigcap_{k \in \mathbb{N}} B\Big(z, r + \frac{1}{k}\Big).$$

We summarize these considerations in the form of a proposition.

Proposition 5.27 *Given $z \in \mathbb{R}^n$ and $r \in \mathbb{R}_+$,*

(a) $\lambda^n\big(B(z,r)\big) = \lambda^n\big(\overline{B(z,r)}\big) = r^n \lambda^n\big(B(0,1)\big)$;

(b) $\overline{B(z,r)} \setminus B(z,r)$ *is a λ^n-null set.*

We will also need Vitali's theorem on the covering of a set by spheres, which is also important in other contexts, such as in the study of differentiability.

Take $A \subset \mathbb{R}^n$. A **Vitali cover** of A is a set $\mathfrak{B}$ of spheres in $\mathbb{R}^n$ such that for each $x \in A$ and each real number $\delta > 0$ there is a sphere $B \in \mathfrak{B}$ whose radius is shorter than δ with $x \in B$.

Theorem 5.28 (Vitali's Covering Theorem) *Let $\mathfrak{B}$ be a Vitali cover of $A \subset \mathbb{R}^n$. Let $E \subset \mathbb{R}^n$ be an open set with $A \subset E$. Then there is a countable family $(B_\iota)_{\iota \in I}$ of pairwise disjoint spheres in $\mathfrak{B}$ such that*

$$A \setminus \bigcup_{\iota \in I} B_\iota \in \mathfrak{N}(\lambda^n)$$

and $B_\iota \subset E$ for every $\iota \in I$. If A is open, then the B_ι may be chosen so that $B_\iota \subset A$ for every $\iota \in I$.

Proof. We begin by assuming that A is bounded. Given a bounded open set $E \subset \mathbb{R}^n$, $E \supset A$, we define

$$\mathfrak{B}_E := \{B \in \mathfrak{B} \mid B \subset E\}.$$

$\mathfrak{B}_E$ is clearly also a Vitali cover of A. If there is a finite disjoint family $(B_\iota)_{\iota \in I}$ of spheres in $\mathfrak{B}_E$ such that $A \setminus \bigcup_{\iota \in I} B_\iota \in \mathfrak{N}(\lambda^n)$, then there is nothing to prove. Assume that there is no such family. Construct sequences $(B_i)_{i \in \mathbb{N}}$ and $(\delta_i)_{i \in \mathbb{N}}$ recursively, with $B_i \in \mathfrak{B}_E$ and $\delta_i \in \mathbb{R}$ for each $i \in \mathbb{N}$ and exhibiting the following.

(a) The sets B_i are pairwise disjoint.

(b) Given $i \in \mathbb{N}$, δ_i is the supremum of the radii of the spheres B in $\mathfrak{B}_E$ for which $B \cap \left(\bigcup_{k<i} \overline{B_k} \right) = \emptyset$.

(c) Given $i \in \mathbb{N}$, the radius of B_i exceeds $\delta_i/2$.

Suppose that B_k and δ_k have been constructed for every $k < i$. It cannot be the case that $A \setminus \bigcup_{k<i} \overline{B_k} = \emptyset$, for otherwise

$$A \setminus \bigcup_{k<i} B_k \subset \bigcup_{k<i} (\overline{B_k} \setminus B_k) \in \mathfrak{N}(\lambda^n),$$

contradicting our hypothesis. Thus there is some $x \in A \setminus \bigcup_{k<i} \overline{B_k}$ and so also a sphere $B' \in \mathfrak{B}_E$ such that $x \in B'$ and $B' \cap \overline{B_k} = \emptyset$ for every $k < i$. Let δ_i be the supremum of the radii of all such spheres B'. Choose a B' whose radius exceeds $\delta_i/2$ and put $B_i := B'$.

We next show that $A \setminus \bigcup_{i \in \mathbb{N}} B_i \in \mathfrak{N}(\lambda^n)$. Given $i \in \mathbb{N}$, let C_i be a sphere with the same centre as B_i, but with five times the radius. Then $\lambda^n(C_i) = 5^n \lambda^n(B_i)$ and so

$$\sum_{i \in \mathbb{N}} \lambda^n(C_i) = 5^n \sum_{i \in \mathbb{N}} \lambda^n(B_i) \le 5^n \lambda^n(E) < \infty.$$

Thus

$$\inf_{i \in \mathbb{N}} \sum_{k \ge i} \lambda^n(C_k) = 0.$$

Choose a fixed $i \in \mathbb{N}$. Let $x \in A \setminus \bigcup_{k<i} \overline{B_k}$ be arbitrary. Then there is a $B' \in \mathfrak{B}_E$ such that $x \in B' \subset E \setminus \bigcup_{k<i} \overline{B_k}$. Let

$$m := \inf\{k \in \mathbb{N} \mid B' \cap B_k \ne \emptyset\}.$$

Clearly $m \ge i$. Now $B' \cap B_m \ne \emptyset$. But

$$r(B') \le \delta_m \le 2r(B_m),$$

where $r(B)$ denotes the radius of the sphere B. Then $B' \subset C_m$ and in particular

$$x \in \bigcup_{k \ge i} C_k.$$

Thus, given $i \in \mathbb{N}$,

$$A \setminus \bigcup_{k<i} \overline{B_k} \subset \bigcup_{k \ge i} C_k,$$

and so

$$A \setminus \bigcup_{k \in \mathbb{N}} B_k \subset \left(\bigcap_{i \in \mathbb{N}} \bigcup_{k \ge i} C_k \right) \cup \left(\bigcup_{k \in \mathbb{N}} (\overline{B_k} \setminus B_k) \right) \in \mathfrak{N}(\lambda^n).$$

Therefore

$$A \setminus \bigcup_{k \in \mathbb{N}} B_k \in \mathfrak{N}(\lambda^n).$$

If A is not bounded, note that

$$\mathbb{R}^n \setminus \bigcup_{(z_1,\ldots,z_n)\in\mathbb{Z}^n} \left(\prod_{k=1}^n \,]z_k, z_k + 1[\right) \in \mathfrak{N}(\lambda^n)$$

and apply the above argument to the bounded sets

$$A \cap \left(\prod_{k=1}^n \,]z_k, z_k + 1[\right), \qquad ((z_1,\ldots,z_n) \in \mathbb{Z}^n).$$

If A is open, then choose $E = A$, proving the final claim of the theorem.

$\square$

The proof of the invariance of λ^n under congruence mappings of $\mathbb{R}^n$ – i.e. under bijective, distance-preserving mappings $\mathbb{R}^n \to \mathbb{R}^n$ – is now quite simple.

Theorem 5.29 *Let τ be a congruence mapping of $\mathbb{R}^n$. Then*

(a) $A \subset \mathbb{R}^n$ is λ^n-integrable if and only if $\tau(A)$ is λ^n-integrable, and

(b) $\lambda^n(\tau(A)) = \lambda^n(A)$ for every λ^n-integrable $A \subset \mathbb{R}^n$.

Proof. Let A be a λ^n-null set and take $\varepsilon > 0$. By Theorem 5.25(b), there is an open set $B \in \mathfrak{L}(\lambda^n)$, $B \supset A$, with $\lambda^n(B) < \varepsilon$. There is a countable disjoint family $(C_\iota)_{\iota \in I}$ of n-dimensional open cubes satisfying $\bigcup_{\iota \in I} C_\iota \subset B \subset \bigcup_{\iota \in I} \overline{C_\iota}$. Each C_ι, of side length r_ι, can be included in a sphere B_ι of radius $r_\iota \sqrt{n}/2$, and each $D_\iota := \overline{B_\iota}$ can be included in a closed cube of side length $r_\iota \sqrt{n}$. Then $\tau(D_\iota) \in \mathfrak{L}(\lambda^n)$ and, by Proposition 5.27,

$$\lambda^n(\tau(D_\iota)) = \lambda^n(D_\iota) \leq (r_\iota \sqrt{n})^n = (\sqrt{n})^n \lambda^n(C_\iota)$$

for each $\iota \in I$. We conclude $\tau(A) \subset \bigcup_{\iota \in I} \tau(D_\iota)$, $\bigcup_{\iota \in I} \tau(D_\iota) \in \mathfrak{L}(\lambda^n)$ by the monotone convergence theorem, and

$$\lambda^n\left(\bigcup_{\iota \in I} \tau(D_\iota) \right) \leq \sum_{\iota \in I} \lambda^n(\tau(D_\iota)) < (\sqrt{n})^n \varepsilon.$$

Since ε is arbitrary, $\tau(A) \in \mathfrak{N}(\lambda^n)$.

Now let A be an open λ^n-integrable subset of $\mathbb{R}^n$. By the Vitali covering theorem, there is a countable family $(B_\iota)_{\iota \in I}$ of paiwise disjoint spheres such that $\bigcup_{\iota \in I} B_\iota \subset A$ and

$$A \setminus \bigcup_{\iota \in I} B_\iota \in \mathfrak{N}(\lambda^n).$$

Note that

$$\mathfrak{B} = \{ B(x,r) \mid x \in A, \ r > 0 \}$$

is a Vitali cover of A. It follows that

$$\tau(A) \setminus \bigcup_{\iota \in I} \tau(B_\iota) \in \mathfrak{N}(\lambda^n).$$

The sets $\bigcup_{\iota \in I} B_{\iota}$ and $\bigcup_{\iota \in I} \tau(B_{\iota})$ are open and by Proposition 5.27

$$\sum_{\iota \in I} \lambda^n\big(\tau(B_{\iota})\big) = \sum_{\iota \in I} \lambda^n(B_{\iota}) = \lambda^n\bigg(\bigcup_{\iota \in I} B_{\iota}\bigg) \le \lambda^n(A) < \infty.$$

It follows in particular that $\bigcup_{\iota \in I} \tau(B_{\iota})$ is λ^n-integrable and so therefore is $\tau(A)$. Moreover,

$$\lambda^n\big(\tau(A)\big) = \lambda^n\bigg(\bigcup_{\iota \in I} \tau(B_{\iota})\bigg) = \lambda^n\bigg(\bigcup_{\iota \in I} B_{\iota}\bigg) = \lambda^n(A).$$

Finally, let A be an arbitrary element of $\mathfrak{L}(\lambda^n)$ and take $\varepsilon > 0$. Then by Theorems 5.25(b) and 5.26(b) we may choose an open set $B \in \mathfrak{L}(\lambda^n)$ and a compact set C such that $C \subset A \subset B$ and

$$\lambda^n(B) - \lambda^n(C) < \varepsilon.$$

Thus

$$\tau(C) \subset \tau(A) \subset \tau(B),$$

and noting that $C = B \setminus (B \setminus C)$ holds and that $B \setminus C$ is open, we see that

$$\lambda^n\big(\tau(B)\big) - \lambda^n\big(\tau(C)\big) < \varepsilon.$$

Since ε was arbitrary, it follows that $\tau(A) \in \mathfrak{L}(\lambda^n)$ and

$$\lambda^n\big(\tau(A)\big) = \lambda^n(A).$$

The rest follows from the fact that τ^{-1} is also a congruence mapping. $\qquad\square$

In other words, n-dimensional Lebesgue measure solves the measure problem for $\mathbb{R}^n$ formulated in Chapter 1: λ^n is defined for a sufficiently large class of sets containing the relatively compact Borel sets, and congruent sets have the same measure.

Exercises

1. Show that the volume of the ball $B(0,r) \subset \mathbb{R}^n$ of radius r with centre 0 is $\dfrac{\pi^{n/2} r^n}{\Gamma(\frac{n}{2} + 1)}$.

Hint: First use induction to prove that there are constants c_n with $\lambda^n(B(0,r)) = c_n r^n$. To do so, observe that by Fubini's theorem,

$$\lambda^n(B(0,r)) = 2^n \int_0^r \int_0^{\sqrt{r^2 - x_1^2}} \cdots \int_0^{\sqrt{r^2 - (x_1^2 + \cdots + x_{n-1}^2)}} 1 \, dx_n \dots dx_2 \, dx_1$$

and that the computation of the inner $n - 1$ integrals yields

$$\frac{1}{2^{n-1}} \lambda^{n-1}\bigg(B\Big(0, \sqrt{r^2 - x_1^2}\Big)\bigg).$$

Use the substitution $x = r\cos\varphi$ to transform the resulting integral $\int_0^r \left(r^2 - x^2\right)^{(n-1)/2} dx$ into one of the form $\int_0^{\pi/2} (\sin\varphi)^n d\varphi$, which can be evaluated recursively. We thus obtain a recursion formula for the c_n – namely $c_n = \frac{2\pi}{n} c_{n-2}$. Now use the fact that $\Gamma(1/2) = \sqrt{\pi}$ together with the functional equation of the gamma function, $\Gamma(\alpha + 1) = \alpha\Gamma(\alpha)$, to complete the proof.

2. Let λ be Lebesgue measure on $\mathfrak{I}$ and take $\nu := \left(\bigotimes_{k=1}^n \lambda\right)^{\mathbb{R}^n}$. Prove that $\lambda^n = \nu$.

Hint: First use the fact that $K = \bigcap_{k\in\mathbb{N}} I_k$, where $(I_k)_{k\in\mathbb{N}}$ is a decreasing sequence of n-dimensional interval forms, to show that each $K \in \mathfrak{K}(\mathbb{R}^n)$ satisfies $K \in \mathfrak{L}(\nu)$ and $\nu(K) = \lambda^n(K)$. Use the regularity of λ^n to show that for each $A \in \mathfrak{L}(\lambda^n)$, $A \in \mathfrak{L}(\nu)$ and $\nu(A) = \lambda^n(A)$.

6

$\mathcal{L}^p$-spaces

$\mathcal{L}^p$-spaces provide one of the most important links between integration theory and its applications. They play a particularly important part in functional analysis. This chapter investigates properties of $\mathcal{L}^p$-spaces. Throughout this chapter $(X, \mathfrak{R}, \mu)$ denotes a positive measure space.

6.1 The structure of $\mathcal{L}^p$-spaces

We denote by $\mathcal{M}_0(\mu)$ the set of all functions $f \in \overline{\mathbb{R}}^X$ for which there is a sequence $(f_n)_{n \in \mathbb{N}}$ in $\mathcal{L}(\mathfrak{R})$ with $f = \lim_{n \to \infty} f_n$ μ-a.e.

Proposition 6.1 $\mathcal{M}_0(\mu)$ has the following properties.

(a) Given $f, g \in \mathcal{M}_0(\mu)$ and $h \in \overline{\mathbb{R}}^X$, if $h(x) = f(x) + g(x)$ μ-a.e., then $h \in \mathcal{M}_0(\mu)$.

(b) If $f \in \mathcal{M}_0(\mu)$ and $\alpha \in \mathbb{R}$, then $\alpha f \in \mathcal{M}_0(\mu)$.

(c) If $f, g \in \mathcal{M}_0(\mu)$, then $f \vee g \in \mathcal{M}_0(\mu)$ and $f \wedge g \in \mathcal{M}_0(\mu)$.

(d) If $f \in \mathcal{M}_0(\mu)_+$ and $p \in \mathbb{R}$, $p > 0$, then $f^p \in \mathcal{M}_0(\mu)$.

(e) $\mathcal{M}_0(\mu) \cap \mathbb{R}^X$ is a vector lattice of functions.

(f) If $f, g \in \mathcal{M}_0(\mu)$ and $\{|f| = \infty\} \cup \{|g| = \infty\} \in \mathfrak{N}(\mu)$, then $fg \in \mathcal{M}_0(\mu)$.

Proof. The proof of (a)–(e) is trivial. We need only note that for arbitrary sequences $(f_n)_{n \in \mathbb{N}}$ and $(g_n)_{n \in \mathbb{N}}$ in $\mathcal{L}(\mathfrak{R})$, $(f_n + g_n)_{n \in \mathbb{N}}$, $(\alpha f_n)_{n \in \mathbb{N}}$, $(f_n \vee g_n)_{n \in \mathbb{N}}$ and $(f_n \wedge g_n)_{n \in \mathbb{N}}$ are also sequences in $\mathcal{L}(\mathfrak{R})$, as is $(f_n^p)_{n \in \mathbb{N}}$ whenever $p > 0$, as long as $f_n \geq 0$ for every $n \in \mathbb{N}$.

(f) Since a function which agrees μ-a.e. with a function in $\mathcal{M}_0(\mu)$ is itself in $\mathcal{M}_0(\mu)$, we may suppose that f and g are real-valued. Choose sequences $(f_n)_{n \in \mathbb{N}}$ and $(g_n)_{n \in \mathbb{N}}$ in $\mathcal{L}(\mathfrak{R})$ with $f = \lim_{n \to \infty} f_n$ μ-a.e. and $g = \lim_{n \to \infty} g_n$ μ-a.e. Then $(f_n g_n)_{n \in \mathbb{N}}$ is a sequence in $\mathcal{L}(\mathfrak{R})$ converging to fg μ-a.e. $\qquad \square$

The useful integrability criterion formulated for $\mathcal{M}(\mu)$ also applies to $\mathcal{M}_0(\mu)$ – simply because every element of $\mathcal{M}_0(\mu)$ is μ-measurable:

Proposition 6.2

(a) $\mathcal{M}_0(\mu) \subset \mathcal{M}(\mu)$.

(b) Take $f \in \mathcal{M}_0(\mu)$ such that $|f| \leq g$ μ-a.e. for some $g \in \mathcal{L}^1(\mu)$. Then $f \in \mathcal{L}^1(\mu)$.

Proof. (a) follows from Corollary 4.34 and Theorems 4.35(c) and 4.33(e), while (b) is a consequence of (a) and Theorem 4.38(a).   □

Given $p \in \mathbb{R}$, $p > 1$, define

$$\mathcal{L}^p(\mu) := \{ f \in \mathcal{M}_0(\mu) \mid |f|^p \in \mathcal{L}^1(\mu) \}.$$

The elements of $\mathcal{L}^p(\mu)$ are called **p-fold μ-integrable functions**. For $f \in \mathcal{L}^p(\mu)$ and $p \geq 1$, put

$$N_p(f) := \left(\int |f|^p d\mu \right)^{1/p}.$$

The definition of $\mathcal{L}^p(\mu)$ also makes sense for $p = 1$. However, we have already defined a space $\mathcal{L}^1(\mu)$ previously. But both definitions lead to the same object. In fact,

$$\mathcal{L}^1(\mu) = \{ f \in \mathcal{M}_0(\mu) \mid |f| \in \mathcal{L}^1(\mu) \},$$

as we prove in Corollary 6.12. But first we turn to the inequalities of Hölder and Minkowski.

Proposition 6.3 *Take $\alpha, \beta \in \overline{\mathbb{R}}_+$ and $p, q \in \mathbb{R}_+ \setminus \{0\}$ such that $\frac{1}{p} + \frac{1}{q} = 1$. Then*

$$\alpha\beta \leq \frac{\alpha^p}{p} + \frac{\beta^q}{q}. \tag{1}$$

Proof. Without loss of generality, assume that α and β are non-zero real numbers, for otherwise the inequality is trivial. Given $\gamma \in \,]0, 1[$, consider the function

$$g : \,]0, \infty[\,\longrightarrow\, \mathbb{R}, \quad x \longmapsto \gamma x - x^\gamma.$$

Then $g'(x) = \gamma(1 - x^{\gamma-1})$. Thus $g'(x) < 0$ if $x \in \,]0, 1[$ and $g'(x) > 0$ if $x > 1$. This implies that $g(x) \geq g(1)$ for every $x \in \,]0, \infty[$, and therefore

$$x^\gamma \leq \gamma x + (1 - \gamma).$$

Put $x := \alpha^p / \beta^q$. Then

$$\alpha^{p\gamma} \beta^{q(1-\gamma)} \leq \gamma \alpha^p + (1 - \gamma)\beta^q.$$

Now put $\gamma := 1/p$. Then $1 - \gamma = 1/q$, and it follows that $\alpha\beta \leq \alpha^p/p + \beta^q/q$.
                                                                                              □

Theorem 6.4 (Hölder's Inequality) *Take $p, q \in \,]1, \infty[$ with $\frac{1}{p} + \frac{1}{q} = 1$. Let $f \in \mathcal{L}^p(\mu)$ and $g \in \mathcal{L}^q(\mu)$. Then $fg \in \mathcal{L}^1(\mu)$ and*

$$N_1(fg) \leq N_p(f)N_q(g).$$

Proof. In (1), put

$$\alpha := \frac{|f(x)|}{N_p(f)} \quad \text{and} \quad \beta := \frac{|g(x)|}{N_q(g)}$$

for $x \in X$, where $N_p(f)$ and $N_q(g)$ are assumed to be non-zero. Otherwise $f \in \mathcal{N}(\mu)$ or $g \in \mathcal{N}(\mu)$, and the inequality is trivial. Then

$$\frac{|f(x)g(x)|}{N_p(f)N_q(g)} \leq \frac{|f(x)|^p}{pN_p(f)^p} + \frac{|g(x)|^q}{qN_q(g)^q}$$

for every $x \in X$. Since

$$\frac{|f|^p}{pN_p(f)^p} + \frac{|g|^q}{qN_q(g)^q} \in \mathcal{L}^1(\mu),$$

Propositions 6.2(b) and 6.1(f) imply that $fg \in \mathcal{L}^1(\mu)$. Integrating, we obtain

$$\frac{N_1(fg)}{N_p(f)N_q(g)} \leq \frac{1}{p} + \frac{1}{q} = 1.$$

$\square$

Minkowski's inequality, which we prove by applying Hölder's inequality, is the key for much of the following, since it enables us to prove the triangle inequality for a natural pseudometric to be introduced on $\mathcal{L}^p(\mu)$, and later the triangle inequality for the corresponding natural norm on $L^p(\mu)$.

Theorem 6.5 (Minkowski's Inequality) *Take $f, g \in \mathcal{L}^p(\mu)$ with $p \geq 1$. Let $h \in \overline{\mathbb{R}}^X$ such that $h(x) = f(x) + g(x)$ μ-a.e. Then $h \in \mathcal{L}^p(\mu)$ and*

$$N_p(h) \leq N_p(f) + N_p(g).$$

Proof. The assertion is trivial for $p = 1$. So take $p > 1$ and choose $q \in \mathbb{R}$ such that $\frac{1}{p} + \frac{1}{q} = 1$. Then $h \in \mathcal{M}_0(\mu)$ and

$$|h|^p \leq \big(2(|f| \vee |g|)\big)^p \leq 2^p\big(|f|^p + |g|^p\big) \quad \mu\text{-a.e.}$$

Therefore $h \in \mathcal{L}^p(\mu)$ by Proposition 6.2(b). Furthermore,

$$|h|^p = |h|\,|h|^{p-1} \leq |f|\,|h|^{p-1} + |g|\,|h|^{p-1} \quad \mu\text{-a.e.} \tag{2}$$

Since

$$\big(|h|^{p-1}\big)^q = |h|^{pq-q} = |h|^p \in \mathcal{L}^1(\mu),$$

it follows that $|h|^{p-1} \in \mathcal{L}^q(\mu)$. Theorem 6.4 now implies that $|f|\,|h|^{p-1} \in \mathcal{L}^1(\mu)$, $|g|\,|h|^{p-1} \in \mathcal{L}^1(\mu)$ and

$$N_1\big(|f|\,|h|^{p-1}\big) \leq N_p(f)N_q\big(|h|^{p-1}\big) = N_p(f)N_p(h)^{p/q}.$$

Dually

$$N_1\left(|g|\,|h|^{p-1}\right) \leq N_p(g)N_p(h)^{p/q}.$$

Integrating (2), we have

$$N_p(h)^p \leq N_p(h)^{p/q}\left(N_p(f) + N_p(g)\right)$$

and therefore

$$N_p(h) \leq N_p(f) + N_p(g).$$

□

We now summarize a first group of properties of the spaces $\mathcal{L}^p(\mu)$.

Theorem 6.6 *Suppose that $p \geq 1$. Then*

(a) $h \in \mathcal{L}^p(\mu)$ *for all $f, g \in \mathcal{L}^p(\mu)$ and $h \in \mathbb{R}^X$ with $h(x) = f(x) + g(x)$*
μ-*a.e.*

(b) $\alpha f \in \mathcal{L}^p(\mu)$ *for every $f \in \mathcal{L}^p(\mu)$ and $\alpha \in \mathbb{R}$.*

(c) $f \vee g \in \mathcal{L}^p(\mu)$ *and $f \wedge g \in \mathcal{L}^p(\mu)$ for all $f, g \in \mathcal{L}^p(\mu)$.*

(d) $\mathcal{L}^p(\mu) \cap \mathbb{R}^X$ *is a vector lattice of functions, and $f \in \mathcal{L}^p(\mu)$ if and only if there is a $g \in \mathcal{L}^p(\mu) \cap \mathbb{R}^X$ with $f = g$ μ-a.e.*

(e) $\mathcal{L}(\mathfrak{R}) \subset \mathcal{L}^p(\mu)$.

Proof. (a) is part of Theorem 6.5. (b) is an immediate result of the definition. (c) follows from $|f \vee g|^p \leq |f|^p + |g|^p$ and $|f \wedge g|^p \leq |f|^p + |g|^p$ by Proposition 6.2(b). (d) results from (a)–(c). Finally, (e) is obvious.   □

Theorem 6.7 *Take $p \geq 1$. Then:*

(a) $$N_p(h) \leq N_p(f) + N_p(g)$$

for all $f, g \in \mathcal{L}^p(\mu)$ and $h \in \overline{\mathbb{R}}^X$ with $h(x) = f(x) + g(x)$ μ-a.e.

(b) $N_p(\alpha f) = |\alpha| N_p(f)$ *for all $f \in \mathcal{L}^p(\mu)$ and $\alpha \in \mathbb{R}$.*

(c) $N_p(f) \geq 0$, *and $N_p(f) = 0$ if and only if $f \in \mathcal{N}(\mu)$.*

(d) *Given $f, g \in \mathcal{L}^p(\mu)$, if $|f| \leq |g|$, then $N_p(f) \leq N_p(g)$.*

(e) *Take $f \in \mathcal{L}^p(\mu)$ and $g \in \overline{\mathbb{R}}^X$ with $f = g$ μ-a.e. Then $g \in \mathcal{L}^p(\mu)$ and $N_p(g) = N_p(f)$.*

Proof. (a) is part of Theorem 6.5. (b) follows immediately from the definition.

(c) The inequality follows from the definitions. If $N_p(f) = 0$, then $\int |f|^p d\mu = 0$. Thus $|f|^p \in \mathcal{N}(\mu)$ and hence f is also in $\mathcal{N}(\mu)$.

(d) and (e) are also immediate consequences of the definitions.   □

Take $f, g \in \mathcal{L}^p(\mu)$. Given any $h_1, h_2 \in \mathcal{L}^p(\mu)$ for which

$$h_1(x) = f(x) - g(x) \ \mu\text{-a.e.} \quad \text{and} \quad h_2(x) = f(x) - g(x) \ \mu\text{-a.e.},$$

we have $N_p(h_1) = N_p(h_2)$ by Theorem 6.7(e). Define

$$d_p(f, g) := N_p(h), \quad \text{where } h \in \mathcal{L}^p(\mu), \ h(x) = f(x) - g(x) \ \mu\text{-a.e.}$$

Note that we used the independence of $d_p(f, g)$ from the choice of h, which was established above.

We now proceed with more significant properties of the $\mathcal{L}^p$-spaces. Their great importance is due to the fact that they are complete (as pseudometric spaces), as we next show. The theory of Fourier series, for instance, benefits from this property.

Theorem 6.8 *Take $p \geq 1$. Then:*

(a) d_p is a pseudometric on $\mathcal{L}^p(\mu)$.

(b) $\mathcal{L}^p(\mu)$ is complete with respect to d_p.

(c) $\mathcal{L}(\mathfrak{R})$ is a dense subspace of $\mathcal{L}^p(\mu)$ with respect to d_p.

Proof. (a) is a simple consequence of Theorem 6.7(a)–(c).

(c) Take $f \in \mathcal{L}^p(\mu)_+$ and define $\bar{f} := f e_{\{f<\infty\}}$. Then $\bar{f} = f$ μ-a.e. and $\bar{f}^p \in \mathcal{L}^1(\mu)$. There is a decreasing sequence $(f_n)_{n\in\mathbb{N}}$ in $\mathcal{L}(\mathfrak{R})^\uparrow \cap \mathcal{L}^1(\mu)_+$ with $f_n \geq \bar{f}^p$ μ-a.e. for every $n \in \mathbb{N}$ and $\int f_n d\mu \downarrow \int \bar{f}^p d\mu$. Thus $\bigwedge_{n\in\mathbb{N}} f_n \geq \bar{f}^p$ μ-a.e. By the monotone convergence theorem

$$\bigwedge_{n\in\mathbb{N}} f_n \in \mathcal{L}^1(\mu) \quad \text{and} \quad \int \left(\bigwedge_{n\in\mathbb{N}} f_n \right) d\mu = \int \bar{f}^p d\mu.$$

It follows that $\bigwedge_{n\in\mathbb{N}} f_n = \bar{f}^p$ μ-a.e. (Theorem 3.30(e)) and so

$$\lim_{n\to\infty} \left(f_n^{1/p} - \bar{f} \right) = 0 \quad \mu\text{-a.e.}$$

Take $n \in \mathbb{N}$. Then $f_n^{1/p} \in \mathcal{L}^p(\mu)$ and thus $f_n^{1/p} - \bar{f} \in \mathcal{L}^p(\mu)$. Moreover, $\left(f_n^{1/p} - \bar{f} \right)^p \leq 2^p f_1$ μ-a.e. By the Lebesgue convergence theorem

$$\lim_{n\to\infty} \int \left(f_n^{1/p} - \bar{f} \right)^p d\mu = 0,$$

whence

$$\lim_{n\to\infty} N_p \left(f_n^{1/p} - \bar{f} \right) = 0.$$

Take $\varepsilon > 0$. By the above, there is an $h \in \mathcal{L}(\mathfrak{R})^\uparrow \cap \mathcal{L}^1(\mu)_+$ with

$$N_p(h^{1/p} - \bar{f}) < \varepsilon/2.$$

Choose a sequence $(g_n)_{n\in\mathbb{N}}$ in $\mathcal{L}(\mathfrak{R})_+$ with $g_n \uparrow h$. Then $g_n^{1/p} \in \mathcal{L}(\mathfrak{R})_+$ and $h^{1/p} - g_n^{1/p} \in \mathcal{L}^p(\mu)$ for each $n \in \mathbb{N}$. By the monotone convergence theorem

$$\lim_{n\to\infty} N_p \left(h^{1/p} - g_n^{1/p} \right) = 0.$$

Hence there is a $g \in \mathcal{L}(\mathfrak{R})_+$ with

$$N_p(h^{1/p} - g) < \varepsilon/2.$$

Thus

$$N_p(f - g) = N_p(\bar{f} - g) \leq N_p(\bar{f} - h^{1/p}) + N_p(h^{1/p} - g) < \varepsilon.$$

Now let f be an arbitrary element of $\mathcal{L}^p(\mu)$. For $\varepsilon > 0$ there are, as already shown, $g_1, g_2 \in \mathcal{L}(\mathfrak{R})$ such that

$$N_p(f^+ - g_1) < \varepsilon/2 \quad \text{and} \quad N_p(f^- - g_2) < \varepsilon/2.$$

$g := g_1 - g_2$ is an $\mathfrak{R}$-step function, and

$$N_p(f - g) \le N_p(f^+ - g_1) + N_p(f^- - g_2) < \varepsilon.$$

This proves (c).

(b) We make a preliminary observation. Let $(f_n)_{n\in\mathbb{N}}$ and $(g_n)_{n\in\mathbb{N}}$ be sequences in $\mathcal{L}^p(\mu) \cap \mathbb{R}^X$ such that

$$d_p(f_n, g_n) < 1/2^n$$

for every $n \in \mathbb{N}$. Then for arbitrary $i, n \in \mathbb{N}$,

$$\int \left(\sum_{k=i}^{i+n} |f_k - g_k| \right)^p d\mu = \left(N_p\left(\sum_{k=i}^{i+n} |f_k - g_k| \right) \right)^p$$
$$\le \left(\sum_{k=i}^{i+n} N_p(f_k - g_k) \right)^p < 1/2^{i-1}$$

and hence

$$\sup_{n\in\mathbb{N}} \int \left(\sum_{k=i}^{i+n} |f_k - g_k| \right)^p d\mu \le 1/2^{i-1}.$$

It follows from the monotone convergence theorem that

$$\left(\sum_{k\ge i} |f_k - g_k| \right)^p \in \mathcal{L}^1(\mu), \tag{3}$$

$$\int \left(\sum_{k\ge i} |f_k - g_k| \right)^p d\mu \le 1/2^{i-1}, \tag{4}$$

and as a result,

$$\sum_{k\ge i} |f_k(x) - g_k(x)| \in \mathbb{R} \quad \mu\text{-a.e.} \tag{5}$$

By Theorems 6.6(d) and 6.7(e), we need only consider d_p-Cauchy sequences in $\mathcal{L}^p(\mu) \cap \mathbb{R}^X$ to prove (b). So let $(f_n)_{n\in\mathbb{N}}$ be such a sequence. Choose a subsequence $(f_{n_k})_{k\in\mathbb{N}}$ such that

$$d_p(f_{n_{k+1}}, f_{n_k}) < 1/2^k$$

for every $k \in \mathbb{N}$. (5) implies that

$$\sum_{k\in\mathbb{N}} |f_{n_{k+1}}(x) - f_{n_k}(x)| \in \mathbb{R} \quad \mu\text{-a.e.},$$

and it follows that

$$\lim_{k\to\infty} f_{n_k}(x) = f_{n_1}(x) + \sum_{k=1}^{\infty} \left(f_{n_{k+1}}(x) - f_{n_k}(x) \right) \tag{6}$$

exists μ-a.e. We define

$$f(x) := \begin{cases} \lim_{k\to\infty} f_{n_k}(x) & \text{if it exists and is real} \\ 0 & \text{otherwise.} \end{cases}$$

We show that $f \in \mathcal{M}_0(\mu)$. Note that by (c) there are functions $h_k \in \mathcal{L}(\mathfrak{R})$ such that

$$d_p\left(f_{n_k}, h_k \right) < 1/2^k.$$

By (5),

$$\lim_{k\to\infty} \left| f_{n_k}(x) - h_k(x) \right| = 0 \quad \mu\text{-a.e.,}$$

and therefore

$$\lim_{k\to\infty} h_k(x) = \lim_{k\to\infty} f_{n_k}(x) - \lim_{k\to\infty} \left(f_{n_k}(x) - h_k(x) \right) = f(x) \quad \mu\text{-a.e.}$$

Thus, $f \in \mathcal{M}_0(\mu)$.

By (6), if $i \in \mathbb{N}$ then

$$\left| f - f_{n_i} \right|^p = \left| \sum_{k=i}^{\infty} \left(f_{n_{k+1}} - f_{n_k} \right) \right|^p = \lim_{n\to\infty} \left| \sum_{k=i}^{i+n} \left(f_{n_{k+1}} - f_{n_k} \right) \right|^p \quad \mu\text{-a.e.,}$$

and by (3),

$$\left| \sum_{k=i}^{i+n} \left(f_{n_{k+1}} - f_{n_k} \right) \right|^p \leq \left(\sum_{k\geq i} \left| f_{n_{k+1}} - f_{n_k} \right| \right)^p \in \mathcal{L}^1(\mu)$$

for every $n \in \mathbb{N}$. The Lebesgue convergence theorem now implies that $\left| f - f_{n_i} \right|^p \in \mathcal{L}^1(\mu)$, and therefore by (4),

$$\int \left| f - f_{n_i} \right|^p d\mu \leq \int \left(\sum_{k\geq i} \left| f_{n_{k+1}} - f_{n_k} \right| \right)^p d\mu \leq 1/2^{i-1}.$$

We conclude that $f - f_{n_i} \in \mathcal{L}^p(\mu)$. Hence $f \in \mathcal{L}^p(\mu)$ and

$$\lim_{i\to\infty} d_p\left(f, f_{n_i} \right) = 0.$$

Take $\varepsilon > 0$ and pick an $m \in \mathbb{N}$ such that $d_p(f_n, f_{n+k}) < \varepsilon/2$ for all $n, k \in \mathbb{N}$, $n \geq m$. For each $n \geq m$ there is an $i \in \mathbb{N}$ such that $n_i \geq n$ and $d_p\left(f, f_{n_i} \right) < \varepsilon/2$. It follows that

$$d_p(f, f_n) \leq d_p\left(f, f_{n_i} \right) + d_p\left(f_{n_i}, f_n \right) < \varepsilon.$$

Thus $\lim_{n\to\infty} d_p(f, f_n) = 0$, which proves (b). $\qquad\square$

Note that in view of Theorem 6.7(c) d_p is not in general a metric.

We have now available four concepts of convergence in the $\mathcal{L}^p$-spaces: convergence with respect to d_p, uniform convergence, pointwise convergence (almost everywhere) and convergence in measure. How are these concepts related? We have already studied several aspects of this problem in Section 4.4. The following considerations provide us with the necessary information, in particular concerning the relation of convergence with respect to d_p – which was not yet available in Section 4.4 – to the other notions of convergence.

Theorem 6.9 *Take $p \geq 1$, and let $(f_n)_{n \in \mathbb{N}}$ be a sequence in $\mathcal{L}^p(\mu)$ which converges to $f \in \mathcal{L}^p(\mu)$ with respect to d_p. Then there is a subsequence $(f_{n_k})_{k \in \mathbb{N}}$ which converges μ-a.e. to f. We may select $(f_{n_k})_{k \in \mathbb{N}}$ in such a way that $|f_{n_k}| \leq g$ for every $k \in \mathbb{N}$, for some suitable $g \in \mathcal{L}^p(\mu)$.*

Proof. Consider a sequence $(f_n)_{n \in \mathbb{N}}$ in $\mathcal{L}^p(\mu) \cap \mathbb{R}^X$ which converges to $f \in \mathcal{L}^p(\mu) \cap \mathbb{R}^X$ with respect to d_p. Then $(f - f_n)_{n \in \mathbb{N}}$ converges to 0 with respect to d_p, and we can choose a subsequence $(f_{n_k})_{k \in \mathbb{N}}$ such that

$$d_p(f - f_{n_k}, 0) < 1/2^k$$

for every $k \in \mathbb{N}$. We appeal to the preliminary observation in the proof of Theorem 6.8(b). Then (5) implies that

$$\sum_{k \in \mathbb{N}} |f(x) - f_{n_k}(x)| \in \mathbb{R} \quad \mu\text{-a.e.,}$$

and we conclude that $f(x) = \lim_{k \to \infty} f_{n_k}(x)$ μ-a.e.

Observe that $d(f_{n_{k+1}}, f_{n_k}) < 2/2^k$ for all $k \in \mathbb{N}$. Thus, putting

$$g := |f_{n_1}| + \sum_{k \in \mathbb{N}} |f_{n_{k+1}} - f_{n_k}|,$$

$g \in \mathcal{L}^p(\mu)$, by (3). Moreover, $|f_{n_k}| \leq g$ for every $k \in \mathbb{N}$.

Since it suffices to consider only real-valued functions (Theorems 6.6(d) and 6.7(e)), the proof of the theorem is complete. $\square$

Note that the converse is not true, i.e. if a sequence $(f_n)_{n \in \mathbb{N}}$ in $\mathcal{L}^p(\mu)$ converges μ-a.e. to $f \in \mathcal{L}^p(\mu)$, then it may happen that no subsequence of $(f_n)_{n \in \mathbb{N}}$ converges to f with respect to d_p. As an example, consider Lebesgue measure on $\mathbb{R}$ and the sequence $\left(\frac{1}{n} e_{[0,n[}\right)_{n \in \mathbb{N}}$. Note that this sequence converges even uniformly to 0. However, if $X \in \mathfrak{L}(\mu)$, then obviously every uniformly convergent sequence in $\mathcal{L}^p(\mu)$ converges with respect to d_p to the same limit.

It is also essential to observe that, under the hypotheses of Theorem 6.9, we cannot expect that the sequence $(f_n)_{n \in \mathbb{N}}$ itself would be convergent μ-a.e. to f. For example, consider the sequence $(h_n)_{n \in \mathbb{N}}$ defined in Exercise 3(c) of Section 3.4.

Applying Theorem 4.46, we immediately obtain:

Corollary 6.10 *Take $p \geq 1$ and let $(f_n)_{n\in\mathbb{N}}$ be a sequence in $\mathcal{L}^p(\mu)$ which converges to $f \in \mathcal{L}^p(\mu)$ with respect to d_p. Then $f_n \overset{\mu}{\to} f$.*

We shall prove a partial converse to Corollary 6.10 in Corollary 6.25.

We can now completely describe the approximation of the functions in $\mathcal{L}^p(\mu)$ by $\mathfrak{R}$-step functions. The following result is an easy consequence of Theorems 6.8(c) and 6.9, and Corollary 6.10.

Corollary 6.11 *Take $f \in \mathcal{L}^p(\mu)$, with $p \geq 1$. Then there is a sequence $(f_n)_{n\in\mathbb{N}}$ of $\mathfrak{R}$-step functions which converges to f with respect to d_p, in measure μ, and μ-a.e.*

Recall that Corollary 4.11 provided us with another approximation of a μ-integrable function f by step functions. There we considered step functions with respect to the ring of sets $\mathfrak{L}(\mu)$ and we obtained a sequence of $\mathfrak{L}(\mu)$-step functions converging to f *everywhere*. The reason why we could prove convergence everywhere is of course that $\mathfrak{L}(\mu)$ contains all μ-null sets. In the context of $\mathfrak{R}$-step functions, however, we can ensure only convergence μ-a.e. Examples of μ-integrable functions which are *not* the pointwise limit everywhere of a sequence of $\mathfrak{R}$-step functions, are given in the exercises.

Our next result is the one heralded above: that our definition here of $\mathcal{L}^p$ reduces in the case $p = 1$ to being equivalent to our earlier one.

Corollary 6.12 $\mathcal{L}^1(\mu) = \{f \in \mathcal{M}_0(\mu) \mid |f| \in \mathcal{L}^1(\mu)\}.$

Proof. If $f \in \mathcal{L}^1(\mu)$, then obviously $|f| \in \mathcal{L}^1(\mu)$, and by Corollary 6.11 also $f \in \mathcal{M}_0(\mu)$. The reverse inclusion follows from Proposition 6.2. $\square$

As a consequence, we see that we could also have defined the $\mathcal{L}^p$-spaces with the aid of μ-measurable functions.

Corollary 6.13 *For $p \geq 1$, $\mathcal{L}^p(\mu) = \{f \in \mathcal{M}(\mu) \mid |f|^p \in \mathcal{L}^1(\mu)\}.$*

Proof. '$\subset$' follows from Proposition 6.2(a). For the proof of '$\supset$', take $f \in \mathcal{M}(\mu)$ such that $|f|^p \in \mathcal{L}^1(\mu)$. Then $f^+ \in \mathcal{M}(\mu)$, and since

$$\{(f^+)^p < \alpha\} = \{f^+ < \alpha^{1/p}\}$$

for every $\alpha > 0$, it follows that $(f^+)^p \in \mathcal{M}(\mu)$. Since $(f^+)^p \leq |f|^p$, we conclude from Theorem 4.38(a) that $(f^+)^p \in \mathcal{L}^1(\mu)$. Then by Corollary 6.12, $(f^+)^p \in \mathcal{M}_0(\mu)$ and hence $f^+ \in \mathcal{M}_0(\mu)$ (Proposition 6.1(d)). Similarly, $f^- \in \mathcal{M}_0(\mu)$. Thus $f = f^+ - f^- \in \mathcal{M}_0(\mu)$. $\square$

This leads to the following result for countable families – in particular for sequences – in $\mathcal{L}^p(\mu)$.

Corollary 6.14 *Take $p \geq 1$ and let $(f_\iota)_{\iota\in I}$ be a non-empty countable family in $\mathcal{L}^p(\mu)$ which is bounded above (below) in $\mathcal{L}^p(\mu)$. Then $\bigvee_{\iota\in I} f_\iota \in \mathcal{L}^p(\mu)$ $(\bigwedge_{\iota\in I} f_\iota \in \mathcal{L}^p(\mu))$.*

Proof. We prove only the assertion for a family which is bounded above. Let g be an upper bound for $(f_\iota)_{\iota\in I}$ in $\mathcal{L}^p(\mu)$. By Theorem 4.35(a), $f := \bigvee_{\iota\in I} f_\iota \in \mathcal{M}(\mu)$. Hence $|f| \in \mathcal{M}(\mu)$ and so $|f|^p \in \mathcal{M}(\mu)$. But since $|f| \leq |f_\kappa| + |g|$, where κ is some element of I, and since $(|f_\kappa| + |g|)^p \in \mathcal{L}^1(\mu)$, $|f|^p$ also belongs to $\mathcal{L}^1(\mu)$ (Theorem 4.38(a)). The assertion now follows from Corollary 6.13. $\square$

Consider for a moment a locally compact space X. In such a space, we are especially interested in the continuous functions with compact support, and we hope or even expect that every function in $\mathcal{L}^p(\mu)$ (for regular μ) can be approximated by functions of this type. This is indeed confirmed by the following result.

Theorem 6.15 *Let X be a locally compact space and μ a regular positive measure on a ring of sets $\mathfrak{R} \supset \mathfrak{R}(X)$, $\mathfrak{R} \subset \mathfrak{P}(X)$. Take $p \geq 1$. Then:*

(a) $\mathcal{K}(X)$ is a dense subset of $\mathcal{L}^p(\mu)$ with respect to d_p.

(b) For every $f \in \mathcal{L}^p(\mu)$ there is a sequence $(f_n)_{n\in\mathbb{N}}$ in $\mathcal{K}(X)$ converging to f with respect to d_p, in measure μ, and μ-a.e. Moreover, the sequence can be chosen such that $|f_n| \leq g$ for every $n \in \mathbb{N}$, for some suitable $g \in \mathcal{L}^p(\mu)$.

Proof. (a) Take $f \in \mathcal{K}(X)$. Then $f \in \mathcal{L}^1(\mu)$ by Proposition 5.19. Corollary 6.12 implies that $f \in \mathcal{M}_0(\mu)$. Obviously $|f|^p \in \mathcal{K}(X)$ and so $|f|^p \in \mathcal{L}^1(\mu)$. Hence $f \in \mathcal{L}^p(\mu)$ and thus $\mathcal{K}(X) \subset \mathcal{L}^p(\mu)$.

We now show that given $f \in \mathcal{L}^p(\mu)$ and $\varepsilon > 0$, there is a $g \in \mathcal{K}(X)$ with $d_p(f, g) < \varepsilon$. In view of Theorem 6.8(c), it suffices to prove this for $f \in \mathcal{L}(\mathfrak{R})$. But as each $f \in \mathcal{L}(\mathfrak{R})$ can be written in the form $f = \sum_{\iota\in I} \alpha_\iota e_{A_\iota}$ with I finite and $A_\iota \in \mathfrak{R}$, it is clearly enough to prove the claim for $f = e_A$, with $A \in \mathfrak{R}$. Then, by regularity, we need prove this only for $f = e_K$, for $K \in \mathfrak{K}(X)$. But this is an easy consequence of Urysohn's theorem and Theorem 5.25(a). Indeed, we can find a $g \in \mathcal{K}(X)_+$, $e_K \leq g \leq e_X$, such that $\int(g - e_K)d\mu < \varepsilon^p$. Then

$$d_p(e_K, g) = \left(\int (g - e_K)^p d\mu\right)^{1/p} \leq \left(\int (g - e_K)d\mu\right)^{1/p} < \varepsilon.$$

(b) follows from (a), Theorem 6.9 and Corollary 6.10. $\square$

We now present another interesting application of the results of this chapter. Recall that in the Introduction we posed the problem of giving a concrete representation of the completion of the pseudometric space of Riemann integrable functions as a function space. This is now easy.

Until now we have been working with the Riemann integral on an arbitrary but fixed interval $[a, b]$ of $\mathbb{R}$. Let us extend the notion slightly. Call a bounded function $f \in \mathbb{R}^{\mathbb{R}}$ **Riemann integrable on $\mathbb{R}$** if there is a closed interval $[a, b]$ such that $f = 0$ outside $[a, b]$ and $f|_{[a,b]}$ is Riemann integrable on $[a, b]$. In this case,

$$\int f(x)\,dx := \int_a^b f|_{[a,b]}(x)\,dx$$

is said to be the **Riemann integral** of f.

Proposition 6.16 *Let f be Riemann integrable on $\mathbb{R}$. Then f is Lebesgue integrable, and*

$$\int f(x)\,dx = \int f\,d\lambda.$$

Proof. Take $a, b \in \mathbb{R}$ such that f vanishes outside $[a, b]$ and that $f|_{[a,b]}$ is Riemann integrable on $[a, b]$. By Theorems 4.20(b) and 4.19, there is a $g \in \mathcal{L}^1(\lambda)$ with $f|_{[a,b]} = g|_{[a,b]}$. Then $f = ge_{[a,b]} \in \mathcal{L}^1(\lambda)$, by Theorem 4.15. Moreover,

$$\int f(x)\,dx = \int g|_{[a,b]}d(\lambda^{\mathbb{R}}|_{[a,b]}) = \int ge_{[a,b]}d\lambda = \int f\,d\lambda,$$

again by Theorems 4.20(b) and 4.19. □

We define a pseudometric d on the space $\mathcal{R}$ of Riemann integrable functions on $\mathbb{R}$ by

$$d(f, g) := \int |f(x) - g(x)|\,dx,$$

i.e. d is just the restriction of the pseudometric d_1 defined on $\mathcal{L}^1(\lambda)$. Letting $\mathfrak{J}$ be the set of all subsets A of $\mathbb{R}$ for which $e_A \in \mathcal{R}$, we can define an analogous pseudometric d' on $\mathfrak{J}$ by setting

$$d'(A, B) := d(e_A, e_B).$$

Then of course, d' is the restriction of the pseudometric d'_1 on $\mathfrak{L}(\lambda)$, given by

$$d'_1(A, B) := d_1(e_A, e_B) = \lambda^{\mathbb{R}}(A \triangle B).$$

Having introduced the notation needed, we can now state our theorem.

Theorem 6.17 $(\mathcal{L}^1(\lambda), d_1)$ *is a completion of the pseudometric space* $(\mathcal{R}, d)$, *and* $(\mathfrak{L}(\lambda), d'_1)$ *is a completion of the pseudometric space* $(\mathfrak{J}, d')$.

Proof. In view of Theorem 6.8, the first assertion follows immediately from the fact that $\mathcal{L}(\mathfrak{J}) \subset \mathcal{R} \subset \mathcal{L}^1(\lambda)$. For the second assertion, we first note that the set $\mathfrak{J}$ of interval forms is dense in $\mathfrak{L}(\lambda)$ with respect to d'_1. (This is a consequence of Theorem 4.25(b).) It remains to show that $\mathfrak{L}(\lambda)$ is complete with respect to d'_1. So let $(A_n)_{n\in\mathbb{N}}$ be a Cauchy sequence in $\mathfrak{L}(\lambda)$. Then $(e_{A_n})_{n\in\mathbb{N}}$ is a Cauchy sequence in $\mathcal{L}^1(\lambda)$ and therefore converges to some $f \in \mathcal{L}^1(\lambda)$ (Theorem 6.8(b)). By Theorem 6.9, some subsequence of $(e_{A_n})_{n\in\mathbb{N}}$ converges to f pointwise λ-a.e. But since the functions e_{A_n} only take the values 0 and 1, f can also take only these values, except possibly on a λ-null set. Thus $f = e_A$ λ-a.e. for some subset A of $\mathbb{R}$, and it follows that $A = \lim_{n\to\infty} A_n$ in the pseudometric space $(\mathfrak{L}(\lambda), d'_1)$. □

Of course the result analogous to Theorem 6.17 holds if we consider the Riemann and Lebesgue integral on a fixed interval $[a, b]$ of $\mathbb{R}$. The reader is invited to verify this in detail.

We also mention that Theorem 6.17 remains valid for the Riemann and Lebesgue integral on $\mathbb{R}^n$. Here, however, some minor technical problems arise. First of all note that we did not define n-dimensional Lebesgue measure λ^n on the ring of n-dimensional interval forms, but on $\mathfrak{L}_1 :=$ $\mathfrak{L}(\bigotimes_{k=1}^n \lambda^{\mathbb{R}})$, so that we can no longer conclude that the step functions on $\mathfrak{L}_1$ are Riemann integrable. However, we can apply Theorem 6.15 instead. But to be precise, it would still remain to show that every Riemann integrable function on $\mathbb{R}^n$ is, in fact, Lebesgue integrable. As we have not explicitly considered the Riemann integral on $\mathbb{R}^n$ in this book, we shall not go into this in detail.

A few remarks are appropriate. The results of this section show clearly how relevant the null sets and null functions are to the investigation of the properties of $\mathcal{L}^p$-spaces. As far as integration is concerned, they may be ignored. Yet they create several technical difficulties. They are the reason why algebraic operations in $\mathcal{L}^p(\mu)$ cannot be carried out unconditionally. They are also the reason why d_p is not, in general, a metric, but only a pseudometric.

There is one simple way of solving this problem: that is, the formation of equivalence classes via the relation $f = g$ μ-a.e. However, this takes us beyond the realm of the treatment of functions into the domain of abstract vector lattices. Since these are also of great importance for integration theory in other respects, the next chapter provides an introduction to them.

Exercises

1. Show that in general $\mathcal{M}_0(\mu) \neq \mathcal{M}(\mu)$.

2. Given $f \in \overline{\mathbb{R}}^X$, let

$$\int^* f \, d\mu := \inf \left\{ \int g \, d\mu \,\Big|\, g \in \mathcal{L}^1(\mu), \, g \geq f \right\}.$$

$\int^* f \, d\mu$ is called the **upper integral** of f with respect to μ.

(a) Prove that for $f \in \overline{\mathbb{R}}^X$

$$\int^* f \, d\mu = \inf \left\{ \int g \, d\mu \,\Big|\, g \in \overline{\mathcal{L}}(\mu), \, g \geq f \text{ } \mu\text{-a.e.} \right\}$$

$$= \inf \left\{ \int h \, d\mu \,\Big|\, h \in \mathcal{L}(\mathfrak{R})^\uparrow \cap \mathcal{L}^1(\mu), \, h \geq f \text{ } \mu\text{-a.e.} \right\}.$$

Hence $\int^* f \, d\mu = \ell_\mu^*(f)$ (cf. Exercise 3 of Section 4.4).

(b) Define the lower integral $\int_* f \, d\mu$ of $f \in \bar{\mathbb{R}}^X$ analogously to the upper integral and prove that

$$f \in \mathcal{L}^1(\mu) \iff \int^* f \, d\mu = \int_* f \, d\mu \in \mathbb{R}.$$

3. Show that for $p \in [1, \infty[$ and $f \in \bar{\mathbb{R}}^X$ the following are equivalent.

   (a) $f \in \mathcal{L}^p(\mu)$.
   (b) $f \in \mathcal{M}_0(\mu)$ and $\int^* |f|^p d\mu < \infty$.
   (c) $f \in \mathcal{M}(\mu)$ and $\int^* |f|^p d\mu < \infty$.
   (d) $(f^+)^p \in \mathcal{L}^1(\mu)$ and $(f^-)^p \in \mathcal{L}^1(\mu)$.

4. Let $\mathfrak{F}$ be the set of finite subsets of $\mathbb{N}$ and $\mu := \mu^{e_{|N}}$ counting measure on $\mathfrak{F}$. Prove:

   (a) If $p, q \in [1, \infty[$ and $p \le q$, then $\mathcal{L}^p(\mu) \subset \mathcal{L}^q(\mu)$.
   (b) If $p < q$, then $\mathcal{L}^p(\mu) \ne \mathcal{L}^q(\mu)$.

   Hint for (b): Consider the function $f : \mathbb{N} \to \mathbb{R}$, $n \mapsto n^{-1/p}$.

5. Let μ be bounded and take $p, q \in [1, \infty[$ with $p \le q$. Show that $\mathcal{L}^q(\mu) \subset \mathcal{L}^p(\mu)$ and

$$N_p(f) \le N_q(f) \left(\int e_X d\mu \right)^{\frac{q-p}{pq}} \qquad \text{for } f \in \mathcal{L}^q(\mu).$$

6. Consider the sequence $(h_n)_{n \in \mathbb{N}}$ defined in Exercise 3(c) of Section 3.4.

   (a) Find a subsequence $(h_{n_k})_{k \in \mathbb{N}}$ which converges to 0 for each x.
   (b) Put $h := e_{\{0, 1/2\}}$. Then $(h_n)_{n \in \mathbb{N}}$ also converges to h with respect to d_1. Find a subsequence $(h_{n_j})_{j \in \mathbb{N}}$ which converges to h λ-a.e., and show that no subsequence of $(h_n)_{n \in \mathbb{N}}$ converges to h everywhere.

7. Take $p \in [1, \infty[$ and let $(f_\iota)_{\iota \in I}$ be a non-empty, countable, upward directed family in $\mathcal{L}^p(\mu)_+$ which is bounded above in $\mathcal{L}^p(\mu)$. Prove that $N_p(\bigvee_{\iota \in I} f_\iota) = \sup_{\iota \in I} N_p(f_\iota)$.

8. Take $p \in [1, \infty[$ and let $(f_n)_{n \in \mathbb{N}}$ be a sequence in $\mathcal{L}^p(\mu) \cap \bar{\mathbb{R}}^X$ with $\sum_{n \in \mathbb{N}} N_p(f_n) < \infty$. Define

$$A := \{x \in X \mid (f_n(x))_{n \in \mathbb{N}} \text{ is summable}\},$$

$$f : X \longrightarrow \mathbb{R}, \quad x \longmapsto \begin{cases} \sum_{n \in \mathbb{N}} f_n(x) & \text{if } x \in A \\ 0 & \text{if } x \in X \setminus A. \end{cases}$$

   Prove that

   (a) $X \setminus A \in \mathfrak{N}(\mu)$.
   (b) $f \in \mathcal{L}^p(\mu)$.
   (c) $\lim_{m \to \infty} N_p(f - \sum_{n=1}^m f_n) = 0$.

(d) $N_p(f) \leq \sum_{n \in \mathbb{N}} N_p(f_n)$.

Hint: Use Minkowski's inequality to show that $\left(\sum_{n \in \mathbb{N}} |f_n| \right)^p \in \mathcal{L}^1(\mu)$.

9. Let $\mathfrak{F}$ be the set of finite subsets of a set X. Take $g \in \mathbb{R}_+^X$ and $p \in [1, \infty[$. Prove the following for $\mu := \mu^g$:

(a)
$$\mathcal{L}^p(\mu) = \left\{ f \in \mathbb{R}^X \;\middle|\; \sum_{x \in X} |f(x)|^p g(x) < \infty \right\},$$

$$N_p(f) = \left(\sum_{x \in X} |f(x)|^p g(x) \right)^{1/p} \qquad \text{for } f \in \mathcal{L}^p(\mu).$$

(b) Suppose $q > p$. Then $\mathcal{L}^p(\mu) \subset \mathcal{L}^q(\mu)$ if and only if $\inf_{x \in \{g > 0\}} g(x) > 0$.

10. Take $f \in \overline{\mathbb{R}}^X$ and define $I := \{p \in [1, \infty[\,|\, f \in \mathcal{L}^p(\mu)\}$. Prove the following:

(a) I is an interval.

(b) The map $I \to \mathbb{R}$, $p \mapsto N_p(f)$ is continuous.

(c) If q is an endpoint of I with $q \notin I$, then
$$\lim_{\substack{p \to q \\ p \in I}} N_p(f) = \infty.$$

11. Take $p, q, r \in [1, \infty[$ with $\frac{1}{p} + \frac{1}{q} = \frac{1}{r}$ and $f \in \mathcal{L}^p(\mu)$, $g \in \mathcal{L}^q(\mu)$. Show that $fg \in \mathcal{L}^r(\mu)$ and $N_r(fg) \leq N_p(f)N_q(g)$.

12. Define
$$\mathcal{L}^\infty(\mu) := \{ f \in \mathcal{M}(\mu) \,|\, \exists \alpha \in \mathbb{R}_+, \{|f| \geq \alpha\} \in \mathfrak{N}(\mu) \},$$
$$N_\infty(f) := \inf \{ \alpha \,|\, \{|f| \geq \alpha\} \in \mathfrak{N}(\mu) \} \qquad \text{for } f \in \mathcal{L}^\infty(\mu)$$

and $\frac{1}{\infty} := 0$.

(a) Show that $|f| \leq N_\infty(f)$ μ-a.e. for $f \in \mathcal{L}^\infty(\mu)$.

(b) Prove that Theorems 6.4–6.7, 6.8(a),(b) and 6.9 and Corollary 6.14 also hold for $p, q \in [1, \infty]$. (Define d_∞ analogously to d_p.)

(c) Show that, in general, $\mathcal{L}(\mathfrak{R})$ is not dense in $\mathcal{L}^\infty(\mu)$ with respect to d_∞.

(d) Solve Exercises 4, 5, 7–11 for $p, q \in [1, \infty]$ (How must 5 and 9(a) be modified?).

(e) Prove that $fg \in \mathcal{L}^\infty(\mu)$ whenever $f, g \in \mathcal{L}^\infty(\mu)$.

13. Let X be a set. Put $\mathfrak{F} := \{A \subset X \mid A \text{ is finite}\}$, $\mu := \mu^{ex}$ and

$$\ell^p(X) := \left\{ f \in \mathbb{R}^X \;\middle|\; \sup_{A \in \mathfrak{F}} \sum_{x \in A} |f(x)|^p < \infty \right\} \qquad \text{for } p \in \,]1, \infty[\,.$$

Show that $\mathcal{L}^p(\mu) = \ell^p(X)$ for every $p \in [1, \infty]$.

14. Take $f \in C(\mathbb{R})_+$. Show that there is an increasing sequence $(f_n)_{n \in \mathbb{N}}$ of $\mathcal{L}(\mathfrak{J})$-step functions such that $f_n \uparrow f$.

15. Let X be a Hausdorff space, and let $\mathfrak{R} \supset \mathfrak{R}(X)$, $\mathfrak{R} \subset \mathfrak{P}(X)$ be a ring of sets. Show that for each $f \in \mathcal{K}(X)$ there is a sequence of $\mathfrak{R}$-step functions converging to f everywhere.

16. Put $X := \mathbb{R}$ and $\mathfrak{R} := \{\emptyset, \{1\}\}$. Define the positive measure μ on $\mathfrak{R}$ by $\mu(\emptyset) := 0$, $\mu(\{1\}) := 1$. Show that $e_X \in \mathcal{L}^1(\mu)$, but that there is no sequence of $\mathfrak{R}$-step functions converging to e_X everywhere.

17. Let λ be Lebesgue measure on $\mathfrak{J}$ and take a λ-integrable set A which is not a Borel set (see Exercise 8 of Section 4.2). Show that e_A is not the limit everywhere of a sequence of $\mathfrak{J}$-step functions.

    Hint: Following the pattern in Theorem 4.35, show that if $(f_n)_{n \in \mathbb{N}}$ is a pointwise convergent sequence of Borel measurable functions, then $\lim_{n \to \infty} f_n$ is Borel measurable. (A function $f \in \overline{\mathbb{R}}^X$ on the topological space X is called **Borel measurable** if $\{f < \alpha\} \in \mathfrak{B}(X)$ for every $\alpha \in \mathbb{R}$.)

18. Let δ_x denote the Dirac measure at the point $x \in X$ (cf. Exercise 5 of Section 2.4), and let $p \in [1, \infty]$.

    (a) Describe $\mathcal{L}^p(\delta_x)$.
    (b) Take $f \in \mathcal{L}^p(\delta_x)$ and $f_n \in \mathcal{L}^p(\delta_x)$ $(n \in \mathbb{N})$. Prove that

$$f = \lim_{n \to \infty} f_n \text{ with respect to } d_p \iff f(x) = \lim_{n \to \infty} f_n(x).$$

6.2 Uniform integrability

Let $p \geq 1$ be a real number.

We introduce the following convention: if $f, g \in \overline{\mathbb{R}}^X$ are finite μ-a.e., then we write $f \overset{\bullet}{-} g$ for a function $h \in \overline{\mathbb{R}}^X$ satisfying $h(x) = f(x) - g(x)$ for every $x \in X$ for which the difference is defined. Thus h is uniquely determined only up to a μ-null set, but since we use this new convention exclusively in connection with integrals with respect to μ, the ambiguity has no material consequences.

We recall the Lebesgue convergence theorem for $\mathcal{L}^1(\mu)$: if $(f_n)_{n \in \mathbb{N}}$ is a sequence in $\mathcal{L}^1(\mu)$ which converges pointwise μ-a.e. to $f \in \overline{\mathbb{R}}^X$, and if an additional condition is satisfied (namely $|f_n| \leq g$ μ-a.e. for some $g \in \mathcal{L}^1(\mu)$), then

$$f \in \mathcal{L}^1(\mu) \quad \text{and} \quad \lim_{n \to \infty} \int |f_n \overset{\bullet}{-} f| \, d\mu = 0.$$

(Note that in our original formulation we did not put the modulus under the integral. At first glance, the present version seems to be slightly stronger. But it is easily seen that both formulations are equivalent in the sense that one formulation immediately implies the other. We leave the verification to the reader.) We saw in Chapter 3 that the additional condition is sufficient but not necessary for Lebesgue's convergence theorem. It is one of the goals of this section to provide conditions which are sufficient *and* necessary. We do this in the context of the $\mathcal{L}^p$-spaces. We begin with some preparatory results, interesting in their own right.

The reader is invited to sketch in a single figure the functions and sets appearing in our first result. This will illustrate the naturality of the assertions of the theorem.

Theorem 6.18 *Take $f \in \mathcal{L}^p(\mu)$ and $\varepsilon > 0$. Then there are an $A \in \mathfrak{R}$ and an $\alpha > 1$ such that*

$$N_p(fe_{\{|f|>\alpha e_A\}}) < \varepsilon. \tag{1}$$

In particular, since $|f|e_{X\setminus A} \leq |f|e_{\{|f|>\alpha e_A\}}$ and $|f|e_{\{|f|>\alpha\}} \leq |f|e_{\{|f|>\alpha e_A\}}$,

$$N_p(fe_{X\setminus A}) < \varepsilon \quad and \quad N_p(fe_{\{|f|>\alpha\}}) < \varepsilon.$$

Proof. Since $|f|^p \in \mathcal{L}^1(\mu)$, we can use Corollary 4.28 to find an $A \in \mathfrak{R}$ with

$$\int_A |f|^p d\mu > \int |f|^p d\mu - (\varepsilon/2)^p,$$

which implies that

$$N_p(fe_{X\setminus A}) = \left(\int |f|^p d\mu - \int_A |f|^p d\mu\right)^{1/p} < \left((\varepsilon/2)^p\right)^{1/p} = \varepsilon/2.$$

On the other hand,

$$|f|^p e_{\{|f|>n\}} \downarrow \infty e_{\{|f|=\infty\}}.$$

But since $\{|f| = \infty\} \in \mathfrak{N}(\mu)$ (Theorem 3.30(a)), the monotone convergence theorem implies that

$$\inf_{n\in\mathbb{N}} \int |f|^p e_{\{|f|>n\}} d\mu = \int \infty e_{\{|f|=\infty\}} d\mu = 0.$$

Hence there is an $\alpha \in \mathbb{N}$, $\alpha > 1$, such that

$$\int |f|^p e_{\{|f|>\alpha\}} d\mu < (\varepsilon/2)^p,$$

i.e.

$$N_p(fe_{\{|f|>\alpha\}}) < \varepsilon/2.$$

Since

$$|f|e_{\{|f|>\alpha e_A\}} \leq |f|e_{X\setminus A} + |f|e_{\{|f|>\alpha\}},$$

Minkowski's inequality implies (1). $\square$

Heuristically speaking, those parts of a p-fold integrable function which lie outside a big set $A \in \mathfrak{R}$ or exceed a large number α contribute only very little to the value of the integral. In the next proposition, we consider a contrary situation: we study the behaviour of a function which is 'cut off' at a level α and outside a set A.

Proposition 6.19 *Take* $A \in \mathfrak{L}(\mu)$ *and* $\alpha \geq 0$. *For* $f \in \overline{\mathbb{R}}^X$, *put*

$$f^* := f^+ \wedge (\alpha e_A) - f^- \wedge (\alpha e_A).$$

Then:

(a) $|f - f^*| \leq |fe_{\{|f| > \alpha e_A\}}|$,
$|f^*| = |f| \wedge (\alpha e_A) \leq |f| \leq |f^*| + |fe_{\{|f| > \alpha e_A\}}|$.

(b) *If* $f \in \mathcal{M}(\mu)$, *then* $f^* \in \mathcal{L}^p(\mu) \cap \mathbb{R}^X$ *(and in particular* $f^* \in \mathcal{M}_0(\mu)$*)*.

(c) *If* $f \in \mathcal{L}^p(\mu)$, *then* $f^* \in \mathcal{L}^p(\mu)$ *and*

$$N_p\big(fe_{\{|f| \geq 2\alpha e_A\}}\big) \leq 2N_p(f - f^*).$$

(d) *If* $(f_n)_{n \in \mathbb{N}}$ *is a sequence in* $\mathcal{L}^p(\mu)$ *converging to* f μ-*a.e., then* $f^* \in \mathcal{L}^p(\mu)$, $f^* = \lim_{n \to \infty} f_n^*$ μ-*a.e. and*

$$\lim_{n \to \infty} N_p(f^* - f_n^*) = 0 \quad \text{and} \quad \lim_{n \to \infty} N_p(f_n^*) = N_p(f^*).$$

(e) *If* $f, g \in \mathcal{L}^p(\mu)$, *then* $N_p(f^* - g^*) \leq N_p(f \overset{\bullet}{-} g)$.

Proof. (a) is left to the reader.

(b) By Theorems 4.33 and 4.35, $f^* \in \mathcal{M}(\mu)$, and in view of $|f^*| \leq \alpha e_A$, we use Theorem 4.38(a) to conclude that $f^* \in \mathcal{L}^p(\mu)$.

(c) The claim follows from

$$\begin{aligned}
|f|e_{\{|f| \geq 2\alpha e_A\}} &\leq |f|e_{X \backslash A} + |f|e_{A \cap \{|f| \geq 2\alpha\}} \\
&\leq |f - f^*|e_{X \backslash A} + |f^*|e_{A \cap \{|f| \geq 2\alpha\}} + |f - f^*|e_{A \cap \{|f| \geq 2\alpha\}} \\
&\leq |f - f^*|e_{X \backslash A} + \alpha e_{A \cap \{|f| \geq 2\alpha\}} + |f - f^*| \\
&\leq |f - f^*|e_{X \backslash A} + |f - f^*|e_A + |f - f^*| \\
&= 2|f - f^*|.
\end{aligned}$$

(d) Since $f \in \mathcal{M}(\mu)$ (Theorems 4.35(c) and 4.33(e)), we infer from (b) that $f^* \in \mathcal{L}^p(\mu)$. It is obvious that $f^* = \lim_{n \to \infty} f_n^*$ μ-a.e. Hence

$$\lim_{n \to \infty} |f^* - f_n^*|^p = 0 \quad \mu\text{-a.e.},$$

and in view of

$$|f^* - f_n^*|^p \leq \big(|f^*| + |f_n^*|\big)^p \leq (2\alpha e_A)^p = 2^p \alpha^p e_A,$$

Lebesgue's convergence theorem implies that $\lim_{n \to \infty} N_p(f^* - f_n^*) = 0$. Since

$$\big|N_p(f^*) - N_p(f_n^*)\big| \leq N_p(f^* - f_n^*),$$

by Minkowski's inequality, it follows immediately that also $\lim_{n\to\infty} N_p(f_n^*)$ $= N_p(f^*)$.

(e) follows from $|f^* - g^*| \le |f \overset{\cdot}{-} g|$ μ-a.e. $\qquad\square$

We also need the following result.

Proposition 6.20 *Let* $(f_n)_{n\in\mathbb{N}}$ *be a sequence in* $\mathcal{L}^p(\mu)$ *converging to* $f \in$ $\mathcal{L}^p(\mu)$ μ-*a.e. If* $N_p(f) = \lim_{n\to\infty} N_p(f_n)$, *then*

$$N_p(fe_A) = \lim_{n\to\infty} N_p(f_n e_A) \qquad \text{for every } A \in \mathfrak{M}(\mu).$$

Proof. We have

$$\liminf_{n\to\infty} \int_A |f_n|^p d\mu \le \liminf_{n\to\infty} \int |f_n|^p d\mu = \liminf_{n\to\infty} \left(N_p(f_n) \right)^p$$
$$= \left(\lim_{n\to\infty} N_p(f_n) \right)^p = (N_p(f))^p$$

and similarly

$$\liminf_{n\to\infty} \int_{X\setminus A} |f_n|^p d\mu \le (N_p(f))^p.$$

Applying Fatou's lemma twice, we see that

$$\liminf_{n\to\infty} \int_A |f_n|^p d\mu \ge \int_A |f|^p d\mu = \int |f|^p d\mu - \int_{X\setminus A} |f|^p d\mu$$
$$\ge \int |f|^p d\mu - \liminf_{n\to\infty} \int_{X\setminus A} |f_n|^p d\mu$$
$$= \limsup_{n\to\infty} \int |f_n|^p d\mu + \limsup_{n\to\infty} \left(- \int_{X\setminus A} |f_n|^p d\mu \right)$$
$$\ge \limsup_{n\to\infty} \left(\int |f_n|^p d\mu - \int_{X\setminus A} |f_n|^p d\mu \right)$$
$$= \limsup_{n\to\infty} \int_A |f_n|^p d\mu.$$

Thus

$$\liminf_{n\to\infty} \int_A |f_n|^p d\mu = \limsup_{n\to\infty} \int_A |f_n|^p d\mu = \int_A |f|^p d\mu$$

which implies the assertion. $\qquad\square$

Subsets of $\mathcal{L}^p(\mu)$ for which the assertion of Theorem 6.18 holds uniformly are of particular interest. We call a non-empty subset $\mathcal{F}$ of $\mathcal{L}^p(\mu)$ **uniformly** p-μ-**integrable** (if $p = 1$, simply **uniformly** μ-**integrable**) if for every $\varepsilon > 0$ there are an $A \in \mathfrak{R}$ and an $\alpha > 1$ satisfying

$$\sup_{f\in\mathcal{F}} N_p\big(fe_{\{|f|>\alpha e_A\}}\big) < \varepsilon.$$

A sequence $(f_n)_{n\in\mathbb{N}}$ in $\mathcal{L}^p(\mu)$ is called **uniformly** p-μ-**integrable** if the set $\{f_n \mid n \in \mathbb{N}\}$ is uniformly p-μ-integrable.

Of course, every non-empty subset of a uniformly p-μ-integrable set is itself uniformly p-μ-integrable. It follows at once from Theorem 6.18 that every finite non-empty subset of $\mathcal{L}^p(\mu)$ is uniformly p-μ-integrable. Another immediate conclusion is that every non-empty subset $\mathcal{F}$ of $\mathcal{M}(\mu)$ satisfying $|f| \leq M e_A$ for some $A \in \mathfrak{L}(\mu)$, some $M > 0$ and each $f \in \mathcal{F}$, is uniformly p-μ-integrable. More generally, we have:

Proposition 6.21 *Let $\mathcal{F}$ be a non-empty subset of $\mathcal{L}^p(\mu)$ and suppose that there is some $g \in \mathcal{L}^p(\mu)$ such that $|f| \leq g$ μ-a.e. for every $f \in \mathcal{F}$. Then $\mathcal{F}$ is uniformly p-μ-integrable.*

Proof. Take $\varepsilon > 0$. By Theorem 6.18, there are an $A \in \mathfrak{R}$ and an $\alpha > 1$ such that $N_p\big(g e_{\{|g| > \alpha e_A\}}\big) < \varepsilon$. We have

$$\big| f e_{\{|f| > \alpha e_A\}} \big| \leq \big| g e_{\{|g| > \alpha e_A\}} \big| \quad \mu\text{-a.e.}$$

for every $f \in \mathcal{F}$, which implies that

$$\sup_{f \in \mathcal{F}} N_p\big(f e_{\{|f| > \alpha e_A\}} \big) < \varepsilon.$$

$\square$

However, putting $f_n := \frac{1}{n} e_{[n, n+1[}$ for $n \in \mathbb{N}$, we see that a uniformly Lebesgue integrable sequence need not have an integrable majorant.

The following characterization of uniformly p-μ-integrable sets is useful.

Theorem 6.22 *Let $\mathcal{F}$ be a non-empty subset of $\mathcal{L}^p(\mu)$. Then the following are equivalent.*

(a) $\mathcal{F}$ is uniformly p-μ-integrable.

(b) $\sup_{f \in \mathcal{F}} N_p(f) < \infty$ and for every $\varepsilon > 0$ there are an $A \in \mathfrak{R}$ and a $\delta > 0$ such that

$$\sup_{f \in \mathcal{F}} N_p(f e_{X \setminus A}) < \varepsilon$$

$$\sup_{f \in \mathcal{F}} N_p(f e_B) < \varepsilon \quad \text{whenever } B \in \mathfrak{L}(\mu) \text{ and } \mu^X(B) < \delta.$$

Proof. (a)$\Rightarrow$(b). Take $\varepsilon > 0$. By assumption, there are an $A \in \mathfrak{R}$ and an $\alpha > 1$ such that

$$\sup_{f \in \mathcal{F}} N_p\big(f e_{\{|f| > \alpha e_A\}} \big) < \frac{\varepsilon}{3}.$$

Hence (Theorem 6.18)

$$\sup_{f \in \mathcal{F}} N_p(f e_{X \setminus A}) < \frac{\varepsilon}{3}.$$

Put $\delta := \left(\frac{\varepsilon}{3\alpha} \right)^p$. Take $B \in \mathfrak{L}(\mu)$ with $\mu^X(B) < \delta$. Then for every $f \in \mathcal{F}$

$$N_p\big(f e_{B \cap A \cap \{|f| \leq \alpha e_A\}} \big) \leq \alpha (\mu^X(B))^{1/p} < \frac{\varepsilon}{3}.$$

Put $B_f := B \cap A \cap \{|f| \le \alpha e_A\}$ and $C_f := B \cap A \cap \{|f| > \alpha e_A\}$. Then, since $B = (B \setminus A) \cup B_f \cup C_f$,

$$\sup_{f \in \mathcal{F}} N_p(fe_B) \le \sup_{f \in \mathcal{F}} N_p(fe_{B \setminus A}) + \sup_{f \in \mathcal{F}} N_p(fe_{B_f}) + \sup_{f \in \mathcal{F}} N_p(fe_{C_f})$$

$$< \frac{\varepsilon}{3} + \frac{\varepsilon}{3} + \frac{\varepsilon}{3} = \varepsilon.$$

Finally,

$$\sup_{f \in \mathcal{F}} N_p(f) \le \sup_{f \in \mathcal{F}} N_p(fe_{\{|f| > \alpha e_A\}}) + \sup_{f \in \mathcal{F}} N_p(fe_{\{|f| \le \alpha e_A\}})$$

$$\le \frac{\varepsilon}{3} + \alpha(\mu(A))^{1/p} < \infty.$$

(b)$\Rightarrow$(a). Take $\varepsilon > 0$. Choose $A \in \mathfrak{R}$ and $\delta > 0$ such that the conditions of (b) are satisfied for $\varepsilon/2$ in place of ε. Put $M := \sup_{f \in \mathcal{F}} N_p(f) + 1$ and

$$\alpha := \sup \left\{ 2, M\left(\frac{2}{\delta}\right)^{1/p} \right\}.$$

Given $f \in \mathcal{F}$,

$$\mu^X(\{|f| > \alpha\}) = \mu^X\left(\left\{\frac{|f|^p}{\alpha^p} > 1\right\}\right) \le \int \frac{|f|^p}{\alpha^p} d\mu$$

$$= \left(\frac{N_p(f)}{\alpha}\right)^p \le \left(\frac{N_p(f)}{M}\right)^p \cdot \frac{\delta}{2} < \delta.$$

Observing that $\{|f| > \alpha e_A\} \subset (X \setminus A) \cup \{|f| > \alpha\}$, we conclude that

$$\sup_{f \in \mathcal{F}} N_p(fe_{\{|f| > \alpha e_A\}}) \le \sup_{f \in \mathcal{F}} N_p(fe_{X \setminus A}) + \sup_{f \in \mathcal{F}} N_p(fe_{\{|f| > \alpha\}}) < \frac{\varepsilon}{2} + \frac{\varepsilon}{2} = \varepsilon.$$

$\square$

Theorem 6.18 furnishes the following interesting continuity property of the integral.

Corollary 6.23 *Take $f \in \mathcal{L}^p(\mu)$. Then for each $\varepsilon > 0$ there is a $\delta > 0$ such that $N_p(fe_B) < \varepsilon$ whenever $B \in \mathfrak{L}(\mu)$ and $\mu^X(B) < \delta$.*

Note that by Proposition 6.21 every sequence satisfying the assumption of the Lebesgue convergence theorem is uniformly integrable. Therefore our following main theorem is in fact a substantial generalization of the Lebesgue convergence theorem.

Theorem 6.24 (Generalized Lebesgue Convergence Theorem)
Take $f \in \overline{\mathbb{R}}^X$ and let $(f_n)_{n \in \mathbb{N}}$ be a sequence in $\mathcal{L}^p(\mu)$ such that

$$f = \lim_{n \to \infty} f_n \quad \mu\text{-a.e.}$$

Then the following are equivalent.

(a) $(f_n)_{n \in \mathbb{N}}$ is uniformly p-μ-integrable.

(b) $f \in \mathcal{L}^p(\mu)$ and $N_p(f) = \lim_{n\to\infty} N_p(f_n)$.

(c) $f \in \mathcal{L}^p(\mu)$ and $\lim_{n\to\infty} N_p(f_n \overset{\cdot}{-} f) = 0$.

The implication (c)$\Rightarrow$(a) holds even without the hypothesis $f = \lim_{n\to\infty} f_n$ μ-a.e.

In the case $p = 1$, the equivalent conditions (a)–(c) imply

$$\int f \, d\mu = \lim_{n\to\infty} \int f_n d\mu.$$

Proof. The last conclusion for $p = 1$ follows from (c) and the fact that

$$\left| \int f_n d\mu - \int f \, d\mu \right| \le \int |f_n \overset{\cdot}{-} f| \, d\mu.$$

(a)$\Rightarrow$(b). Take $\varepsilon > 0$. By (a) there are an $A \in \mathfrak{R}$ and an $\alpha > 1$ such that

$$\sup_{n\in\mathbb{N}} N_p\big(f_n e_{\{|f_n|>\alpha e_A\}}\big) < \varepsilon.$$

Given $g \in \overline{\mathbb{R}}^X$, put

$$g^* := g^+ \wedge (\alpha e_A) - g^- \wedge (\alpha e_A).$$

Then, by Proposition 6.19(a),

$$|f_n^*| \le \alpha e_A \quad \text{and} \quad |f_n| \le |f_n^*| + \big|f_n e_{\{|f_n|>\alpha e_A\}}\big|$$

for every $n \in \mathbb{N}$ and thus

$$N_p(f_n) \le N_p(f_n^*) + \varepsilon \le \alpha(\mu(A))^{1/p} + \varepsilon. \tag{2}$$

Using Fatou's lemma we conclude that $\liminf_{n\to\infty} |f_n|^p \in \mathcal{L}^1(\mu)$ (hence $f \in \mathcal{L}^p(\mu)$, by Corollary 6.13 and Theorems 4.35(c) and 4.33(e)) and that

$$N_p(f) = \left(\int \Big(\liminf_{n\to\infty} |f_n|^p \Big) d\mu \right)^{1/p} \le \liminf_{n\to\infty} N_p(f_n).$$

Moreover, using first (2) and then Proposition 6.19(d),

$$\limsup_{n\to\infty} N_p(f_n) \le \limsup_{n\to\infty} N_p(f_n^*) + \varepsilon = N_p(f^*) + \varepsilon \le N_p(f) + \varepsilon.$$

Since ε is arbitrary,

$$\limsup_{n\to\infty} N_p(f_n) \le N_p(f) \le \liminf_{n\to\infty} N_p(f_n),$$

which proves the claim.

(b)$\Rightarrow$(c). Take $\varepsilon > 0$. By Theorem 6.18 and Corollary 6.23, there are an $A \in \mathfrak{R}$ and a $\delta > 0$ such that $N_p(f e_{X\backslash A}) < \varepsilon$ and $N_p(f e_B) < \varepsilon$ whenever $B \in \mathfrak{L}(\mu)$ with $\mu^X(B) < \delta$. Egoroff's theorem now enables us to find a $C \in \mathfrak{L}(\mu)$ contained in A and satisfying $\mu^X(A \backslash C) < \delta$ such that $f = \lim_{n\to\infty} f_n$ uniformly on C. This means that we can find an $n_0 \in \mathbb{N}$ such that

$$|f_n(x) - f(x)| < \frac{\varepsilon}{(\mu^X(C) + 1)^{1/p}}$$

for every $n > n_0$ and every $x \in C$. Thus given $n > n_0$,

$$N_p(f_n e_C - f e_C) = \left(\int_C |f_n \overset{\cdot}{-} f|^p d\mu \right)^{1/p} < \varepsilon.$$

Using this result and Proposition 6.20 together with assumption (b), we obtain

$$\limsup_{n\to\infty} N_p(f_n \overset{\cdot}{-} f)$$

$$\leq \limsup_{n\to\infty} \left(N_p((f_n \overset{\cdot}{-} f)e_{X\setminus A}) + N_p((f_n \overset{\cdot}{-} f)e_{A\setminus C}) + N_p((f_n \overset{\cdot}{-} f)e_C) \right)$$

$$\leq \limsup_{n\to\infty} \left(N_p(f_n e_{X\setminus A}) + N_p(f e_{X\setminus A}) + N_p(f_n e_{A\setminus C}) + N_p(f e_{A\setminus C}) \right) + \varepsilon$$

$$= 2N_p(f e_{X\setminus A}) + 2N_p(f e_{A\setminus C}) + \varepsilon < 5\varepsilon.$$

Hence $\limsup_{n\to\infty} N_p(f_n \overset{\cdot}{-} f) = 0$ and thus $\lim_{n\to\infty} N_p(f_n \overset{\cdot}{-} f) = 0$.

(c)$\Rightarrow$(a). Take $\varepsilon > 0$. By Theorem 6.18, we can find a $B \in \mathfrak{R}$ and a $\beta > 1$ such that

$$N_p(f e_{\{|f|>\beta e_B\}}) < \frac{\varepsilon}{4}.$$

Given $g \in \overline{\mathbb{R}}^X$, put

$$g^* := g^+ \wedge (\beta e_B) - g^- \wedge (\beta e_B).$$

Then, by Proposition 6.19, f^* and each f_n^* belong to $\mathcal{L}^p(\mu)$ and $|f - f^*| \leq |f e_{\{|f|>\beta e_B\}}|$, whence $N_p(f - f^*) < \varepsilon/4$. Thus for every $n \in \mathbb{N}$

$$N_p(f_n - f_n^*) \leq N_p(f_n \overset{\cdot}{-} f) + N_p(f - f^*) + N_p(f^* \overset{\cdot}{-} f_n^*)$$

$$\leq N_p(f_n \overset{\cdot}{-} f) + \frac{\varepsilon}{4} + N_p(f \overset{\cdot}{-} f_n)$$

(Proposition 6.19(e)). From (c) we conclude that

$$\limsup_{n\to\infty} N_p(f_n - f_n^*) \leq \frac{\varepsilon}{4}.$$

Hence there is an $n_0 \in \mathbb{N}$ with

$$\sup_{n>n_0} N_p(f_n - f_n^*) < \frac{\varepsilon}{3}.$$

Thus, applying Proposition 6.19(c), given $n > n_0$,

$$N_p(f_n e_{\{|f_n| \geq 2\beta e_B\}}) \leq 2N_p(f_n - f_n^*) < \frac{2\varepsilon}{3}.$$

Now take $n \leq n_0$. By Theorem 6.18, there are an $A_n \in \mathfrak{R}$ and an $\alpha_n > 1$ such that

$$N_p(f_n e_{\{|f_n|>\alpha_n e_{A_n}\}}) < \varepsilon.$$

Then, putting $A := B \cup \left(\bigcup_{n \leq n_0} A_n \right)$ and $\alpha := 2\beta + \sum_{n \leq n_0} \alpha_n$, we see that

$$\sup_{n \in \mathbb{N}} N_p(f_n e_{\{|f_n|>\alpha e_A\}}) < \varepsilon.$$

Note that we did not use the hypothesis $f = \lim_{n\to\infty} f_n$ μ-a.e. in the proof of (c)$\Rightarrow$(a). $\qquad\square$

Of course the statement that $\lim_{n\to\infty} N_p(f_n \overset{\bullet}{-} f) = 0$ in (c) is equivalent to the statement that $(f_n)_{n\in\mathbb{N}}$ converges to f with respect to d_p.

Note that the condition $N_p(f) = \lim_{n\to\infty} N_p(f_n)$ in (b) is considerably weaker than the condition $\lim_{n\to\infty} N_p(f_n \overset{\bullet}{-} f) = 0$ in (c). However, if it is already known that $(f_n)_{n\in\mathbb{N}}$ converges pointwise μ-a.e. to $f \in \mathcal{L}^p(\mu)$, then the weaker condition implies the stronger one. Another result of this type is contained in the following corollary, which is a partial converse to Corollary 6.10.

Corollary 6.25 *Let f and f_n $(n \in \mathbb{N})$ be elements of $\mathcal{L}^p(\mu)$ and suppose that $f_n \overset{\mu}{\to} f$ and $N_p(f) = \lim_{n\to\infty} N_p(f_n)$. Then*

$$\lim_{n\to\infty} N_p(f_n \overset{\bullet}{-} f) = 0.$$

Proof. Assume that there is some $\varepsilon > 0$ such that for every $n_0 \in \mathbb{N}$ there is an $n \geq n_0$ with $N_p(f_n \overset{\bullet}{-} f) \geq \varepsilon$. Then we can construct a subsequence $(f_{n_k})_{k\in\mathbb{N}}$ satisfying

$$N_p(f_{n_k} \overset{\bullet}{-} f) \geq \varepsilon \qquad \text{for every } k \in \mathbb{N}. \tag{3}$$

Since $f_{n_k} \overset{\mu}{\to} f$, the sequence $(f_{n_k})_{k\in\mathbb{N}}$ contains a subsequence $(f_{n_{k(m)}})_{m\in\mathbb{N}}$ converging pointwise μ-a.e. to f (Corollary 4.45). Theorem 6.24 (b)$\Rightarrow$(c) now implies that

$$\lim_{m\to\infty} N_p(f_{n_{k(m)}} \overset{\bullet}{-} f) = 0,$$

which contradicts (3). $\qquad\square$

Another very useful result follows, giving necessary and sufficient conditions for the statement that $f \in \mathcal{L}^p(\mu)$ and $f = \lim_{n\to\infty} f_n$ with respect to d_p. Note that we do not require $(f_n)_{n\in\mathbb{N}}$ to converge pointwise μ-a.e.

Theorem 6.26 *Given a μ-measurable function $f \in \overline{\mathbb{R}}^X$ and a sequence $(f_n)_{n\in\mathbb{N}}$ in $\mathcal{L}^p(\mu)$, the following are equivalent.*

(a) $(f_n)_{n\in\mathbb{N}}$ is uniformly p-μ-integrable and $f_n \overset{\mu}{\to} f$.

(b) $f \in \mathcal{L}^p(\mu)$ and $\lim_{n\to\infty} N_p(f_n \overset{\bullet}{-} f) = 0$.

(c) Each subsequence of $(f_n)_{n\in\mathbb{N}}$ has a subsequence which is bounded in $\mathcal{L}^p(\mu)$ and converges to f μ-a.e.

Proof. (a)$\Rightarrow$(b). By Corollary 4.45, $(f_n)_{n\in\mathbb{N}}$ has a subsequence converging to f μ-a.e. Applying Theorem 6.24 (a)$\Rightarrow$(b), we see that $f \in \mathcal{L}^p(\mu)$. Now assume that there is an $\varepsilon > 0$ such that for every $n_0 \in \mathbb{N}$ we can find an $n \geq n_0$ with $N_p(f_n \overset{\bullet}{-} f) \geq \varepsilon$. Then, using Corollary 4.45 again, we can construct a subsequence $(f_{n_k})_{k\in\mathbb{N}}$ such that

$$f = \lim_{k\to\infty} f_{n_k} \ \mu\text{-a.e.} \quad \text{and} \quad N_p(f_{n_k} \overset{\bullet}{-} f) \geq \varepsilon \text{ for every } k \in \mathbb{N}.$$

This obviously contradicts Theorem 6.24 (a)⇒(c). Hence

$$\lim_{n\to\infty} N_p(f_n \overset{\bullet}{-} f) = 0.$$

(b)⇒(c) follows from Theorem 6.9.

(c)⇒(a). By the assumption of boundedness, we can apply Theorem 4.38(a) to conclude that $|f|^p \in \mathcal{L}^1(\mu)$ and hence that $f \in \mathcal{L}^p(\mu)$ (Corollary 6.13). Then by Theorem 4.46, $f_n \overset{\mu}{\to} f$. Assume that there are an $\varepsilon > 0$ and a subsequence $(f_{n_k})_{k\in\mathbb{N}}$ satisfying

$$N_p(f_{n_k} \overset{\bullet}{-} f) \geq \varepsilon \qquad \text{for every } k \in \mathbb{N}. \tag{4}$$

By (c), we may suppose that $(f_{n_k})_{k\in\mathbb{N}}$ is bounded in $\mathcal{L}^p(\mu)$ and converges to f μ-a.e. Proposition 6.21 implies that $(f_{n_k})_{k\in\mathbb{N}}$ is uniformly p-μ-integrable. But then Theorem 6.24 (a)⇒(c) shows that (4) is impossible. Hence $\lim_{n\to\infty} N_p(f_n \overset{\bullet}{-} f) = 0$. Theorem 6.24 (c)⇒(a) now proves that $(f_n)_{n\in\mathbb{N}}$ is uniformly p-μ-integrable. □

Of course the boundedness condition is the crux of (c).

Finally we remark that the two conditions appearing in (a), namely the uniform p-μ-integrability and the convergence in measure μ of a sequence, are independent properties. Let λ be Lebesgue measure on $\mathbb{R}$. Then the sequence $(ne_{[n,n+1[})_{n\in\mathbb{N}}$ converges in measure λ to 0 but it is not uniformly λ-integrable. On the other hand, if $g_{2m} := 0$ and $g_{2m-1} := e_{[0,1]}$ for every $m \in \mathbb{N}$, then the sequence $(g_n)_{n\in\mathbb{N}}$ is uniformly λ-integrable but does not converge in measure λ.

Exercises

1. Decide which of the following sequences $(f_n)_{n\in\mathbb{N}}$ are uniformly p-λ-integrable with respect to Lebesgue measure λ on $\mathbb{R}$ ($p = 1, 2$):

   (a) $f_{2n} = e_{[0,\frac{1}{n}]}, \ f_{2n-1} = e_{[n-\frac{1}{n},n]}$

   (b) $f_n(t) = \begin{cases} t^{\frac{1-n}{n}} & \text{if } 0 < t < 1, \\ 0 & \text{otherwise.} \end{cases}$

2. Let $\mu := \mu^{e_{||}}$ be counting measure on $\mathbb{N}$. Decide whether the sequences $(\frac{1}{n}e_{\{n\}})_{n\in\mathbb{N}}$ and $(\frac{1}{n}e_{\{1,\ldots,n\}})_{n\in\mathbb{N}}$ are uniformly p-μ-integrable ($p = 1, 2$).

3. Determine when the sets $\mathcal{L}(\mathfrak{R})$ and $\{e_A \mid A \in \mathfrak{L}(\mu)\}$ are uniformly p-μ-integrable.

4. Let λ be Lebesgue measure on $\mathbb{R}$ and for $n \in \mathbb{N}$ put

$$f_n := \sum_{k=0}^{2^n-1} (-1)^k e_{[k2^{-n}, (k+1)2^{-n}[}.$$

   Show that $(f_n)_{n\in\mathbb{N}}$ is uniformly λ-integrable, but no subsequence of $(f_n)_{n\in\mathbb{N}}$ converges pointwise λ-a.e.

5. Let $\mathcal{F}$ be a non-empty subset of $\mathcal{L}^p(\mu)$. Show that $\mathcal{F}$ is uniformly p-μ-integrable if and only if for every $\varepsilon > 0$ there is a $g \in \mathcal{L}^p(\mu)$ such that
$$\sup_{f \in \mathcal{F}} N_p(f e_{\{|f| > |g|\}}) < \varepsilon.$$
   Hint: $\{|f| > \alpha e_A\} \subset \{|f| > |g|\} \cup \{\alpha e_A < |f| \le |g|\}$.

6. Let $(f_n)_{n \in \mathbb{N}}$ be a sequence in $\mathcal{L}^p(\mu)$ converging μ-a.e. to $f \in \mathcal{L}^0(\mu)$ and suppose that the two 'ε-conditions' of Theorem 6.22(b) are satisfied. Prove that $(f_n)_{n \in \mathbb{N}}$ is uniformly p-μ-integrable.
   Hint: Take $\varepsilon > 0$, and choose $A \in \mathfrak{R}$ and $\delta > 0$ as in the hypothesis. Show that there is an $m \in \mathbb{N}$ with $\mu^X(A \cap \{|f| > m\}) < \delta$ and put $B := A \cap \{|f| > m\}$. Thus $\sup_{n \in \mathbb{N}} N_p(f_n e_B)$ is finite and the same is true of $\sup_{n \in \mathbb{N}} N_p(f_n e_{X \setminus A})$. To show that $\sup_{n \in \mathbb{N}} N_p(f_n e_{A \setminus B}) < \infty$, use Egoroff's theorem. It follows that $\sup_{n \in \mathbb{N}} N_p(f_n) < \infty$.

7. Let δ_x denote Dirac measure at the point $x \in X$. Find a sequence $(f_n)_{n \in \mathbb{N}}$ of δ_x-integrable functions converging pointwise to some $f \in \overline{\mathbb{R}}^X$ and satisfying the two 'ε-conditions' of Theorem 6.22(b), but satisfying neither $\sup_{n \in \mathbb{N}} N_1(f_n) < \infty$ nor $f \in \mathcal{L}^0(\delta_x)$. (Hence $(f_n)_{n \in \mathbb{N}}$ is not uniformly δ_x-integrable.)

8. Let μ be bounded and $\mathcal{F}$ a non-empty subset of $\mathcal{L}^p(\mu)$. Suppose that for every $\varepsilon > 0$ there is a $\delta > 0$ such that $\sup_{f \in \mathcal{F}} N_p(f e_B) < \varepsilon$ whenever $B \in \mathfrak{R}$ and $\mu(B) < \delta$. Prove that for every $\varepsilon > 0$ there is an $A \in \mathfrak{R}$ with $\sup_{f \in \mathcal{F}} N_p(f e_{X \setminus A}) < \varepsilon$. Is the assumption that μ is bounded essential? (Consider, for example, counting measure on $\mathbb{N}$.)
   Hint: Assume that there is an $\varepsilon > 0$ such that $\sup_{f \in \mathcal{F}} N_p(f e_{X \setminus A}) > \varepsilon$ for every $A \in \mathfrak{R}$. Choose $\delta > 0$ such that $N_p(f e_B) < \varepsilon/2$ whenever $f \in \mathcal{F}$ and $B \in \mathfrak{R}$ with $\mu(B) < \delta$. Use recursion to construct a disjoint sequence $(A_n)_{n \in \mathbb{N}}$ in $\mathfrak{R}$ with $\mu(A_n) \ge \delta$ for every $n \in \mathbb{N}$.

9. Let μ be bounded and take $p > 1$. Take a non-empty subset $\mathcal{F}$ of $\mathcal{L}^p(\mu)$ with $\sup_{f \in \mathcal{F}} N_p(f) < \infty$. Show that $\mathcal{F} \subset \mathcal{L}^1(\mu)$ and that $\mathcal{F}$ is uniformly μ-integrable. Is the assertion also true for $p = 1$?
   Hint: To show that $\sup_{f \in \mathcal{F}} N_1(f) < \infty$, note that
$$X = \{|f| \le 1\} \cup \{|f| > 1\}.$$
   To verify the condition in Exercise 8, use Hölder's inequality.

10. Suppose that $(X, \mathfrak{R}, \mu)$ is σ-finite. Let $\mathcal{F}$ be a non-empty subset of $\mathcal{L}^p(\mu)$ such that every sequence in $\mathcal{F}$ is uniformly p-μ-integrable. Show that $\mathcal{F}$ itself is uniformly p-μ-integrable.

7

Vector lattices, L^p-spaces

7.1 Vector lattices

This section is devoted to a brief discussion of the structure of vector lattices, the abstract structure which is the cornerstone of integration theory. We have already made use of important vector lattices, namely vector lattices of real functions. They provide excellent examples to illustrate the following discussion of abstract vector lattices.

Let E be a real vector space. E is called an **ordered vector space** if there is an ordering $\leq$ defined on E such that

(i) $x + z \leq y + z$ for all $x, y, z \in E$ with $x \leq y$.

(ii) $\alpha x \leq \alpha y$ for all $x, y \in E$ with $x \leq y$ and every $\alpha \in \mathbb{R}_+$.

For the ordered vector space E, put

$$E_+ := \{x \in E \mid x \geq 0\}.$$

The elements of E_+ are called **positive**.

Proposition 7.1 *Let E be an ordered vector space. Then:*

(a) $-x \leq 0$ whenever $x \geq 0$.

(b) $-y \leq -x$ for all $x, y \in E$ with $x \leq y$.

(c) Given a family $(x_\iota)_{\iota \in I}$ in E, $x \in E$ and $\alpha \in \mathbb{R}_+ \setminus \{0\}$:

(c1) $\bigvee_{\iota \in I} x_\iota$ exists if and only if $\bigvee_{\iota \in I}(x + x_\iota)$ exists, and in this case

$$\bigvee_{\iota \in I}(x + x_\iota) = x + \bigvee_{\iota \in I} x_\iota.$$

The dual statement holds for infima.

(c2) $\bigvee_{\iota \in I} x_\iota$ exists if and only if $\bigvee_{\iota \in I}(\alpha x_\iota)$ exists, and in this case

$$\bigvee_{\iota \in I}(\alpha x_\iota) = \alpha \bigvee_{\iota \in I} x_\iota.$$

The dual statement holds for infima.

(c3) $\bigvee_{\iota \in I} x_\iota$ *exists if and only if* $\bigwedge_{\iota \in I}(-x_\iota)$ *exists, and in this case*

$$\bigvee_{\iota \in I} x_\iota = - \bigwedge_{\iota \in I}(-x_\iota).$$

Here, too, the dual statement holds.

Proof. (a) follows immediately from (i) by adding $(-x)$ to both sides of the inequality $x \geq 0$.

(b) If $x \leq y$, then by (i) $y - x \geq 0$, and hence by (a)

$$(-y) - (-x) = -(y - x) \leq 0.$$

Thus $-y \leq -x$.

(c1) Assume that $\bigvee_{\iota \in I} x_\iota$ exists. Then $x + x_\iota \leq x + \bigvee_{\iota \in I} x_\iota$ for every $\iota \in I$. If z is an upper bound for $(x + x_\iota)_{\iota \in I}$, then $x + x_\iota \leq z$ and therefore $x_\iota \leq z - x$ for every $\iota \in I$. It follows that $\bigvee_{\iota \in I} x_\iota \leq z - x$ and hence $x + \bigvee_{\iota \in I} x_\iota \leq z$. Thus $\bigvee_{\iota \in I}(x + x_\iota)$ exists and

$$\bigvee_{\iota \in I}(x + x_\iota) = x + \bigvee_{\iota \in I} x_\iota.$$

Conversely, if $\bigvee_{\iota \in I}(x + x_\iota)$ exists, then by what we have just shown,

$$\bigvee_{\iota \in I} x_\iota = \bigvee_{\iota \in I}\left((x + x_\iota) - x\right)$$

exists.

(c2) is left to the reader.

(c3) follows from the fact that z is an upper bound for $(x_\iota)_{\iota \in I}$ if and only if $-z$ is a lower bound for $(-x_\iota)_{\iota \in I}$ (cf. (b)).   □

An ordered vector space E is called a **vector lattice** if E is a lattice with respect to the ordering $\leq$. If E is a vector lattice and $x \in E$, then we define

$$x^+ := x \vee 0, \quad x^- := (-x) \vee 0, \quad \text{and} \quad |x| := x \vee (-x).$$

We call x^+, x^- and $|x|$ the **positive part**, the **negative part** and the **absolute value** of x, respectively. Note that the negative part x^- is a positive element of E. While this might seem strange, it means that the structure of a vector lattice is fully determined by its positive elements – see Theorem 7.2(b) and Proposition 7.4.

Many well-known results for functions correspond to formulae valid in vector lattices. This richness is one of the reasons for the fruitfulness of the theory of vector lattices. We list a few of these formulae in the following theorem. Note that, for example, (b) says that each element of a vector lattice is the difference between two positive elements, namely between its positive and its negative part. Moreover, (d) asserts that these are ortho-gonal. (Orthogonality is discussed below.) In addition, the absolute value of x is, by (e), just the sum of these elements.

Theorem 7.2 *Let x, y, z be elements of the vector lattice E. Then:*

(a) $x + y = x \vee y + x \wedge y$.

(b) $x = x^+ - x^-$.

(c) *If $x = y - z$ with $y, z \in E_+$, then $x^+ = y - y \wedge z$, $x^- = z - y \wedge z$ and $x^+ \leq y$ and $x^- \leq z$.*

(d) $x^+ \wedge x^- = 0$.

(e) $|x| = x^+ + x^-$.

(f) $x \vee y = \frac{1}{2}(x + y + |x - y|)$,
$\quad x \wedge y = \frac{1}{2}(x + y - |x - y|)$

(g) $|x + y| \leq |x| + |y|$, $\quad (x + y)^+ \leq x^+ + y^+$, $\quad (x + y)^- \leq x^- + y^-$.

(h) *If $x, y, z \in E_+$, then $x \wedge (y + z) \leq x \wedge y + x \wedge z$.*

(i) $\big| |x| - |y| \big| \leq |x - y|$.

(j) *If $\alpha \in \mathbb{R}$, then $|\alpha x| = |\alpha| |x|$.*

Proof. (a) By Proposition 7.1(c1),(c3),

$$x \vee y - x = 0 \vee (y - x) = -(0 \wedge (x - y)) = -(x \wedge y) + y.$$

(b) It follows from (a) and Proposition 7.1(c3) that

$$x = x + 0 = x \vee 0 + x \wedge 0 = x^+ - ((-x) \vee 0) = x^+ - x^-.$$

(c) By (b) and Proposition 7.1(c1),(c3),

$$y - y \wedge z = x + z - y \wedge z = x + z + (-y) \vee (-z)$$
$$= x + (z - y) \vee 0 = x + (-x) \vee 0 = x^+ - x^- + x^- = x^+.$$

Dually $z - y \wedge z = x^-$. It follows immediately from these formulae that $x^+ \leq y$ and $x^- \leq z$, since $y \wedge z \geq 0$.

(d) By (b), there are $y, z \in E_+$ such that $x = y - z$. Using (c), it follows that

$$x^+ \wedge x^- = (y - y \wedge z) \wedge (z - y \wedge z) = y \wedge z - y \wedge z = 0.$$

(e) results from

$$|x| = x \vee (-x) = (x + x) \vee 0 - x = (2x) \vee 0 - x$$
$$= 2(x \vee 0) - x = 2x^+ - (x^+ - x^-) = x^+ + x^-.$$

(f) By (a) and (e),

$$\frac{1}{2}(x + y + |x - y|)$$

$$= \frac{1}{2}(x \vee y + x \wedge y + (x - y) \vee 0 + (y - x) \vee 0)$$

$$= \frac{1}{2}(x \vee y + x \wedge y + (x \vee y - y) + ((-x) \vee (-y) + y))$$

$$= \frac{1}{2}(x \vee y + x \wedge y + x \vee y - x \wedge y)$$

$$= x \vee y.$$

The second relation follows similarly.

(g) Since $x + y = (x^+ + y^+) - (x^- + y^-)$, it follows by (c) that

$$(x + y)^+ \leq x^+ + y^+ \quad \text{and} \quad (x + y)^- \leq x^- + y^-.$$

Therefore

$$|x + y| = (x + y)^+ + (x + y)^- \leq x^+ + y^+ + x^- + y^- = |x| + |y|.$$

(h) First,

$$x \wedge (y + z) - x \wedge y \leq x - x \wedge y \leq x.$$

Then since

$$x \wedge (y + z) - x \wedge y = z + (x - z) \wedge y - x \wedge y \leq z,$$

we have that

$$x \wedge (y + z) - x \wedge y \leq x \wedge z.$$

(i) By (g), $|x| \leq |x - y| + |y|$ and $|y| \leq |y - x| + |x|$. Hence

$$\big||x| - |y|\big| = (|x| - |y|) \vee (|y| - |x|) \leq |x - y|.$$

(j) For $\alpha \geq 0$, the claim follows from Proposition 7.1(c2). $|-x| = |x|$ is immediate from the definition. For $\alpha < 0$,

$$|\alpha x| = \big||\alpha|(-x)\big| = |\alpha| \, |-x| = |\alpha| \, |x|.$$

$\square$

A particularly important property of vector lattices is the **distributivity** presented in the following theorem. General lattices need not be distributive, which is somewhat surprising since the formulation of distributivity uses only order-theoretic concepts. However, as one might now guess, the proof of distributivity in vector lattices uses their algebraic properties.

Theorem 7.3 *Let $(x_\iota)_{\iota \in I}$ be a family in the vector lattice E. If $\bigvee_{\iota \in I} x_\iota$ exists, then $\bigvee_{\iota \in I}(x_\iota \wedge x)$ also exists for every $x \in E$ and*

$$\bigvee_{\iota \in I}(x_\iota \wedge x) = \left(\bigvee_{\iota \in I} x_\iota\right) \wedge x.$$

The dual statement also holds.

Proof. Put $y := \bigvee_{\iota \in I} x_\iota$. Given $\iota \in I$, $y \wedge x \geq x_\iota \wedge x$. If z is an upper bound for $(x_\iota \wedge x)_{\iota \in I}$, then for each $\iota \in I$

$$z \geq x_\iota \wedge x = x_\iota + x - x_\iota \vee x$$

and hence

$$z + x_\iota \vee x \geq x_\iota + x.$$

Therefore

$$z + y \vee x = z + \bigvee_{\iota \in I} (x_\iota \vee x) = \bigvee_{\iota \in I} (z + x_\iota \vee x) \geq \bigvee_{\iota \in I} (x_\iota + x)$$

$$= \left(\bigvee_{\iota \in I} x_\iota \right) + x = y + x = y \vee x + y \wedge x.$$

Thus $z \geq y \wedge x$, and the conclusion follows. □

Significantly, to show that an ordered vector space E is a vector lattice, we do not need to establish the existence of $x \vee y$ and $x \wedge y$ for all $x, y \in E$. The following proposition presents a simpler criterion.

Proposition 7.4 *Let E be an ordered vector space. Then the following are equivalent.*

(a) E is a vector lattice.

(b) $E = \{y - z \mid y, z \in E_+\}$, and $x \vee y$ exists for all $x, y \in E_+$.

Proof. (a)$\Rightarrow$(b) is trivial.

(b)$\Rightarrow$(a). Take $x_1, x_2 \in E$. Then there are $y_1, z_1, y_2, z_2 \in E_+$ such that $x_1 = y_1 - z_1$ and $x_2 = y_2 - z_2$. It follows that

$$x_1 \vee x_2 = (y_1 - z_1) \vee (y_2 - z_2) = (y_1 + z_2) \vee (y_2 + z_1) - (z_1 + z_2)$$

exists, as does $(-x_1) \wedge (-x_2)$ and therefore also

$$x_1 \wedge x_2 = \left(-x_2 + (x_1 + x_2) \right) \wedge \left(-x_1 + (x_1 + x_2) \right).$$

□

A vector lattice E is said to be **complete** if $\bigvee_{\iota \in I} x_\iota$ exists for each non-empty family $(x_\iota)_{\iota \in I}$ in E which is bounded above and if $\bigwedge_{\lambda \in L} y_\lambda$ exists for each non-empty family $(y_\lambda)_{\lambda \in L}$ in E which is bounded below. If there is any danger of confusion with other notions of completeness (such as completeness with respect to a norm or a metric), then we refer more precisely to the **order completeness** of E.

Note that the definition of completeness is somewhat different for vector lattices from that for lattices. Completeness in the sense of lattice theory would be meaningless here. (Recall that a lattice is complete if every subset of it has both a supremum and an infimum, which implies that a complete lattice always possesses a largest and a smallest element.) In fact, a vector lattice E can only contain a largest or smallest element if $E = \{0\}$, since

$|x| + |x| > x$ and $x > -|x| - |x|$ for all $x \neq 0$. Concerning the notion of completeness of vector lattices, we also speak at times of conditional completeness. However, we have no need for this notion here.

There is a simple criterion for the completeness of a vector lattice.

Proposition 7.5 *For each vector lattice E, the following are equivalent.*

(a) E is complete.

(b) $\bigvee_{\iota \in I} x_\iota$ exists for each non-empty, upward directed family $(x_\iota)_{\iota \in I}$ in E_+ which is bounded above.

Proof. (a)$\Rightarrow$(b) is trivial.

(b)$\Rightarrow$(a). Suppose that $(x_\iota)_{\iota \in I}$ is non-empty and bounded above in E. Let $\mathfrak{J}$ be the set of all finite non-empty subsets of I, and for each $J \in \mathfrak{J}$ put $x_J := \bigvee_{\iota \in J} x_\iota$. Then $(x_J)_{J \in \mathfrak{J}}$ is non-empty and directed up. It has the same upper bounds in E as $(x_\iota)_{\iota \in I}$.

Take $J_0 \in \mathfrak{J}$. Then $\left(x_J - x_{J_0} \right)_{J \in \mathfrak{J}, J \supset J_0}$ satisfies the hypotheses of (b), and hence

$$\bigvee_{\substack{J \in \mathfrak{J} \\ J \supset J_0}} (x_J - x_{J_0}) = \left(\bigvee_{\substack{J \in \mathfrak{J} \\ J \supset J_0}} x_J \right) - x_{J_0}$$

exists. Then $\bigvee_{J \in \mathfrak{J}} x_J$ also exists and we conclude that $\bigvee_{\iota \in I} x_\iota$ exists.

If $(y_\lambda)_{\lambda \in L}$ is a non-empty family in E which is bounded below, then by the above argument $\bigvee_{\lambda \in L}(-y_\lambda)$ exists, and therefore so does $\bigwedge_{\lambda \in L} y_\lambda = -\bigvee_{\lambda \in L}(-y_\lambda)$. $\qquad \square$

The vector lattice $\mathbb{R}^X$ is complete. The vector lattice $E := \mathcal{C}([0,1])$, however, is not. To see this, consider the sequence $(f_n)_{n \in \mathbb{N}}$ in E, defined as follows:

$$f_n|_{[0,\frac{1}{2}]} = 0, \quad f_n|_{[\frac{1}{2}+\frac{1}{2n}, 1]} = 1, \quad f_n \text{ is linear in } [\tfrac{1}{2} + \tfrac{1}{2n}, 1].$$

The sequence $(f_n)_{n \in \mathbb{N}}$ is obviously bounded above in E. Suppose that $(f_n)_{n \in \mathbb{N}}$ has a supremum f in E. Then, being the *smallest* upper bound of $(f_n)_{n \in \mathbb{N}}$, the function f must satisfy $f|_{[0,\frac{1}{2}[} = 0$ and $f|_{]\frac{1}{2},1]} = 1$, which is impossible for a continuous function. (Observe that $(f_n)_{n \in \mathbb{N}}$ *has* a supremum in $\mathbb{R}^{[0,1]}$, namely the function $e_{]\frac{1}{2},1]}$.)

The following substructures of a vector lattice E will prove important to us.

- A **vector sublattice** of E is a vector subspace F of E with the property that, for all $x, y \in F$, $x \wedge y \in F$ and $x \vee y \in F$.

- A **solid subspace** of E is a vector subspace F of E with the property that given $x \in F$ and $y \in E$ with $|y| \leq |x|$, it follows that $y \in F$.

- A **band** in E is a solid subspace F of E with the property that for each family $(x_\iota)_{\iota \in I}$ in F, the existence of $\bigvee_{\iota \in I} x_\iota$ in E implies that $\bigvee_{\iota \in I} x_\iota \in F$.

Before presenting examples, we enunciate some basic properties of our new substructures.

Theorem 7.6 *Let E be a vector lattice.*

(a) If F is a vector sublattice of E, then F is a vector lattice in its own right, with structures induced by E. If $x, y \in F$, then the supremum of x and y in F is identical to that in E and dually the infimum of x and y in F is precisely that in E.

(b) If F is a solid subspace of E and $x \in E$, then $x \in F$ if and only if $|x| \in F$.

(c) Each solid subspace of E, and therefore each band in E, is a vector sublattice of E.

(d) If F is a band in E and $(x_\iota)_{\iota \in I}$ is a family in F for which the supremum $\bigvee_{\iota \in I} x_\iota$ exists in E, then the supremum in F also exists, and it is equal to $\bigvee_{\iota \in I} x_\iota$. The dual holds for infima.

Proof. (a) Take $x, y \in F$. Since $x \vee y$ belongs to F by definition of a vector sublattice, and since it is the smallest upper bound of x and y in E, it is a fortiori the smallest upper bound of x and y in F. Thus the supremum of x and y in F exists and coincides with the supremum in E. The dual holds for the infimum. This proves (a).

We leave the remaining proofs to the reader as an exercise. For (c), use Theorem 7.2(f). □

Since the pointwise supremum and infimum of two continuous real-valued functions is continuous, $\mathcal{C}(X)$ is always a vector sublattice of $\mathbb{R}^X$. On the other hand $E := \mathcal{C}([0,1])$ is clearly not a solid subspace of $\mathbb{R}^{[0,1]}$. Note also that the supremum of a sequence in E need not coincide with the supremum of the same sequence in $\mathbb{R}^{[0,1]}$, contrary to the situation for a finite collection of functions. To verify this, consider the sequence $(f_n)_{n \in \mathbb{N}}$ in E defined by

$$f_n(x) := \begin{cases} nx & \text{if } 0 \leq x \leq \frac{1}{n} \\ 1 & \text{if } \frac{1}{n} < x \leq 1. \end{cases}$$

Then the supremum of $(f_n)_{n \in \mathbb{N}}$ in E is $e_{[0,1]}$, while its supremum in $\mathbb{R}^{[0,1]}$ is $e_{]0,1]}$.

The set $\{f \in E \mid f(0) = 0\}$ is an example of a solid subspace of E which is not a band in E. (To verify this, use the sequence $(f_n)_{n \in \mathbb{N}}$ just defined.) However, the set $\{f \in E \mid f|_{[0,\frac{1}{2}]} = 0\}$ is a band in E. We leave the easy proof of this fact to the reader.

The verification of the band property can also be somewhat simplified, as we next show.

Proposition 7.7 *Let F be a solid subspace of the vector lattice E. Then the following are equivalent.*

(a) F is a band in E.

(b) $\bigvee_{\iota \in I} x_\iota \in F$ whenever the family $(x_\iota)_{\iota \in I}$ is directed up in F_+ and $\bigvee_{\iota \in I} x_\iota$ exists in E.

The proof uses the same arguments as in Proposition 7.5.

Knowing that vector sublattices, solid subspaces and bands are themselves vector lattices, we can study the concept of completeness for them. Solid subspaces behave particularly well, in the sense of our next theorem.

Theorem 7.8 *If E is a complete vector lattice, then each solid subspace, and hence each band in E, is also complete.*

Proof. Let F be a solid subspace of E. Take a non-empty family in F_+, $(x_\iota)_{\iota \in I}$, which is directed up and bounded above in F by, say, y. Then, taking the supremum in E, $0 \leq \bigvee_{\iota \in I} x_\iota \leq y$ and hence $\bigvee_{\iota \in I} x_\iota \in F$. This implies that $\bigvee_{\iota \in I} x_\iota$ must be the smallest upper bound of $(x_\iota)_{\iota \in I}$ in F. By Proposition 7.5, F is complete. $\square$

Of course, the preceding theorem does not generalize to vector sublattices, as is shown by the vector sublattice $\mathcal{C}([0,1])$ of $\mathbb{R}^{[0,1]}$.

Let E be a vector lattice. If $(F_\iota)_{\iota \in I}$ is a non-empty family of solid subspaces (bands) of E, then it is easy to see that $\bigcap_{\iota \in I} F_\iota$ is also a solid subspace (band) of E. Thus if $A \subset E$, then there is a smallest solid subspace S_A (smallest band B_A) of E for which $A \subset S_A$ ($A \subset B_A$), namely,

$$S_A = \bigcap \{F' \mid F' \text{ is a solid subspace of } E \text{ and } A \subset F'\}$$

$$(B_A = \bigcap \{F' \mid F' \text{ is a band in } E \text{ and } A \subset F'\}).$$

We call S_A the **solid subspace of E generated by A** and B_A the **band in E generated by A**.

Since our concern in this text is focused on complete vector lattices, we are, of course, primarily interested in their properties. In order to formulate one of the most important ones given in the bands theorem (Theorem 7.11) below, we need the notion of orthogonality.

Two elements x, y of a vector lattice E are said to be **orthogonal** if $|x| \wedge |y| = 0$. We write $x \perp y$ to denote this. Given $A \subset E$, define

$$A^\perp := \{x \in E \mid x \perp y \text{ for all } y \in A\}.$$

$A^\perp$ is known as the **orthogonal complement** of A in E.

Note that $f, g \in \mathbb{R}^X$ are orthogonal if and only if $\{f \neq 0\} \cap \{g \neq 0\} = \emptyset$. This simple example is a good illustration of the idea of orthogonality.

Theorem 7.9 *Let E be a vector lattice.*

(a) Take $x, y, z \in E$. If $x \perp z$ and $y \perp z$, then $(x+y) \perp z$.

(b) Take $x, y \in E$. If $x \perp y$, then $\alpha x \perp y$ for every $\alpha \in \mathbb{R}$.

(c) Take $x, y, z \in E$. If $x \perp y$ and $|z| \le |x|$, then $z \perp y$.

(d) Let $(x_\iota)_{\iota \in I}$ be a family in E for which $\bigvee_{\iota \in I} x_\iota$ exists. Take $x \in E$ such that $x \perp x_\iota$ for every $\iota \in I$. Then $x \perp \bigvee_{\iota \in I} x_\iota$.

(e) If $x, y \in E$ such that $x \perp y$, then $|x + y| = |x| + |y|$.

Proof. (a) follows from

$$0 \le |x + y| \wedge |z| \le (|x| + |y|) \wedge |z| \le |x| \wedge |z| + |y| \wedge |z| = 0.$$

Note that we have used Proposition 7.2(h).

(b) is a result of

$$0 \le |\alpha x| \wedge |y| \le ((|\alpha| + 1)|x|) \wedge |((|\alpha| + 1)|y|) = (|\alpha| + 1)(|x| \wedge |y|) = 0.$$

(c) follows from

$$0 \le |z| \wedge |y| \le |x| \wedge |y| = 0.$$

(d) Put $u := \bigvee_{\iota \in I} x_\iota$. Then by Theorem 7.3,

$$0 \le u^+ \wedge |x| = \left(\bigvee_{\iota \in I} x_\iota{}^+\right) \wedge |x| = \bigvee_{\iota \in I} (x_\iota{}^+ \wedge |x|) \le \bigvee_{\iota \in I} (|x_\iota| \wedge |x|) = 0.$$

Thus $u^+ \perp x$. Fixing $\lambda \in I$, we have $x_\lambda \le u$ and hence $-u \le -x_\lambda$. Therefore

$$u^- = (-u) \vee 0 \le (-x_\lambda) \vee 0 = x_\lambda{}^-.$$

It follows that

$$0 \le u^- \wedge |x| \le x_\lambda{}^- \wedge |x| \le |x_\lambda| \wedge |x| = 0,$$

and thus $u^- \perp x$. Thus, it follows from (a) that $|u| = (u^+ + u^-)$ is orthogonal to x. Now (c) implies that $u \perp x$.

(e) By Theorem 7.2(f),

$$0 = |x| \wedge |y| = \frac{1}{2}(|x| + |y| - ||x| - |y||).$$

It follows from Theorem 7.2(i),(g) that

$$|x| + |y| = ||x| - |y|| = ||x| - |-y|| \le |x + y| \le |x| + |y|.$$

$\square$

Corollary 7.10 *Let E be a vector lattice. Then for each $A \subset E$, $A^\perp$ is a band in E and $A \subset (A^\perp)^\perp$.*

Let E be a vector lattice. Let F and G be solid subspaces of E. We say that E is the **direct sum** of F and G and write $E = F \oplus G$ if and only if

(i) for each $x \in E$ there are a $y \in F$ and a $z \in G$ such that $x = y + z$;

(ii) $y \perp z$ for every $y \in F$ and every $z \in G$.

If $E = F \oplus G$, then by (ii) $F \cap G = \{0\}$. If $x \in E$ and if $y', y'' \in F$, $z', z'' \in G$ satisfy

$$x = y' + z' = y'' + z'',$$

then $y' - y'' = z'' - z' \in F \cap G$. Thus $y' - y'' = z'' - z' = 0$, and therefore $y' = y''$ and $z' = z''$. The elements y, z of representation (i) are thus uniquely determined. We call y the **component of x in F** and z the **component of x in G**.

If $x \geq 0$, then by Theorem 7.9(e), $x = |x| = |y + z| = |y| + |z|$, and the uniqueness of the decomposition demands that both components of x be positive.

Our next theorem is one of the highlights of the theory of complete vector lattices.

Theorem 7.11 (Bands Theorem) *Let E be a complete vector lattice.*

(a) If F is a band in E, then $E = F \oplus F^{\perp}$.

(b) Given $A \subset E$, $(A^{\perp})^{\perp}$ is the band in E generated by A.

Proof. (a) Take $x \in E_+$. Since E is complete,

$$u := \bigvee \{y \in F \mid y \leq x\}$$

exists. Since F is a band, $u \in F$. We have $0 \leq u \leq x$ and hence $x = u + (x - u)$, where $x - u \geq 0$. Take $v \in F$. Since F is a band, it is a solid subspace. Hence $|v| \wedge (x - u) \in F$. But

$$x \geq (u + |v|) \wedge x = u + \big(|v| \wedge (x - u)\big),$$

and by the definition of u,

$$|v| \wedge (x - u) = 0.$$

Thus, $x - u \in F^{\perp}$, verifying property (i) of direct sums for positive elements. For arbitrary $x \in E$, the property holds since $x = x^+ - x^-$. Property (ii) is trivial.

(b) By Corollary 7.10, $(A^{\perp})^{\perp}$ is a band in E and $A \subset (A^{\perp})^{\perp}$. Let F be an arbitrary band in E with $A \subset F$. Then $A^{\perp} \supset F^{\perp}$ and $(A^{\perp})^{\perp} \subset (F^{\perp})^{\perp}$. From (a) it follows that

$$E = F \oplus F^{\perp} = (F^{\perp})^{\perp} \oplus F^{\perp}.$$

Take $x \in (F^{\perp})^{\perp}$ with $x \geq 0$. Then there is a $y \in F_+$ and a $z \in (F^{\perp})_+$ with $x = y + z$. We have $z = z \wedge x = 0$ and therefore $x = y \in F$. It follows that $x = x^+ - x^- \in F$ for any $x \in (F^{\perp})^{\perp}$ and hence $F = (F^{\perp})^{\perp}$. But then $(A^{\perp})^{\perp} \subset F$ and therefore $(A^{\perp})^{\perp}$ is the smallest band in E containing A. $\square$

Finally, we consider morphisms of vector lattices briefly. Let E and F be vector lattices. The map $\varphi : E \to F$ is

- a **homomorphism of vector lattices** if φ is linear and given $x, y \in E$,
  $\varphi(x \vee y) = \varphi(x) \vee \varphi(y)$ and $\varphi(x \wedge y) = \varphi(x) \wedge \varphi(y)$.
- an **isomorphism of vector lattices** if φ is bijective and if φ and φ^{-1}
  are vector lattice homomorphisms.

E is said to be **isomorphic as vector lattice** to F if there is a vector lattice isomorphism $\varphi : E \to F$.

Proposition 7.12 *Let $\varphi : E \to F$ be a linear map of vector lattices. Then the following are equivalent.*

(a) *φ is a vector lattice homomorphism.*

(b) *$\varphi(|x|) = |\varphi(x)|$ for every $x \in E$.*

(c) *$\varphi(x \vee y) = \varphi(x) \vee \varphi(y)$ for every $x, y \in E$.*

Proof. (a)$\Rightarrow$(b) follows from

$$\varphi(|x|) = \varphi(x \vee (-x)) = \varphi(x) \vee \varphi(-x) = \varphi(x) \vee (-\varphi(x)) = |\varphi(x)|.$$

(b)$\Rightarrow$(c) follows from

$$\varphi(x \vee y) = \varphi\left(\frac{1}{2}(x + y + |x - y|)\right) = \frac{1}{2}(\varphi(x) + \varphi(y) + \varphi(|x - y|))$$
$$= \frac{1}{2}(\varphi(x) + \varphi(y) + |\varphi(x) - \varphi(y)|) = \varphi(x) \vee \varphi(y).$$

(c)$\Rightarrow$(a) follows from

$$\varphi(x \wedge y) = \varphi(x + y - x \vee y) = \varphi(x) + \varphi(y) - \varphi(x) \vee \varphi(y) = \varphi(x) \wedge \varphi(y).$$

$\square$

It is immediate that a vector lattice homomorphism is increasing (see the proof of (a)$\Rightarrow$(b) in the next proposition). But an increasing linear map between two vector lattices need not be a homomorphism of vector lattices! However, the situation is different in the case of isomorphisms. Our next proposition shows that a sufficient condition for a bijective linear map to be a vector lattice isomorphism is that both it and its inverse be increasing.

Proposition 7.13 *Let $\varphi : E \to F$ be a linear map of vector lattices. Then the following are equivalent.*

(a) *φ is a vector lattice isomorphism.*

(b) *φ is bijective, and φ and φ^{-1} are increasing.*

Proof. (a)$\Rightarrow$(b). Given $x, y \in E$, if $x \leq y$, then $x = x \wedge y$ and therefore $\varphi(x) = \varphi(x) \wedge \varphi(y) \leq \varphi(y)$. A similar argument shows that φ^{-1} is increasing.

(b)$\Rightarrow$(a). Take $x, y \in E$. Then $\varphi(x) \leq \varphi(x \vee y)$ and $\varphi(y) \leq \varphi(x \vee y)$. Thus

$$\varphi(x) \vee \varphi(y) \leq \varphi(x \vee y).$$

Similarly, if $u, v \in F$, then

$$\varphi^{-1}(u) \vee \varphi^{-1}(v) \leq \varphi^{-1}(u \vee v).$$

Putting $u := \varphi(x)$ and $v := \varphi(y)$, we obtain

$$x \vee y \leq \varphi^{-1}\big(\varphi(x) \vee \varphi(y)\big),$$

and hence

$$\varphi(x \vee y) \leq \varphi(x) \vee \varphi(y).$$

Thus

$$\varphi(x \vee y) = \varphi(x) \vee \varphi(y).$$

By Proposition 7.12 (c)$\Rightarrow$(a), φ is a vector lattice homomorphism. A similar argument applies to φ^{-1}. ☐

We shall use the following consequence in several places.

Corollary 7.14 *Let E be a vector lattice, F an ordered vector space and $\varphi : E \to F$ a linear, bijective map such that both φ and φ^{-1} are increasing. Then F is a vector lattice and φ is a vector lattice isomorphism. If E is complete, then so is F.*

Proof. Take $u, v \in F_+$. Then $\varphi\big(\varphi^{-1}(u) \vee \varphi^{-1}(v)\big)$ is an upper bound for u and for v. If $w \in F$ is an upper bound for u and for v, then $\varphi^{-1}(w) \geq \varphi^{-1}(u)$ and $\varphi^{-1}(w) \geq \varphi^{-1}(v)$. Therefore $\varphi^{-1}(w) \geq \varphi^{-1}(u) \vee \varphi^{-1}(v)$ and so

$$w = \varphi\big(\varphi^{-1}(w)\big) \geq \varphi\big(\varphi^{-1}(u) \vee \varphi^{-1}(v)\big).$$

Thus

$$u \vee v = \varphi\big(\varphi^{-1}(u) \vee \varphi^{-1}(v)\big)$$

exists.

Since φ is linear, $\varphi(0) = 0$. Take $u \in F$. Then there are elements $x, y \in E_+$ such that $\varphi^{-1}(u) = x - y$. It follows that

$$u = \varphi\big(\varphi^{-1}(u)\big) = \varphi(x) - \varphi(y), \quad \varphi(x) \geq 0, \quad \varphi(y) \geq 0.$$

Thus the hypotheses of Proposition 7.4 are satisfied and F is therefore a vector lattice. The isomorphism property is a result of Proposition 7.13. We leave the remainder of the proof to the reader. ☐

We will shortly look at L^p-spaces as examples of vector lattices. However, the significance of the vector lattice structure for integration theory extends far beyond this example. Vector lattices represent the basic structure of integration theory. Accordingly, a deeper study of integration theory requires a deeper study of the theory of vector lattices. However, the treatment here is adequate for our more limited purposes.

Exercises

1. Let X be a Hausdorff space and put

$$\mathcal{C}_b(X) := \{f \in \mathcal{C}(X) \mid f \text{ is bounded}\}.$$

Prove the following.

(a) $\mathcal{C}(X)$ is a solid subspace of $\mathbb{R}^X$ if and only if X is discrete.

(b) $\{f \in \mathcal{C}([0,1]) \mid f \text{ is piecewise linear}\}$ is a vector sublattice, but not a solid subspace of $\mathcal{C}([0,1])$.

(c) $\mathcal{C}_b(X)$ is a solid subspace, but in general not a band in $\mathcal{C}(X)$.

(d) Determine $(\mathcal{C}_b(X)^\perp)^\perp$ and the band generated in $\mathcal{C}(X)$ by $\mathcal{C}_b(X)$.

(e) $\mathcal{K}(X)$ is a solid subspace of $\mathcal{C}_b(X)$.

(f) If X is locally compact, then $\mathcal{K}(X)$ is a band in $\mathcal{C}_b(X)$ if and only if X is compact.

(g) $F := \{f \in \mathcal{C}([0,1]) \mid f|_{[0,\frac{1}{2}]} = 0\}$ is a band in $\mathcal{C}([0,1])$, but $\mathcal{C}([0,1])$ does not decompose as the direct sum of F and $F^\perp$.

2. Verify that for any $A \subset X$, $\{f \in \mathbb{R}^X \mid \{f \neq 0\} \subset A\}$ is a band in $\mathbb{R}^X$.

3. Determine the solid subspace and the band generated by $\mathcal{F}(X)$ in $\mathbb{R}^X$.

4. Determine the solid subspace and the band generated by $e_{[0,1]}$ in $\mathbb{R}^{\mathbb{R}}$.

5. Let F be a solid subspace of the vector lattice E. Prove the following.

(a) If $x, y \in F$ and $z \in E$ with $x \leq z \leq y$, then $z \in F$.

(b) Let $(x_\iota)_{\iota \in I}$ be a family in F which is bounded above in F. Then $\bigvee_{\iota \in I} x_\iota$ exists in E if and only if $\bigvee_{\iota \in I} x_\iota$ exists in F, and if they exist, then they coincide.

6. Let E be a vector lattice. Take $A \subset E$ and let S_A (B_A) be the solid subspace (band) of E generated by A. Show that B_A is the band in E generated by S_A.

7. Let E be a vector lattice. Prove the following.

(a) For each $A \subset E$

$$\left\{x \in E \,\middle|\, \text{there is a finite family } (x_\iota)_{\iota \in I} \text{ in } A \text{ with } |x| \leq \sum_{\iota \in I} |x_\iota|\right\}$$

is the solid subspace of E generated by A.

(b) For each $x \in E$

$$\{y \in E \mid \exists n \in \mathbb{N}, \ |y| \leq n|x|\}$$

is the solid subspace of E generated by x.

8. Let E be a vector lattice and take $x \in E_+$. Prove the following.

   (a) If F is a solid subspace of E and if $u := \bigvee\{y \in F \mid y \leq x\}$ exists, then $x - u \in F^{\perp}$.

   (b) If E is complete and F a solid subspace of E, then $\bigvee\{y \in F \mid y \leq x\}$ is the component of x in the band $(F^{\perp})^{\perp}$ of E generated by F.

   (c) If E is complete and $y \in E_+$, then $\bigvee_{n \in \mathbb{N}} (ny \wedge x)$ is the component of x in the band of E generated by y.

   Hint for (a): Adapt the proof of Theorem 7.11(a).

9. Let E be a vector lattice. Take $A \subset E$ and define $|A| := \{|x| \mid x \in A\}$. Show that A and $|A|$ generate the same solid subspace and the same band in E.

10. Let E be a vector lattice, F a solid subspace (band) of E and take $A \subset F$. Prove that the solid subspace (band) of E generated by A coincides with the solid subspace (band) of F generated by A.

11. Let E be a vector lattice and define

$$\Phi := \{F \mid F \text{ is a solid subspace of } E\},$$
$$\Psi := \{F \mid F \text{ is a band in } E\}.$$

Show that Φ and Ψ are complete lattices with respect to inclusion and determine the largest and smallest elements in Φ and Ψ as well as $\bigvee_{\iota \in I} F_{\iota}$ and $\bigwedge_{\iota \in I} F_{\iota}$ for a non-empty family $(F_{\iota})_{\iota \in I}$ from Φ (from Ψ).

12. Let $(E_{\iota})_{\iota \in I}$ be a non-empty family of vector lattices. Endow $E := \prod_{\iota \in I} E_{\iota}$ with the operations

$$(x_{\iota})_{\iota \in I} + (y_{\iota})_{\iota \in I} := (x_{\iota} + y_{\iota})_{\iota \in I}$$
$$\alpha(x_{\iota})_{\iota \in I} := (\alpha x_{\iota})_{\iota \in I} \quad \text{for } \alpha \in \mathbb{R}$$

and the order relation

$$(x_{\iota})_{\iota \in I} \leq (y_{\iota})_{\iota \in I} \quad :\Longleftrightarrow \quad x_{\iota} \leq y_{\iota} \text{ for all } \iota \in I.$$

Prove that E is a vector lattice that is complete if and only if each of the E_{ι} is complete. An important example of the structure introduced here is furnished by $\mathbb{R}^X$.

13. Define the **lexicographic order** on $\mathbb{R}^2$ by

$$(x, y) \preccurlyeq (u, v) \quad :\Longleftrightarrow \quad x < u \text{ or } (x = u \text{ and } y \leq v).$$

Prove the following.

   (a) $\mathbb{R}^2$ is a vector lattice with respect to $\preccurlyeq$.

   (b) There is a pair $(x, y) \succcurlyeq (0, 0)$ such that $\{\frac{1}{n}(x, y) \mid n \in \mathbb{N}\}$ has no infimum.

14. Show that if E is a complete vector lattice and $x \in E_+$, then $\bigwedge_{n\in\mathbb{N}} \frac{1}{n}x = 0$.

15. Verify that $\mathbb{R}^2$ with the lexicographic order is not isomorphic as vector lattice to $\mathbb{R}^2$ with the usual order.

16. Let $\varphi : E \to F$ be a homomorphism of vector lattices. Prove the following.

   (a) If G is a vector sublattice of E, then φG is a vector sublattice of F.
   (b) If H is a vector sublattice of F, then $\varphi^{-1}H$ is a vector sublattice of E.
   (c) If G is a solid subspace of E, then φG is a solid subspace of φE.
   (d) If H is a solid subspace of F, then $\varphi^{-1}H$ is a solid subspace of E.
   (e) $\ker \varphi := \varphi^{-1}(\{0\})$ is a solid subspace of E.

17. Let F and G be solid subspaces of the vector lattice E. Show that if $E = F \oplus G$, then F and G are bands in E.

7.2 L^p-spaces

We now return to the spaces $\mathcal{L}^p(\mu)$ where $(X, \mathfrak{R}, \mu)$ is a positive measure space and $p \in [1, \infty[$. By Proposition 3.25 '$f = g$ μ-a.e.' defines an equivalence relation on each $\mathcal{L}^p(\mu)$. For $f \in \mathcal{L}^p(\mu)$, let $\dot{f}$ denote the equivalence class of f. We define

$$L^p(\mu) := \{\dot{f} \mid f \in \mathcal{L}^p(\mu)\}.$$

In other words, we pass from $\mathcal{L}^p(\mu)$ to $L^p(\mu)$ simply by identifying functions which are μ-a.e. equal.

The following proposition will enable us to define operations on $L^p(\mu)$.

Proposition 7.15 *Take* $\mathcal{F}, \mathcal{G} \in L^p(\mu)$ *and* $\alpha \in \mathbb{R}$.

   (a) *Given any* $f_1, f_2 \in \mathcal{F} \cap \mathbb{R}^X$ *and* $g_1, g_2 \in \mathcal{G} \cap \mathbb{R}^X$, $f_1 + g_1 = f_2 + g_2$ μ-a.e.
   (b) *Given any* $f_1, f_2 \in \mathcal{F}$, $\alpha f_1 = \alpha f_2$ μ-a.e.
   (c) *Given any* $f_1, f_2 \in \mathcal{F}$ *and* $g_1, g_2 \in \mathcal{G}$, $f_1 \leq g_1$ μ-a.e. *if and only if* $f_2 \leq g_2$ μ-a.e.
   (d) *Given any* $f_1, f_2 \in \mathcal{F}$, $N_p(f_1) = N_p(f_2)$.

The proofs are trivial.

Given $\mathcal{F}, \mathcal{G} \in L^p(\mu)$ and $\alpha \in \mathbb{R}$, Proposition 7.15 allows us to define:

   (i) $\mathcal{F} + \mathcal{G} := (f + g)^{\cdot}$ for $f \in \mathcal{F} \cap \mathbb{R}^X$ and $g \in \mathcal{G} \cap \mathbb{R}^X$.
   (ii) $\alpha\mathcal{F} := (\alpha f)^{\cdot}$ for $f \in \mathcal{F}$.
   (iii) $\mathcal{F} \leq \mathcal{G} :\iff f \leq g$ μ-a.e. for $f \in \mathcal{F}$ and $g \in \mathcal{G}$.
   (iv) $\|\mathcal{F}\|_p := N_p(f)$ for $f \in \mathcal{F}$.

We put $-\mathcal{F} := (-1)\mathcal{F}$ and $\mathcal{F} - \mathcal{G} := \mathcal{F} + (-1)\mathcal{G}$.

The following theorem shows that $L^p(\mu)$ enjoys all the pleasing properties which make life easy for a mathematician: $L^p(\mu)$ is complete in two senses, namely as a vector lattice and as a normed space, and there is a powerful connection between order and norm, as shown in (c) and (d). Recall that $\mathcal{L}^p(\mu)$ has substantially weaker properties: in general, N_p is not a norm on $\mathcal{L}^p(\mu) \cap \mathbb{R}^X$, due to the fact that $N_p(f) = 0$ for every $f \in \mathcal{N}(\mu)$, and $\mathcal{L}^p(\mu) \cap \mathbb{R}^X$ is not a complete vector lattice. To verify the last statement (for Lebesgue measure), consider the family $(e_A)_{A \in \mathfrak{A}}$, where $\mathfrak{A}$ is the set of all finite subsets of a non-Lebesgue integrable subset of $[0, 1]$.

Theorem 7.16

(a) $L^p(\mu)$ *is a complete vector lattice with respect to the operations (i)–(iii).*

(b) $\|\cdot\|_p$ *is a norm on $L^p(\mu)$, and $L^p(\mu)$ is a Banach space with respect to this norm.*

(c) *Take $\mathcal{F}, \mathcal{G} \in L^p(\mu)$ with $0 \leq \mathcal{F} \leq \mathcal{G}$. Then $\|\mathcal{F}\|_p \leq \|\mathcal{G}\|_p$, and if $\|\mathcal{F}\|_p = \|\mathcal{G}\|_p$ then $\mathcal{F} = \mathcal{G}$.*

(d) *If $(\mathcal{F}_n)_{n\in\mathbb{N}}$ is an increasing sequence in $L^p(\mu)_+$ such that*

$$\sup_{n\in\mathbb{N}} \|\mathcal{F}_n\|_p < \infty,$$

then $\mathcal{F} := \bigvee_{n\in\mathbb{N}} \mathcal{F}_n$ exists and

$$\|\mathcal{F}\|_p = \sup_{n\in\mathbb{N}} \|\mathcal{F}_n\|_p.$$

Proof. It is easy to verify that $L^p(\mu)$ is an ordered vector space with respect to relations (i)–(iii) and that

$$L^p(\mu) = \{\mathcal{F} - \mathcal{G} \mid \mathcal{F}, \mathcal{G} \in L^p(\mu)_+\}.$$

Take $\mathcal{F}, \mathcal{G} \in L^p(\mu)_+$. Choose $f \in \mathcal{F}, g \in \mathcal{G}$ and put $\mathcal{H} := (f \vee g)^{\cdot}$. By definition, $\mathcal{H} \geq \mathcal{F}$ and $\mathcal{H} \geq \mathcal{G}$. If $\mathcal{H}'$ is an arbitrary upper bound for both $\mathcal{F}$ and $\mathcal{G}$, then we can choose $h' \in \mathcal{H}'$ and then $h' \geq f$ μ-a.e. and $h' \geq g$ μ-a.e. Thus $h' \geq f \vee g$ μ-a.e. and therefore $\mathcal{H}' \geq \mathcal{H}$. Thus $\mathcal{F} \vee \mathcal{G} = \mathcal{H}$ exists and so by Proposition 7.4, $L^p(\mu)$ is a vector lattice.

We note that (b) and (c) are trivial results of Theorems 6.7 and 6.8(b) and proceed to prove (d).

Let $(\mathcal{F}_n)_{n\in\mathbb{N}}$ be an increasing sequence in $L^p(\mu)_+$ such that

$$\sup_{n\in\mathbb{N}} \|\mathcal{F}_n\|_p < \infty.$$

Given $n \in \mathbb{N}$, choose $f_n \in (\mathcal{F}_n)_+$ such that $(f_n)_{n\in\mathbb{N}}$ is increasing. The functions f_n are in $\mathcal{L}^p(\mu)$. That is, $f_n \in \mathcal{M}_0(\mu)$ and $f_n^p \in \mathcal{L}^1(\mu)$ for each $n \in \mathbb{N}$. We put $g := \uparrow f_n^p$. Note that $(f_n^p)_{n\in\mathbb{N}}$ is increasing and

$$\sup_{n\in\mathbb{N}} \int f_n^p d\mu = \sup_{n\in\mathbb{N}} N_p(f_n)^p = \sup_{n\in\mathbb{N}} (\|\mathcal{F}_n\|_p)^p < \infty.$$

Thus $g \in \mathcal{L}^1(\mu)_+$ and

$$\int g \, d\mu = \sup_{n \in \mathbb{N}} \int f_n^p d\mu.$$

By Corollary 6.12, $g \in \mathcal{M}_0(\mu)$ and therefore $f := g^{1/p} \in \mathcal{M}_0(\mu)$ as well (Proposition 6.1(d)). We conclude that $f \in \mathcal{L}^p(\mu)$, $f = \bigvee_{n \in \mathbb{N}} f_n$ and

$$\|\dot{f}\|_p = N_p(f) = \left(\int f^p d\mu \right)^{1/p} = \sup_{n \in \mathbb{N}} \left(\int f_n^p d\mu \right)^{1/p} = \sup_{n \in \mathbb{N}} \|\mathcal{F}_n\|_p.$$

It remains to show that $\dot{f} = \bigvee_{n \in \mathbb{N}} \mathcal{F}_n$. But $f \geq f_n$ for every $n \in \mathbb{N}$ and therefore $\dot{f} \geq \mathcal{F}_n$. Given $\mathcal{G} \in L^p(\mu)$ such that $\mathcal{G} \geq \mathcal{F}_n$ for every $n \in \mathbb{N}$, if $g \in \mathcal{G}$, then $g \geq f_n$ μ-a.e. for every $n \in \mathbb{N}$ and hence $g \geq f$ μ-a.e. Thus $\mathcal{G} \geq \dot{f}$ and hence $\dot{f} = \bigvee_{n \in \mathbb{N}} \mathcal{F}_n$.

We now prove that the vector lattice $L^p(\mu)$ is complete. We use Proposition 7.5.

Let $(\mathcal{F}_\iota)_{\iota \in I}$ be a non-empty family in $L^p(\mu)_+$ which is directed up and bounded above. Put

$$\alpha := \sup_{\iota \in I} \|\mathcal{F}_\iota\|_p \in \mathbb{R}.$$

Then there is an increasing sequence $(\mathcal{F}_{\iota_n})_{n \in \mathbb{N}}$ $(\iota_n \in I$ for every $n \in \mathbb{N})$ such that

$$\alpha = \sup_{n \in \mathbb{N}} \|\mathcal{F}_{\iota_n}\|_p.$$

By (d), $\mathcal{F} := \bigvee_{n \in \mathbb{N}} \mathcal{F}_{\iota_n}$ exists and $\|\mathcal{F}\|_p = \alpha$. We show that, for each $\iota \in I$, $\mathcal{F} \geq \mathcal{F}_\iota$. Fix $\iota \in I$. Since $(\mathcal{F}_\iota)_{\iota \in I}$ is directed up, there is a sequence $(\lambda_n)_{n \in \mathbb{N}}$ in I such that $\mathcal{F}_{\lambda_n} \geq \mathcal{F}_{\iota_n}$ and $\mathcal{F}_{\lambda_n} \geq \mathcal{F}_\iota$ for every $n \in \mathbb{N}$ and such that $(\mathcal{F}_{\lambda_n})_{n \in \mathbb{N}}$ is increasing. As a result,

$$\alpha = \sup_{n \in \mathbb{N}} \|\mathcal{F}_{\lambda_n}\|_p,$$

and by (d), $\mathcal{G} := \bigvee_{n \in \mathbb{N}} \mathcal{F}_{\lambda_n}$ exists, with $\|\mathcal{G}\|_p = \alpha$. But then $\mathcal{G} \geq \mathcal{F}$ and $\|\mathcal{G}\|_p = \|\mathcal{F}\|_p$, which implies that $\mathcal{F} = \mathcal{G} \geq \mathcal{F}_\iota$. Hence $\mathcal{F}$ is an upper bound for $(\mathcal{F}_\iota)_{\iota \in I}$, and, trivially, it is the smallest upper bound. This proves that $L^p(\mu)$ is complete. □

Note that the zero element of $L^p(\mu)$ is precisely the set $\mathcal{N}(\mu)$ of μ-null functions.

For countable families in $\mathcal{L}^p(\mu)$ which are bounded above, the formation of equivalence classes and the taking of suprema can be interchanged, as the following result shows. This does not generalize to uncountable families, as shown in the exercises.

Corollary 7.17 Let $(\mathcal{F}_\iota)_{\iota \in I}$ be a non-empty countable family in $L^p(\mu)$ which is bounded above. Then

$$\bigvee_{\iota \in I} \mathcal{F}_\iota = \left(\bigvee_{\iota \in I} f_\iota \right)^{\cdot}$$

where, for each $\iota \in I$, f_ι is chosen arbitrarily in $\mathcal{F}_\iota$. If $(\mathcal{F}_\iota)_{\iota \in I}$ is bounded below, then

$$\bigwedge_{\iota \in I} \mathcal{F}_\iota = \left(\bigwedge_{\iota \in I} f_\iota\right)^\cdot$$

where, for each $\iota \in I$, f_ι is chosen arbitrarily in $\mathcal{F}_\iota$.

Proof. We only prove the first assertion. By Theorem 7.16(a), $\bigvee_{\iota \in I} \mathcal{F}_\iota$ exists in $L^p(\mu)$, and by Corollary 6.14, $\bigvee_{\iota \in I} f_\iota \in \mathcal{L}^p(\mu)$ (so that $\left(\bigvee_{\iota \in I} f_\iota\right)^\cdot$ is well defined). Using the fact that a countable union of null sets is a null set, it is not difficult to see that every upper bound for $(\mathcal{F}_\iota)_{\iota \in I}$ is an upper bound for $\left(\bigvee_{\iota \in I} f_\iota\right)^\cdot$ in $L^p(\mu)$. On the other hand, $\left(\bigvee_{\iota \in I} f_\iota\right)^\cdot$ is obviously an upper bound for $(\mathcal{F}_\iota)_{\iota \in I}$, which proves the claim. $\qquad\square$

The L^p-spaces are thus complete vector lattices. These spaces play an important part in functional analysis. This is especially true of the space $L^2(\mu)$, as we have:

Theorem 7.18 $L^2(\mu)$ *is a Hilbert space with respect to the inner product*

$$L^2(\mu) \times L^2(\mu) \longrightarrow \mathbb{R}, \quad (\mathcal{F}, \mathcal{G}) \longmapsto \int fg\, d\mu \quad (f \in \mathcal{F}, g \in \mathcal{G}).$$

Proof. By Theorem 6.4, the map is well defined. The verification of the inner product properties is simple and left to the reader. $\qquad\square$

Exercises

1. Let λ be Lebesgue measure and define $\mathfrak{F} := \{A \subset [0,1] \,|\, A \text{ is finite}\}$. Show that for $p \in [1, \infty[$:

   (a) $(e_A)_{A \in \mathfrak{F}}$ is directed up in $\mathcal{L}^p(\lambda)$;

   (b) $\bigvee_{A \in \mathfrak{F}} e_A = e_{[0,1]} \in \mathcal{L}^p(\lambda)$;

   (c) $\bigvee_{A \in \mathfrak{F}} \dot{e}_A = 0 < \dot{e}_{[0,1]}$.

   Hence Corollary 7.17 cannot be generalized to uncountable families.

2. Verify the following for $p \in [1, \infty[$. If $L^p(\mu)$ has a countable base with respect to $\|\cdot\|_p$, then there is a sequence $(A_n)_{n \in \mathbb{N}}$ in $\mathfrak{R}$ with $X \setminus \bigcup_{n \in \mathbb{N}} A_n \in \mathfrak{N}(\mu)$.

3. Take $p \in [1, \infty[$. Prove that the following are equivalent.

   (a) There is an $\mathcal{F} \in L^p(\mu)$ such that $L^p(\mu)$ is the band generated by $\mathcal{F}$ in $L^p(\mu)$.

   (b) There is a countable subset M of $L^p(\mu)$ such that $L^p(\mu)$ is the band generated by M in $L^p(\mu)$.

   (c) There is a sequence $(A_n)_{n \in \mathbb{N}}$ in $\mathfrak{R}$ with $X \setminus \bigcup_{n \in \mathbb{N}} A_n \in \mathfrak{N}(\mu)$.

4. Take $p \in [1, \infty[$, $f \in \overline{\mathbb{R}}^X$ and let $(f_n)_{n \in \mathbb{N}}$ be a sequence in $\mathcal{L}^p(\mu)$ such that $(f_n)_{n \in \mathbb{N}}$ converges to f μ-a.e. and that $(\dot{f}_n)_{n \in \mathbb{N}}$ is a Cauchy sequence. Show that $f \in \mathcal{L}^p(\mu)$ and $\dot{f} = \lim_{n \to \infty} \dot{f}_n$.

5. Let μ be bounded and take $p, q \in [1, \infty[$ with $p \le q$. Prove the following.

   (a) $L^q(\mu)$ is a solid subspace of $L^p(\mu)$.
   (b) The inclusion map $L^q(\mu) \to L^p(\mu)$ is continuous.

6. Define $L^\infty(\mu) := \{\dot{f} \mid f \in \mathcal{L}^\infty(\mu)\}$.

   (a) Verify that Proposition 7.15 also holds for $p = \infty$.

   Hence the same operations can be introduced on $L^\infty(\mu)$ as on $L^p(\mu)$. Show, further, the following.

   (b) $L^\infty(\mu)$ is a vector lattice and Theorem 7.16(b)–(d) and Corollary 7.17 remain valid for $p = \infty$.
   (c) The results of Exercises 1, 2, 4, 5 also hold for $p, q \in [1, \infty]$.
   (d) If $\mathfrak{R}$ is a σ-ring, then $\{\dot{f} \mid f \in \mathcal{L}(\mathfrak{R})\}$ is dense in $L^\infty(\mu)$.
   (e) Find a positive measure space $(X, \mathfrak{R}, \mu)$ and a decreasing sequence $(\mathcal{F}_n)_{n \in \mathbb{N}}$ in $L^\infty(\mu)$ such that $0 \le \mathcal{F}_n \le \dot{e}_X$ and $\bigwedge_{n \in \mathbb{N}} \mathcal{F}_n = 0$, but $\inf_{n \in \mathbb{N}} \|\mathcal{F}_n\|_\infty = 1$.

7. Take $\mathcal{F}, \mathcal{G} \in L^\infty(\mu)$.

   (a) Show that $f_1 g_1 = f_2 g_2$ μ-a.e. whenever $f_1, f_2 \in \mathcal{F}$ and $g_1, g_2 \in \mathcal{G}$.

   Define $\mathcal{FG} := (fg)^\cdot$ for $f \in \mathcal{F}$ and $g \in \mathcal{G}$. Prove each of the statements below.

   (b) The following are equivalent.

   (b1) $\mathcal{F} \ge 0$.
   (b2) There is an $\mathcal{H} \in L^\infty(\mu)$ with $\mathcal{F} = \mathcal{H}^2$.

   (c) The following are equivalent.

   (c1) $\mathcal{F}$ is invertible (i.e. there is an $\mathcal{H} \in L^\infty(\mu)$ with $\mathcal{FH} = \dot{e}_X$).
   (c2) There is an $\alpha > 0$ with $|\mathcal{F}| \ge \alpha \dot{e}_X$.

   (d) The following are equivalent.

   (d1) $\mathcal{F}^2 = \mathcal{F}$.
   (d2) $\mathcal{F} \wedge (\dot{e}_X - \mathcal{F}) = 0$.
   (d3) $\mathcal{F} \vee (\dot{e}_X - \mathcal{F}) = \dot{e}_X$.
   (d4) There is an $A \in \mathfrak{M}(\mu)$ with $\mathcal{F} = \dot{e}_A$.

8

Spaces of measures

This chapter deals with another example of the influence of the theory of vector lattices in integration theory. We do not consider individual measure spaces but rather the set of all measures on a ring of sets. At this point, we deliberately do *not* refer to positive measures. We deal with an extension of this concept which is a natural consequence of our study of vector lattices.

8.1 The vector lattice structure and Hahn's theorem

Let $\mathfrak{R}$ be a ring of sets. Let $M(\mathfrak{R})$ be the set of all real-valued functions λ on $\mathfrak{R}$ for which there are two positive contents μ, ν on $\mathfrak{R}$ with $\lambda = \mu - \nu$. We call the elements of $M(\mathfrak{R})$ **(real) contents** on $\mathfrak{R}$. It is clear that each content λ on $\mathfrak{R}$ is **additive**, i.e.

$$\lambda(A \cup B) = \lambda(A) + \lambda(B) \qquad \text{for all } A, B \in \mathfrak{R} \text{ with } A \cap B = \emptyset.$$

Note also that it is essential for the preceding definition that positive contents take only finite values. This is one of the reasons why we do not assign content or measure ∞ to 'large' measurable sets.

The definition above is indeed very natural, taking into account that every element of a vector lattice is the difference of two positive elements. Before speaking of a vector lattice structure on $M(\mathfrak{R})$, however, we need to introduce an appropriate order relation $\leq$ on $M(\mathfrak{R})$ which makes the positive contents, in fact, positive with respect to $\leq$. This is done in the obvious manner.

For two maps $\mu, \nu \in \mathbb{R}^{\mathfrak{R}}$, define $\mu \leq \nu$ if and only if $\mu(A) \leq \nu(A)$ for every $A \in \mathfrak{R}$. This is precisely the order relation defined for real-valued functions.

We begin immediately with the fundamental result describing the structure of $M(\mathfrak{R})$.

Theorem 8.1 $M(\mathfrak{R})$ *is a complete vector lattice with respect to the ordering* $\leq$, *and* $M(\mathfrak{R})_+$ *is the set of all positive contents on* $\mathfrak{R}$.

Proof. It follows from the definitions that $M(\mathfrak{R})$ is an ordered vector space, that

$$M(\mathfrak{R})_+ = \{\mu \mid \mu \text{ is a positive content on } \mathfrak{R}\},$$

and that

$$M(\mathfrak{R}) = \{\mu - \nu \mid \mu, \nu \in M(\mathfrak{R})_+\}.$$

Take $\mu, \nu \in M(\mathfrak{R})$. Given $A \in \mathfrak{R}$, let

$$\lambda(A) := \sup\{\mu(B) + \nu(A \setminus B) \mid B \in \mathfrak{R}, B \subset A\}.$$

We show that λ is additive. Take $A_1, A_2 \in \mathfrak{R}$ with $A_1 \cap A_2 = \emptyset$. Then for $B_1, B_2 \in \mathfrak{R}$ with $B_1 \subset A_1$ and $B_2 \subset A_2$,

$$\big(\mu(B_1) + \nu(A_1 \setminus B_1)\big) + \big(\mu(B_2) + \nu(A_2 \setminus B_2)\big)$$
$$= \mu(B_1 \cup B_2) + \nu\big((A_1 \cup A_2) \setminus (B_1 \cup B_2)\big) \leq \lambda(A_1 \cup A_2).$$

As a result,

$$\lambda(A_1) + \lambda(A_2) \leq \lambda(A_1 \cup A_2).$$

Conversely, given $C \in \mathfrak{R}$ with $C \subset A_1 \cup A_2$,

$$\mu(C) + \nu\big((A_1 \cup A_2) \setminus C\big) = \big(\mu(A_1 \cap C) + \nu(A_1 \setminus (A_1 \cap C))\big)$$
$$+ \big(\mu(A_2 \cap C) + \nu(A_2 \setminus (A_2 \cap C))\big) \leq \lambda(A_1) + \lambda(A_2).$$

It follows that

$$\lambda(A_1 \cup A_2) \leq \lambda(A_1) + \lambda(A_2),$$

and λ is thus additive.

Let λ' be additive and suppose that $\lambda' \geq \mu$ and $\lambda' \geq \nu$. Then for all $A, B \in \mathfrak{R}$ with $B \subset A$,

$$\mu(B) + \nu(A \setminus B) \leq \lambda'(B) + \lambda'(A \setminus B) = \lambda'(A).$$

Hence $\lambda \leq \lambda'$.

Now if μ and ν are positive, then so is λ and hence λ is in $M(\mathfrak{R})_+$. This implies that $\lambda = \mu \vee \nu$. Thus, by Proposition 7.4, $M(\mathfrak{R})$ is a vector lattice.

We show that the vector lattice $M(\mathfrak{R})$ is complete. Let $(\lambda_\iota)_{\iota \in I}$ be a nonempty, upward directed family in $M(\mathfrak{R})_+$ which is bounded above. Then for each $A \in \mathfrak{R}$,

$$\lambda(A) := \sup_{\iota \in I} \lambda_\iota(A) \in \mathbb{R}.$$

λ is positive and additive, since for all $A, B \in \mathfrak{R}$ with $A \cap B = \emptyset$,

$$\lambda(A \cup B) = \sup_{\iota \in I} \lambda_\iota(A \cup B) = \sup_{\iota \in I} \big(\lambda_\iota(A) + \lambda_\iota(B)\big)$$
$$= \sup_{\iota \in I} \lambda_\iota(A) + \sup_{\iota \in I} \lambda_\iota(B) = \lambda(A) + \lambda(B).$$

(Note that the third step relies on the fact that $(\lambda_\iota)_{\iota \in I}$ is directed up.) Thus $\lambda \in M(\mathfrak{R})_+$. But then by the definition of λ, we have $\lambda = \bigvee_{\iota \in I} \lambda_\iota$. The claim then follows by Proposition 7.5. $\qquad\square$

The following criterion is very useful for deciding whether an additive map on $\mathfrak{R}$ is a content.

Corollary 8.2 *Given any additive mapping μ of $\mathfrak{R}$ into $\mathbb{R}$, the following are equivalent.*

(a) $\mu \in M(\mathfrak{R})$.

(b) *Given any $A \in \mathfrak{R}$, $\sup\{|\mu(B)| \,|\, B \in \mathfrak{R}, B \subset A\} < \infty$.*

Proof. (a)$\Rightarrow$(b). If $\mu \in M(\mathfrak{R})$, then for all $A, B \in \mathfrak{R}$ with $B \subset A$,

$$-|\mu|(A) \le -|\mu|(B) \le \mu(B) \le |\mu|(B) \le |\mu|(A),$$

and hence

$$\sup\{|\mu(B)| \,|\, B \in \mathfrak{R}, B \subset A\} \le |\mu|(A) < \infty.$$

(b)$\Rightarrow$(a). (b) implies the existence of

$$\lambda' : \mathfrak{R} \longrightarrow \mathbb{R}, \quad A \longmapsto \sup\{\mu(B) \,|\, B \in \mathfrak{R}, B \subset A\}$$

and

$$\lambda'' : \mathfrak{R} \longrightarrow \mathbb{R}, \quad A \longmapsto -\inf\{\mu(B) \,|\, B \in \mathfrak{R}, B \subset A\}.$$

It can easily be shown (cf. the proof of Theorem 8.1) that λ' and λ'' are positive contents on $\mathfrak{R}$. Then, since

$$
\begin{aligned}
\sup\{\mu(B) \,|\, B \in \mathfrak{R}, B \subset A\} &= \sup\{\mu(A) - \mu(A \setminus B) \,|\, B \in \mathfrak{R}, B \subset A\} \\
&= \mu(A) - \inf\{\mu(A \setminus B) \,|\, B \in \mathfrak{R}, B \subset A\} \\
&= \mu(A) - \inf\{\mu(C) \,|\, C \in \mathfrak{R}, C \subset A\}
\end{aligned}
$$

for every $A \in \mathfrak{R}$, we see that $\mu = \lambda' - \lambda'' \in M(\mathfrak{R})$. $\square$

Another important characterization of the elements of $M(\mathfrak{R})$ uses the concept of variation on the sets of $\mathfrak{R}$ and is given below in Theorem 8.5.

We now describe suprema and infima in $M(\mathfrak{R})$. The reader may wonder why this is necessary. Should we not expect, for example, that $(\mu \vee \nu)(A)$ is the pointwise supremum of $\mu(A)$ and $\nu(A)$, bearing in mind that the order structure is defined pointwise on the 'points' of $\mathfrak{R}$? No, we should not! This would only hold true if $M(\mathfrak{R})$ were a vector sublattice of $\mathbb{R}^{\mathfrak{R}}$. But this is in general not the case, as the following theorem (implicitly) reveals.

Theorem 8.3 *Take arbitrary $\mu, \nu \in M(\mathfrak{R})$ and $A \in \mathfrak{R}$. Then:*

(a) $(\mu \vee \nu)(A) = \sup\{\mu(B) + \nu(A \setminus B) \,|\, B \in \mathfrak{R}, B \subset A\}$.

(b) $(\mu \wedge \nu)(A) = \inf\{\mu(B) + \nu(A \setminus B) \,|\, B \in \mathfrak{R}, B \subset A\}$.

(c) $\mu^+(A) = \sup\{\mu(B) \,|\, B \in \mathfrak{R}, B \subset A\}$.

(d) $\mu^-(A) = -\inf\{\mu(B) \,|\, B \in \mathfrak{R}, B \subset A\}$.

(e) $|\mu|(A) = \sup\{\mu(B) - \mu(A \setminus B) \,|\, B \in \mathfrak{R}, B \subset A\}$.

Proof. (a) Let λ be defined as in the proof of Theorem 8.1. It was shown there that λ is the smallest upper bound of μ and ν in the set of additive real-valued mappings on $\mathfrak{R}$. It remains to show that λ belongs to $M(\mathfrak{R})$. But this is an easy consequence of Corollary 8.2, since condition (b) is satisfied for μ and ν and therefore also for λ.

Assertions (c),(d) and (e) are immediate consequences of (a), since $\mu^+ = \mu \vee 0$, $\mu^- = (-\mu) \vee 0$, and $|\mu| = \mu \vee (-\mu)$. (b) follows from (a) since $\mu \wedge \nu = \mu + \nu - \mu \vee \nu$. $\square$

The next result shows that $M(\mathfrak{R})$ is actually a little bit 'more than complete', since the existence of $\bigvee_{\iota \in I} \mu_\iota$ for an upward directed family $(\mu_\iota)_{\iota \in I}$ is already guaranteed by the pointwise boundedness of this family.

Theorem 8.4 *Let $(\mu_\iota)_{\iota \in I}$ be a non-empty family in $M(\mathfrak{R})$.*

(a) If $\sup_{\iota \in I} \mu_\iota(A) \in \mathbb{R}$ for every $A \in \mathfrak{R}$ (in particular if $(\mu_\iota)_{\iota \in I}$ is bounded above in $M(\mathfrak{R})$), and if $(\mu_\iota)_{\iota \in I}$ is directed up, then $\bigvee_{\iota \in I} \mu_\iota$ exists in $M(\mathfrak{R})$, and

$$\left(\bigvee_{\iota \in I} \mu_\iota \right)(A) = \sup_{\iota \in I} \mu_\iota(A) \qquad \text{for every } A \in \mathfrak{R}.$$

(b) If $\inf_{\iota \in I} \mu_\iota(A) \in \mathbb{R}$ for every $A \in \mathfrak{R}$ (in particular if $(\mu_\iota)_{\iota \in I}$ is bounded below in $M(\mathfrak{R})$), and if $(\mu_\iota)_{\iota \in I}$ is directed down, then $\bigwedge_{\iota \in I} \mu_\iota$ exists in $M(\mathfrak{R})$, and

$$\left(\bigwedge_{\iota \in I} \mu_\iota \right)(A) = \inf_{\iota \in I} \mu_\iota(A) \qquad \text{for every } A \in \mathfrak{R}.$$

Proof. Suppose first that $\mu_\iota \geq 0$ for every $\iota \in I$. Then the proof given in Theorem 8.1 also applies here. Note that in that proof only our present hypothesis $\sup_{\iota \in I} \mu_\iota(A) \in \mathbb{R}$ was used, rather than the boundedness of $(\mu_\iota)_{\iota \in I}$ in $M(\mathfrak{R})$. The general case follows from the first by fixing $\iota_0 \in I$ and considering the family $(\mu_\iota - \mu_{\iota_0})_{\iota \in J}$, where $J := \{\iota \in I \mid \mu_\iota \geq \mu_{\iota_0}\}$.

(b) follows from (a) because $\bigwedge_{\iota \in I} \mu_\iota = -\bigvee_{\iota \in I}(-\mu_\iota)$. $\square$

We have seen that suprema and infima in $M(\mathfrak{R})$ are completely different from those in $\mathbb{R}^{\mathfrak{R}}$, although the ordering of $M(\mathfrak{R})$ is induced by that of $\mathbb{R}^{\mathfrak{R}}$. Thus $M(\mathfrak{R})$ is not a vector lattice of functions in the sense of Chapter 2. Theorem 8.4 shows, however, that suprema and infima in the two vector lattices agree in the case of directed families.

As promised, we now show how to characterize $M(\mathfrak{R})$ yet in another manner. Let $\mathfrak{R}$ be a ring of sets. For $A \in \mathfrak{R}$, let $\Delta(A)$ denote the set of all finite disjoint families in $\mathfrak{R}$ whose union is A.

Consider an additive, real-valued function μ on $\mathfrak{R}$. Given $A \in \mathfrak{R}$, put

$$V\mu(A) := \sup \left\{ \sum_{\iota \in I} |\mu(A_\iota)| \,\Big|\, (A_\iota)_{\iota \in I} \in \Delta(A) \right\}.$$

$V\mu(A)$ is called the **variation of** μ **on** A. μ is said to be of **locally finite variation** on $\Re$ if $V\mu(A) < \infty$ for all $A \in \Re$.

Theorem 8.5 *Given an additive map* $\mu : \Re \to \mathbb{R}$, *the following are equivalent.*

(a) $\mu \in M(\Re)$.

(b) $V\mu(A) < \infty$ *for all* $A \in \Re$.

If these conditions are satisfied, then $V\mu$ *is a positive content and* $V\mu = |\mu|$.

Proof. (a)$\Rightarrow$(b). Take $\mu \in M(\Re)$. Then for each $A \in \Re$,

$$V\mu(A) \leq \sup \left\{ \sum_{\iota \in I} |\mu|(A_\iota) \,\middle|\, (A_\iota)_{\iota \in I} \in \Delta(A) \right\} = |\mu|(A) < \infty.$$

In particular, $V\mu \leq |\mu|$.

(b)$\Rightarrow$(a) is a result of Corollary 8.2 (b)$\Rightarrow$(a).

We now prove the additional claims. Take $A, B \in \Re$ with $A \cap B = \emptyset$. Then for arbitrary $(A_\iota)_{\iota \in I} \in \Delta(A)$ and $(B_\lambda)_{\lambda \in L} \in \Delta(B)$,

$$\sum_{\iota \in I} |\mu(A_\iota)| + \sum_{\lambda \in L} |\mu(B_\lambda)| \leq V\mu(A \cup B)$$

and hence

$$V\mu(A) + V\mu(B) \leq V\mu(A \cup B).$$

Conversely, if $(C_\iota)_{\iota \in I} \in \Delta(A \cup B)$, then

$$(C_\iota \cap A)_{\iota \in I} \in \Delta(A), \quad (C_\iota \cap B)_{\iota \in I} \in \Delta(B),$$

and therefore

$$\sum_{\iota \in I} |\mu(C_\iota)| \leq \sum_{\iota \in I} |\mu(C_\iota \cap A)| + \sum_{\iota \in I} |\mu(C_\iota \cap B)| \leq V\mu(A) + V\mu(B).$$

Thus,

$$V\mu(A \cup B) \leq V\mu(A) + V\mu(B).$$

Hence $V\mu$ is a positive content.

Clearly $\mu \leq V\mu$ and $-\mu \leq V\mu$, so $|\mu| = \mu \vee (-\mu) \leq V\mu$. Combining inequalities, we conclude that $V\mu = |\mu|$. $\qquad\square$

Thus the elements of $M(\Re)$ are precisely the additive real-valued functions of locally finite variation on $\Re$.

As we will see in the exercises, not every additive map on a ring of sets is a content according to our definition. Theorem 8.1 shows, however, that the definition of a content given in this section leads to interesting structures. We define measures in a similar manner.

A mapping μ of a ring of sets $\Re$ into $\mathbb{R}$ is called a **(real) measure** on $\Re$ if it is representable as the difference of two positive measures on $\Re$. Let $M^\sigma(\Re)$ denote the set of all measures on $\Re$. $(X, \Re, \mu)$ is called a **(real) measure space** if X is a set, $\Re \subset \mathfrak{P}(X)$ is a ring of sets and $\mu \in M^\sigma(\Re)$.

$M^\sigma(\mathfrak{R})$ is not merely a subset of $M(\mathfrak{R})$! It is rather a vector sublattice of 'the best kind', in the sense of our next theorem.

Theorem 8.6 *$M^\sigma(\mathfrak{R})$ is a band in $M(\mathfrak{R})$ and thus a complete vector lattice. Moreover, suprema and infima in $M^\sigma(\mathfrak{R})$ are formed in accordance with Theorems 8.3 and 8.4.*

Proof. Clearly $M^\sigma(\mathfrak{R})$ is a vector subspace of $M(\mathfrak{R})$. Take $\mu \in M^\sigma(\mathfrak{R})$ and $\nu \in M(\mathfrak{R})$ such that $|\nu| \le |\mu|$. By definition, there are positive measures μ_1, μ_2 on $\mathfrak{R}$ such that $\mu = \mu_1 - \mu_2$. Then by Theorem 7.2(c), $\mu^+ \le \mu_1$ and $\mu^- \le \mu_2$. By (e) of the same theorem,

$$0 \le \nu^+ + \nu^- = |\nu| \le |\mu| = \mu^+ + \mu^- \le \mu_1 + \mu_2,$$

and so $\nu^+ \le \mu_1 + \mu_2$ and $\nu^- \le \mu_1 + \mu_2$. Moreover, for each sequence $(A_n)_{n\in\mathbb{N}}$ in $\mathfrak{R}$ for which $A_n \downarrow \emptyset$,

$$0 \le \inf_{n\in\mathbb{N}} \nu^+(A_n) \le \inf_{n\in\mathbb{N}} (\mu_1 + \mu_2)(A_n) = 0,$$

and similarly $\inf_{n\in\mathbb{N}} \nu^-(A_n) = 0$. Therefore ν^+ and ν^- are positive measures. Then $\nu = \nu^+ - \nu^- \in M^\sigma(\mathfrak{R})$ and we conclude that $M^\sigma(\mathfrak{R})$ is a solid subspace of $M(\mathfrak{R})$.

Now let $(\nu_\iota)_{\iota\in I}$ be an upward directed family in $M^\sigma(\mathfrak{R})_+$ such that $\nu := \bigvee_{\iota\in I} \nu_\iota \in M(\mathfrak{R})_+$ exists. We show that ν is null-continuous. Let $(A_n)_{n\in\mathbb{N}}$ be a sequence in $\mathfrak{R}$ such that $A_n \downarrow \emptyset$. Let $\varepsilon > 0$. By Theorem 8.4(a), there is a $\lambda \in I$ with

$$\nu(A_1) - \nu_\lambda(A_1) < \varepsilon$$

and therefore

$$(\nu - \nu_\lambda)(A_n) \le (\nu - \nu_\lambda)(A_1) < \varepsilon$$

for all $n \in \mathbb{N}$. It follows from

$$\nu(A_n) = \nu_\lambda(A_n) + (\nu - \nu_\lambda)(A_n) \le \nu_\lambda(A_n) + \varepsilon$$

and the null-continuity of ν_λ that

$$\inf_{n\in\mathbb{N}} \nu(A_n) \le \varepsilon.$$

Then since ε was arbitrary,

$$\inf_{n\in\mathbb{N}} \nu(A_n) = 0.$$

Thus, in view of Proposition 7.7, $M^\sigma(\mathfrak{R})$ is a band in $M(\mathfrak{R})$. By Theorems 8.1 and 7.8, $M^\sigma(\mathfrak{R})$ is complete.

That Theorem 8.3 holds for $M^\sigma(\mathfrak{R})$, follows from the vector sublattice property of $M^\sigma(\mathfrak{R})$ and Theorem 7.6(a). Similarly, the band property of $M^\sigma(\mathfrak{R})$ and Theorem 7.6(d) establish Theorem 8.4 for $M^\sigma(\mathfrak{R})$.   □

A characterization of the real measures, similar to that in Theorem 8.5 for real contents, is provided in the exercises.

Consider a Hausdorff space X. Let $M_r(X)$ denote the set of all real-valued functions on $\mathfrak{R}(X)$ which are representable as the difference of two regular positive measures on $\mathfrak{R}(X)$. We call the elements of $M_r(X)$ **regular measures** on $\mathfrak{R}(X)$. Then $M_r(X) \subset M^\sigma(\mathfrak{R}(X)) \subset M(\mathfrak{R}(X))$, and the same strong subspace properties hold for $M_r(X)$ as for $M^\sigma(\mathfrak{R})$:

Theorem 8.7 $M_r(X)$ *is a band in* $M^\sigma(\mathfrak{R}(X))$ *and therefore also a band in* $M(\mathfrak{R}(X))$. *Thus* $M_r(X)$ *is a complete vector lattice, and the formulae of Theorems 8.3 and 8.4 also apply in* $M_r(X)$.

Proof. It is easy to show that $M_r(X)$ is a vector subspace of $M^\sigma(\mathfrak{R}(X))$. Take $\mu \in M_r(X)_+$ and $\nu \in M^\sigma(\mathfrak{R}(X))_+$ with $\nu \leq \mu$. Take $A \in \mathfrak{R}(X)$ and $\varepsilon > 0$. Then there is a set $K \in \mathfrak{K}(X)$, $K \subset A$, such that $\mu(A \setminus K) < \varepsilon$. It follows that

$$\nu(A \setminus K) \leq \mu(A \setminus K) < \varepsilon.$$

This shows that ν is regular and hence $\nu \in M_r(X)$. Now if $\mu \in M_r(X)$ and $\nu \in M^\sigma(\mathfrak{R}(X))$ satisfy $|\nu| \leq |\mu|$, then by decomposing ν into ν^+ and ν^- as in the proof of Theorem 8.6, we see that $\nu \in M_r(X)$. $M_r(X)$ is thus a solid subspace of $M^\sigma(\mathfrak{R}(X))$.

Let $(\nu_\iota)_{\iota \in I}$ be an upward directed family in $M_r(X)_+$ for which $\nu := \bigvee_{\iota \in I} \nu_\iota$ exists in $M^\sigma(\mathfrak{R}(X))$. Then given $A \in \mathfrak{R}(X)$ and $\varepsilon > 0$, Theorem 8.4(a) ensures the existence of an $\iota \in I$ with

$$\nu(A) - \nu_\iota(A) < \varepsilon/2.$$

Moreover, there is a $K \in \mathfrak{K}(X)$ with $K \subset A$, for which

$$\nu_\iota(A) - \nu_\iota(K) < \varepsilon/2.$$

Then

$$\nu(A) - \nu(K) \leq \nu(A) - \nu_\iota(K) = \big(\nu(A) - \nu_\iota(A)\big) + \big(\nu_\iota(A) - \nu_\iota(K)\big) < \varepsilon,$$

and we conclude that ν is regular. This proves the first part of the theorem. For the remaining assertions, use the same arguments as in the preceding theorem. $\square$

We again consider the particularly important special case of a locally compact space X. In the Riesz representation theorem we assigned to each positive linear functional ℓ on $\mathcal{K}(X)$ a regular positive measure μ_ℓ on $\mathfrak{R}(X)$. Let $\mathcal{K}(X)^\delta$ denote the set of all linear functionals on $\mathcal{K}(X)$ which may be expressed as the difference of two positive linear functionals. $\mathcal{K}(X)^\delta$ is called the **order dual** of $\mathcal{K}(X)$. It follows immediately that to each element of $\mathcal{K}(X)^\delta$ there corresponds a regular measure on $\mathfrak{R}(X)$. The next result, which may be considered an extension of the Riesz representation theorem, states that actually more is true: the correspondence is obtained in such a way that it establishes a vector lattice isomorphism between $\mathcal{K}(X)^\delta$ and $M_r(X)$ and, in particular, every regular measure on $\mathfrak{R}(X)$ arises from an element of $K(X)^\delta$ in this way. In brief, $M_r(X)$ is (isomorphic as vector lattice to) the order dual of $K(X)$.

Theorem 8.8

(a) $K(X)^\delta$ *is a complete vector lattice with respect to the order relation*

$$\ell_1 \le \ell_2 \quad :\Longleftrightarrow \quad \ell_2 - \ell_1 \text{ is a positive linear functional.}$$

(b) *There is a unique linear map* $\varphi : K(X)^\delta \to M_r(X)$ *such that* $\varphi(\ell) = \mu_\ell$ *for each positive linear functional* ℓ. φ *is an isomorphism of the vector lattices* $K(X)^\delta$ *and* $M_r(X)$.

Proof. It is easy to see that $K(X)^\delta$ is an ordered vector space with respect to the given ordering.

Given a positive linear functional ℓ, put $\varphi(\ell) := \mu_\ell$. If ℓ_1 and ℓ_2 are positive and linear on $K(X)$, then for each $K \in \mathfrak{R}(X)$,

$$
\begin{aligned}
\mu_{\ell_1+\ell_2}(K) &= \inf\{(\ell_1 + \ell_2)(f) \mid f \in K(X),\, f \ge e_K\} \\
&= \inf\{\ell_1(f) + \ell_2(f) \mid f \in K(X),\, f \ge e_K\} \\
&= \inf\{\ell_1(f) \mid f \in K(X),\, f \ge e_K\} + \inf\{\ell_2(f) \mid f \in K(X),\, f \ge e_K\} \\
&= \mu_{\ell_1}(K) + \mu_{\ell_2}(K).
\end{aligned}
$$

As a result,

$$\mu_{\ell_1+\ell_2}\big|_{\mathfrak{R}(X)} = \left(\mu_{\ell_1} + \mu_{\ell_2}\right)\big|_{\mathfrak{R}(X)},$$

and because the extension to $\mathfrak{R}(X)$ is unique,

$$\varphi(\ell_1 + \ell_2) = \mu_{\ell_1+\ell_2} = \mu_{\ell_1} + \mu_{\ell_2} = \varphi(\ell_1) + \varphi(\ell_2).$$

Take $\ell \in K(X)^\delta$ and let $\ell_1, \ell_2, \ell_1', \ell_2'$ be positive linear functionals such that

$$\ell = \ell_1 - \ell_2 = \ell_1' - \ell_2'.$$

Then $\ell_1 + \ell_2' = \ell_1' + \ell_2$ and hence $\varphi(\ell_1) + \varphi(\ell_2') = \varphi(\ell_1') + \varphi(\ell_2)$. Therefore

$$\varphi(\ell_1) - \varphi(\ell_2) = \varphi(\ell_1') - \varphi(\ell_2').$$

Consequently, we define

$$\varphi : K(X)^\delta \longrightarrow M_r(X), \quad \ell \longmapsto \varphi(\ell_1) - \varphi(\ell_2),$$

where $\ell = \ell_1 - \ell_2$ and ℓ_1, ℓ_2 are positive linear functionals. We leave it to the reader to show that φ is linear. The uniqueness of φ is obvious.

Take $\ell \in K(X)^\delta$ with $\varphi(\ell) = 0$. Let ℓ_1, ℓ_2 be positive linear functionals with $\ell = \ell_1 - \ell_2$. Then $\varphi(\ell_1) - \varphi(\ell_2) = 0$, and therefore $\varphi(\ell_1) = \varphi(\ell_2)$. Then for all $f \in K(X)$,

$$\ell_1(f) = \int f \, d\mu_{\ell_1} = \int f \, d\mu_{\ell_2} = \ell_2(f),$$

by Theorem 5.20. Thus $\ell_1 = \ell_2$, and $\ell = 0$. This shows that φ is injective.

Take $\mu \in M_r(X)_+$. Then, by Theorem 5.21(a), there is a positive linear functional ℓ on $\mathcal{K}(X)$ such that $\mu = \mu_\ell$. We conclude that φ is also surjective and thus bijective. φ^{-1} is clearly linear.

Take $\ell_1, \ell_2 \in \mathcal{K}(X)^\delta$ with $\ell_1 \leq \ell_2$. Then $\ell_2 - \ell_1$ is a positive linear functional and hence

$$\varphi(\ell_2) - \varphi(\ell_1) = \varphi(\ell_2 - \ell_1) = \mu_{\ell_2 - \ell_1} \geq 0.$$

Thus $\varphi(\ell_1) \leq \varphi(\ell_2)$ and so φ is increasing.

If $\mu, \nu \in M_r(X)$, $\mu \leq \nu$, then $\nu - \mu \geq 0$ and by Theorem 5.20

$$\left(\varphi^{-1}(\nu - \mu)\right)(f) = \int f \, d(\nu - \mu) \geq 0$$

for every $f \in \mathcal{K}(X)_+$. Thus

$$\varphi^{-1}(\nu) - \varphi^{-1}(\mu) = \varphi^{-1}(\nu - \mu) \geq 0$$

and $\varphi^{-1}(\nu) \geq \varphi^{-1}(\mu)$. Thus φ^{-1} is also increasing. The conclusion of the theorem now follows from Corollary 7.14. □

The fact that $\mathcal{K}(X)^\delta$ is a complete vector lattice can be generalized substantially. Indeed, if E is an arbitrary vector lattice and E^δ denotes the set of all differences of two positive linear functionals on E, then it can be shown that E^δ is a complete vector lattice with respect to the order relation described in Theorem 8.8.

An obvious question is whether the space $\mathcal{K}(X)'$ of all continuous linear functionals on $\mathcal{K}(X)$ (where $\mathcal{K}(X)$ is equipped with the supremum norm) can be characterized in a manner similar to $\mathcal{K}(X)^\delta$. Indeed, this space can be identified with the complete vector lattice of bounded regular measures on $\mathfrak{R}(X)$, as we shall see in the exercises. In particular, it turns out that $\mathcal{K}(X)'$ is a solid subspace of $\mathcal{K}(X)^\delta$.

Theorem 4.1 implies that each positive measure on a ring of sets $\mathfrak{R}$ can be extended to a positive measure on a δ-ring containing $\mathfrak{R}$. Then, in particular, each positive measure μ on $\mathfrak{R}$ has an extension to $\mathfrak{R}_\delta$, and, by Theorem 4.16, it is uniquely determined. We write $\bar{\mu}$ to denote this extension. For arbitrary $\mu \in M^\sigma(\mathfrak{R})$, we define $\bar{\mu} := \overline{\mu^+} - \overline{\mu^-}$. We now turn our attention to the map

$$\psi : M^\sigma(\mathfrak{R}) \longrightarrow M^\sigma(\mathfrak{R}_\delta), \quad \mu \longmapsto \bar{\mu}.$$

We have:

Theorem 8.9 ψ *is a vector lattice isomorphism and* $\psi^{-1}(\nu) = \nu|_{\mathfrak{R}}$ *for every* $\nu \in M^\sigma(\mathfrak{R}_\delta)$.

Proof. We first show that ψ is additive. Take $\mu, \nu \in M^\sigma(\mathfrak{R})$. Then

$$(\mu + \nu)^+ - (\mu + \nu)^- = \mu + \nu = \mu^+ - \mu^- + \nu^+ - \nu^-,$$

and therefore

$$(\mu + \nu)^+ + \mu^- + \nu^- = (\mu + \nu)^- + \mu^+ + \nu^+.$$

Then

$$\overline{(\mu+\nu)^+ + \mu^- + \nu^-} = \overline{(\mu+\nu)^- + \mu^+ + \nu^+}.$$

But clearly $\overline{(\mu+\nu)^+} + \overline{\mu^-} + \overline{\nu^-}$ is also an extension of $(\mu+\nu)^+ + \mu^- + \nu^-$. Since the extension is unique by Theorem 4.16, it follows that

$$\overline{(\mu+\nu)^+ + \mu^- + \nu^-} = \overline{(\mu+\nu)^+} + \overline{\mu^-} + \overline{\nu^-}.$$

Similarly,

$$\overline{(\mu+\nu)^- + \mu^+ + \nu^+} = \overline{(\mu+\nu)^-} + \overline{\mu^+} + \overline{\nu^+}.$$

Thus we obtain

$$\overline{\mu+\nu} = \overline{(\mu+\nu)^+} - \overline{(\mu+\nu)^-} = \overline{\mu^+} - \overline{\mu^-} + \overline{\nu^+} - \overline{\nu^-} = \overline{\mu} + \overline{\nu}.$$

We leave it to the reader to show that $\overline{\alpha\mu} = \alpha\overline{\mu}$ for all $\alpha \in \mathbb{R}$. Thus ψ is a linear map.

ψ is injective: Take $\mu \in M^\sigma(\mathfrak{R})$ with $\overline{\mu} = 0$. Then $\overline{\mu^+} - \overline{\mu^-} = 0$; that is, $\overline{\mu^+} = \overline{\mu^-}$. It follows that $\mu^+ = \mu^-$, and hence $\mu = \mu^+ - \mu^- = 0$.

ψ is surjective: Take $\nu \in M^\sigma(\mathfrak{R}_\delta)$. Put $\mu_1 := \nu^+|_\mathfrak{R}$, $\mu_2 := \nu^-|_\mathfrak{R}$, and $\mu := \mu_1 - \mu_2$. Since ψ is linear, it follows that

$$\overline{\mu} = \overline{\mu_1} - \overline{\mu_2} = \nu^+ - \nu^- = \nu.$$

Thus ψ is a bijective map and

$$\psi^{-1}(\nu) = \mu = \nu|_\mathfrak{R}.$$

It immediately follows that ψ^{-1} is increasing. But ψ is also increasing, as $\mu \geq 0$ implies that $\overline{\mu} \geq 0$. Thus, by Proposition 7.13, ψ is a vector lattice isomorphism. $\qquad\square$

Hence every measure on $\mathfrak{R}_\delta$ is obtained by extending to $\mathfrak{R}_\delta$ an appropriate measure defined on $\mathfrak{R}$. The preceding theorem tells us that this process of extension obeys several natural rules. For instance, if $\mu \leq \nu$ for $\mu,\nu \in M^\sigma(\mathfrak{R})$, then $\overline{\mu} \leq \overline{\nu}$. Moreover, the extension of $|\mu|$ is the absolute value of the extension of μ, and so on.

Using Theorem 8.9, we now prove an important proposition about the structure of real-valued measures. We may decompose these measures into positive and negative parts in the sense of the theory of vector lattices. We shall see that this leads to a decomposition of the sets in $\mathfrak{R}_\delta$.

Theorem 8.10 (Hahn) *Let $(X, \mathfrak{R}, \mu)$ be a measure space. For each set $A \in \mathfrak{R}_\delta$ there are disjoint sets $B, C \in \mathfrak{R}_\delta$ such that*

(i) $A = B \cup C$;

(ii) $B \in \mathfrak{N}(\mu^+)$ *and therefore $\overline{\mu}(D) \leq 0$ for all $D \in \mathfrak{R}_\delta$ with $D \subset B$;*

(iii) $C \in \mathfrak{N}(\mu^-)$ *and therefore $\overline{\mu}(D) \geq 0$ for all $D \in \mathfrak{R}_\delta$ with $D \subset C$.*

Proof. Put $\nu_1 := \overline{\mu^+}$ and $\nu_2 := \overline{\mu^-}$. By Theorem 8.9, $\nu_1 \wedge \nu_2 = 0$. Take $A \in \mathfrak{R}_\delta$. Then, by Theorem 8.3(b), for each $n \in \mathbb{N}$ there is a set $B_n \in \mathfrak{R}_\delta$ such that $B_n \subset A$ and

$$\nu_1(B_n) + \nu_2(A \setminus B_n) < 1/2^n.$$

Put

$$B := \bigcap_{n \in \mathbb{N}} \bigcup_{m \geq n} B_m$$

and

$$C := A \setminus B = A \setminus \bigcap_{n \in \mathbb{N}} \bigcup_{m \geq n} B_m = \bigcup_{n \in \mathbb{N}} \bigcap_{m \geq n} (A \setminus B_m).$$

B and C are contained in $\mathfrak{R}_\delta$. Furthermore, $B \cap C = \emptyset$ and $B \cup C = A$. For each $n \in \mathbb{N}$, we have $\nu_1(B_n) < 1/2^n$ and $\nu_2(A \setminus B_n) < 1/2^n$. Therefore

$$\nu_1\left(\bigcup_{m \geq n} B_m \right) \leq \sum_{m \geq n} \nu_1(B_m) \leq 1/2^{n-1}$$

and

$$\nu_2\left(\bigcap_{m \geq n} (A \setminus B_m) \right) = 0.$$

By the first relation

$$\nu_1(B) = \nu_1\left(\bigcap_{n \in \mathbb{N}} \bigcup_{m \geq n} B_m \right) = 0,$$

and by the second

$$\nu_2(A \setminus B) = \nu_2\left(\bigcup_{n \in \mathbb{N}} \bigcap_{m \geq n} (A \setminus B_m) \right) = 0.$$

Thus $B \in \mathfrak{N}(\mu^+)$ and $C \in \mathfrak{N}(\mu^-)$.

Finally, take $D \in \mathfrak{R}_\delta$. If $D \subset B$, then

$$\overline{\mu}(D) = \overline{\mu^+}(D) - \overline{\mu^-}(D) = -\overline{\mu^-}(D) \leq 0.$$

If $D \subset C$, then

$$\overline{\mu}(D) = \overline{\mu^+}(D) - \overline{\mu^-}(D) = \overline{\mu^+}(D) \geq 0.$$

$\square$

Thus, restricted to A, μ^+ 'lives' only on the set C, while μ^- 'lives' only on its complement (in A) B. We call a decomposition of a set $A \in \mathfrak{R}_\delta$ with the properties listed in Hahn's theorem a **Hahn decomposition** of A with respect to μ. It is left as an exercise for the reader to prove that the Hahn decomposition of $A \in \mathfrak{R}_\delta$ is unique up to μ-null sets.

We may state the theorem in a more general way. In fact, it can be shown that each μ-integrable set can be decomposed into a μ^+-null set and a μ^--null set. This, too, is postponed to the exercises.

We do not discuss integration with respect to a real measure in this book. But for the sake of completeness, we at least mention *how* such an integral is defined. It is defined simply by means of integration with respect to positive measures. Thus, take $\mu \in M^\sigma(\mathfrak{R})$. Then $f \in \bar{\mathbb{R}}^X$ is called **μ-integrable** if $f \in \mathcal{L}^1(\mu^+) \cap \mathcal{L}^1(\mu^-)$ (or equivalently, if $f \in \mathcal{L}^1(|\mu|)$), and in this case one defines

$$\int f \, d\mu := \int f \, d\mu^+ - \int f \, d\mu^-.$$

For example, consider $\varphi : \mathcal{K}(X)^\delta \to M_r(X)$, the isomorphism of vector lattices in Theorem 8.8, take $\ell \in \mathcal{K}(X)^\delta$ and put $\mu := \varphi(\ell)$. Then

$$\ell(f) = \int f \, d\mu$$

for every $f \in \mathcal{K}(X)$.

Exercises

1. Let λ be Lebesgue measure and δ_x Dirac measure at the point $x \in \mathbb{R}$. Describe $\lambda \wedge \delta_x$, $\lambda \vee \delta_x$, $(\lambda - \delta_x)^+$, $(\lambda - \delta_x)^-$ and $|\lambda - \delta_x|$.

2. Define

$$M_b(\mathfrak{R}) := \{\mu \in M(\mathfrak{R}) \,|\, \sup_{A \in \mathfrak{R}} |\mu|(A) < \infty\},$$

$$M_b^\sigma(\mathfrak{R}) := M_b(\mathfrak{R}) \cap M^\sigma(\mathfrak{R})$$

and

$$\|\mu\| := \sup_{A \in \mathfrak{R}} |\mu|(A) \qquad \text{for all } \mu \in M_b(\mathfrak{R}).$$

Prove the following.

(a) $M_b(\mathfrak{R})$ is a solid subspace of $M(\mathfrak{R})$. $M_b^\sigma(\mathfrak{R})$ is a band in $M_b(\mathfrak{R})$ and a solid subspace of $M^\sigma(\mathfrak{R})$. Hence $M_b(\mathfrak{R})$ and $M_b^\sigma(\mathfrak{R})$ are complete vector lattices.

(b) $\|\cdot\|$ is a norm on $M_b(\mathfrak{R})$.

(c) $\|\mu + \nu\| = \|\mu\| + \|\nu\|$ for all $\mu, \nu \in M_b(\mathfrak{R})$ with $\mu, \nu \geq 0$.

(d) If $(\mu_n)_{n \in \mathbb{N}}$ is an increasing sequence in $(M_b(\mathfrak{R}))_+$ $((M_b^\sigma(\mathfrak{R}))_+)$ with $\sup_{n \in \mathbb{N}} \|\mu_n\| < \infty$, then $\bigvee_{n \in \mathbb{N}} \mu_n = \lim_{n \to \infty} \mu_n$ exists in $M_b(\mathfrak{R})$ $(M_b^\sigma(\mathfrak{R}))$.

(e) $M_b(\mathfrak{R})$ and $M_b^\sigma(\mathfrak{R})$ are norm complete; hence $M_b^\sigma(\mathfrak{R})$ is a closed subspace of $M_b(\mathfrak{R})$.

(f) If $\mathfrak{R}$ is a σ-ring, then $M^\sigma(\mathfrak{R}) = M_b^\sigma(\mathfrak{R})$.

Hint for (e): Let $(\mu_n)_{n \in \mathbb{N}}$ be a Cauchy sequence in $M_b(\mathfrak{R})$. For each $A \in \mathfrak{R}$, put $\mu(A) := \lim_{n \to \infty} \mu_n(A)$. Then μ is additive and satisfies condition (b) of Corollary 8.2. Hence $\mu \in M_b(\mathfrak{R})$. Furthermore, $|\mu|(A) =$

$\lim_{n\to\infty} |\mu_n|(A)$ for each $A \in \mathfrak{R}$. If all μ_n belong to $M_b^\sigma(\mathfrak{R})$, conclude from the last statement that $|\mu| \in M_b^\sigma(\mathfrak{R})$ and hence $\mu \in M_b^\sigma(\mathfrak{R})$ as well. Finally, show that $\mu = \lim_{n\to\infty} \mu_n$.

3. For $\mathfrak{R} := \mathfrak{P}(\mathbb{N})$ and $\mathfrak{S} := \{A \subset \mathbb{N} \mid A \text{ is finite}\}$ describe $M(\mathfrak{R})$, $M^\sigma(\mathfrak{R})$, $M_b(\mathfrak{R})$, $M_b^\sigma(\mathfrak{R})$, $M(\mathfrak{S})$, $M^\sigma(\mathfrak{S})$, $M_b(\mathfrak{S})$, $M_b^\sigma(\mathfrak{S})$.

4. A map $\mu : \mathfrak{R} \to \mathbb{R}$ is called **σ-additive** if $(\mu(A_n))_{n\in\mathbb{N}}$ is summable and $\mu\left(\bigcup_{n\in\mathbb{N}} A_n\right) = \sum_{n\in\mathbb{N}} \mu(A_n)$ whenever $(A_n)_{n\in\mathbb{N}}$ is a disjoint sequence in $\mathfrak{R}$ with $\bigcup_{n\in\mathbb{N}} A_n \in \mathfrak{R}$. Show that for an additive map $\mu : \mathfrak{R} \to \mathbb{R}$ the following are equivalent.

(a) μ is σ-additive.
(b) $\mu\left(\bigcup_{n\in\mathbb{N}} A_n\right) = \lim_{n\to\infty} \mu(A_n)$ for each increasing sequence $(A_n)_{n\in\mathbb{N}}$ in $\mathfrak{R}$ with $\bigcup_{n\in\mathbb{N}} A_n \in \mathfrak{R}$.
(c) $\mu\left(\bigcap_{n\in\mathbb{N}} A_n\right) = \lim_{n\to\infty} \mu(A_n)$ for each decreasing sequence $(A_n)_{n\in\mathbb{N}}$ in $\mathfrak{R}$ with $\bigcap_{n\in\mathbb{N}} A_n \in \mathfrak{R}$.
(d) $\lim_{n\to\infty} \mu(A_n) = 0$ for each decreasing sequence $(A_n)_{n\in\mathbb{N}}$ in $\mathfrak{R}$ with $\bigcap_{n\in\mathbb{N}} A_n = \emptyset$.

Show further that each $\mu \in M^\sigma(\mathfrak{R})$ is σ-additive.

5. Let X be an uncountable set and define
$$\mathfrak{R} := \{A \subset X \mid A \text{ is finite or } X \setminus A \text{ is finite}\},$$
$$\mu : \mathfrak{R} \longrightarrow \mathbb{R}, \quad A \longmapsto \begin{cases} \operatorname{card} A & \text{if } A \text{ is finite} \\ -\operatorname{card} X \setminus A & \text{if } X \setminus A \text{ is finite.} \end{cases}$$

Verify that μ is σ-additive (and in particular additive), but not a content.

6. For $A \in \mathfrak{R}$ define
$$\Delta^\sigma(A) := \Big\{ (A_\iota)_{\iota\in I} \,\Big|\, I \text{ countable, } A_\iota \in \mathfrak{R} \text{ for all } \iota \in I,$$
$$A = \bigcup_{\iota\in I} A_\iota, \ A_\iota \cap A_\lambda = \emptyset \text{ if } \iota \neq \lambda \Big\}.$$

Prove that
$$V\mu(A) = \sup\Big\{ \sum_{\iota\in I} |\mu(A_\iota)| \,\Big|\, (A_\iota)_{\iota\in I} \in \Delta^\sigma(A) \Big\}$$
for every $\mu \in M(\mathfrak{R})$ and every $A \in \mathfrak{R}$.

7. Prove the following statements.

(a) If μ is a σ-additive real map, then $V\mu$ is also σ-additive.
(b) Given $\mu \in \mathbb{R}^\mathfrak{R}$, the following are equivalent.

(b1) $\mu \in M^\sigma(\mathfrak{R})$.
(b2) μ is σ-additive and of locally bounded variation.

8. Let μ be an additive real map on $\mathfrak{R}$. Take $A \in \mathfrak{R}$ and $\alpha \in \mathbb{R}$ with $2\alpha < V\mu(A) - |\mu(A)|$. Show that there are disjoint sets $B, C \in \mathfrak{R}$ such that $B \cup C = A$ and $|\mu(B)| > \alpha$, $|\mu(C)| > \alpha$.
   Hint: Take $(A_\iota)_{\iota \in I} \in \Delta(A)$ with $\sum_{\iota \in I} |\mu(A_\iota)| > 2\alpha + |\mu(A)|$. Put $B := \bigcup_{\iota \in I, \mu(A_\iota) \geq 0} A_\iota$ and $C := A \setminus B$.

9. Suppose that $\mathfrak{R}$ is a δ-ring. Prove the following.

   (a) If $\mu \in \mathbb{R}^{\mathfrak{R}}$ is σ-additive, then μ is of locally bounded variation.
   (b) Given $\mu \in \mathbb{R}^{\mathfrak{R}}$, the following are equivalent.

   (b1) $\mu \in M^\sigma(\mathfrak{R})$.
   (b2) μ is σ-additive.

   Hint for (a): Assume the existence of $A \in \mathfrak{R}$ with $V\mu(A) = \infty$. Apply Exercise 8 to construct recursively a sequence $(A_n)_{n \in \mathbb{N}}$ in $\mathfrak{R}$ such that $A_n \subset A \setminus \bigcup_{m<n} A_m$, $V\mu(A \setminus \bigcup_{m \leq n} A_m) = \infty$ and $|\mu(A_n)| > 1$ for each $n \in \mathbb{N}$.

10. Let X be a locally compact space. Define
$$\|f\| := \sup_{x \in X} |f(x)| \qquad \text{for } f \in \mathcal{K}(X).$$

   (a) Show that $\|\cdot\|$ is a norm on $\mathcal{K}(X)$.

   Define
$$\mathcal{K}(X)' := \{\ell \in \mathcal{K}(X)^\delta \,|\, \ell \text{ is continuous with respect to } \|\cdot\|\}$$
   and
$$M_{rb}(X) := M_r(X) \cap M_b^\sigma(\mathfrak{R}(X)).$$
   Prove the following.

   (b) $M_{rb}(X)$ is a solid subspace of $M_r(X)$ and a band in $M_b^\sigma(\mathfrak{R}(X))$. It is thus a complete vector lattice.
   (c) Take $\mu \in M_r(X)$, $L \in \mathfrak{K}(X)$ and $\varepsilon > 0$. Then there is an $f \in \mathcal{K}(X)$ such that $e_L \leq f \leq e_X$ and
$$\left| \int f \, d\mu - \mu(L) \right| < \varepsilon.$$

   (d) If φ denotes the map in Theorem 8.8, then $\varphi(\mathcal{K}(X)') = M_{rb}(X)$ and the map
$$\mathcal{K}(X)' \longrightarrow M_{rb}(X), \quad \ell \longmapsto \varphi(\ell)$$
   is an isomorphism of vector lattices. In particular, $\mathcal{K}(X)'$ is a solid subspace of $\mathcal{K}(X)^\delta$ and hence a complete vector lattice.

   Hint for (d): Take $\mu \in M_{rb}(X)$ and $f \in \mathcal{K}(X)$. It is easy to see that $|\int f \, d\mu| \leq \|f\| \|\mu\|$. Hence $\ell := \varphi^{-1}(\mu)$ is continuous at 0 and so it is continuous.

For the converse, take $\mu \in M_r(X) \setminus M_{rb}(X)$ and construct a sequence $(K_n)_{n\in\mathbb{N}}$ in $\mathfrak{K}(X)$ with $|\mu|(K_n) > 2n$ for each $n \in \mathbb{N}$. Given $n \in \mathbb{N}$, there is an $L_n \in \mathfrak{K}(X)$ with $L_n \subset K_n$ and $|\mu(L_n)| > n$ (Theorem 8.3(e)). By (c), there is an $f_n \in \mathcal{K}(X)$ with $\|f_n\| \leq 1/n$ and $|\int f_n d\mu| > 1$. Hence $\ell := \varphi^{-1}(\mu)$ is not continuous.

11. Take $\mu \in M^\sigma(\mathfrak{R})$. Show that in general the Hahn decomposition is not unique, but that given two Hahn decompositions (B_1, C_1) and (B_2, C_2) of a set $A \in \mathfrak{R}_\delta$, $\bar{\mu}(B_1 \triangle B_2) = 0 = \bar{\mu}(C_1 \triangle C_2)$.
    Hint: $B_1 \setminus B_2 \subset B_1 \cap C_2$.

12. Given $\mu \in M^\sigma(\mathfrak{R})$, prove that for each $A \in \mathfrak{R}_\sigma$ there are disjoint sets $B, C \in \mathfrak{R}_\sigma$ such that

    (i) $A = B \cup C$;
    (ii) $B \in \mathfrak{N}(\mu^+)$ and hence $\bar{\mu}(D) \leq 0$ for every $D \in \mathfrak{R}_\delta$ with $D \subset B$;
    (iii) $C \in \mathfrak{N}(\mu^-)$ and hence $\bar{\mu}(D) \geq 0$ for every $D \in \mathfrak{R}_\delta$ with $D \subset C$.

13. Given $\mu \in M^\sigma(\mathfrak{R})$, put $\mathfrak{L} := \mathfrak{L}(\mu) := \mathfrak{L}(\mu^+) \cap \mathfrak{L}(\mu^-)$ and $\mu^X := (\mu^+)^X|_\mathfrak{L} - (\mu^-)^X|_\mathfrak{L} \in M^\sigma(\mathfrak{L})$. Prove the following.

    (a) $(\mu^+)^X|_\mathfrak{L} \wedge (\mu^-)^X|_\mathfrak{L} = 0$.
    (b) $(\mu^X)^+ = (\mu^+)^X|_\mathfrak{L}$ and $(\mu^X)^- = (\mu^-)^X|_\mathfrak{L}$.
    (c) For each $A \in \mathfrak{L}$ there are disjoint sets $B, C \in \mathfrak{L}$ with

       (i) $A = B \cup C$;
       (ii) $B \in \mathfrak{N}(\mu^+)$ and hence $\mu^X(D) \leq 0$ for every $D \in \mathfrak{L}$ with $D \subset B$;
       (iii) $C \in \mathfrak{N}(\mu^-)$ and hence $\mu^X(D) \geq 0$ for every $D \in \mathfrak{L}$ with $D \subset C$.

    Hint for (a): Theorem 8.3(b) and Hahn decomposition.

8.2 Absolute continuity and the Radon–Nikodym theorem

Let μ and ν be measures on the ring of sets $\mathfrak{R}$. We say that ν is **absolutely continuous with respect to μ** and write $\nu \ll \mu$ if for each set $A \in \mathfrak{R}$ and for each number $\varepsilon > 0$ there is a $\delta > 0$ such that $|\nu|(B) < \varepsilon$ for every $B \in \mathfrak{R}$ with $B \subset A$ and $|\mu|(B) < \delta$. The following result shows that absolute continuity can be described very elegantly in the language of vector lattices.

For each measure μ on $\mathfrak{R}$, let $B(\mu)$ denote the set of all measures on $\mathfrak{R}$ which are absolutely continuous with respect to μ.

Theorem 8.11 *Let μ be a measure on $\mathfrak{R}$. Then $B(\mu)$ is the band in $M^\sigma(\mathfrak{R})$ generated by μ.*

Proof. It follows immediately from the definition that $B(\mu)$ is a solid subspace of $M^\sigma(\mathfrak{R})$ and $\mu \in B(\mu)$. We show that it is a band. Let $(\nu_\iota)_{\iota\in I}$ be a non-empty upward directed family in $B(\mu)_+$ such that $\nu := \bigvee_{\iota\in I} \nu_\iota \in$

$M^\sigma(\mathfrak{R})$ exists. Take $A \in \mathfrak{R}$ and $\varepsilon > 0$. Then there is a $\lambda \in I$ such that $\nu(A) - \nu_\lambda(A) < \varepsilon/2$, and hence

$$\nu(B) - \nu_\lambda(B) = (\nu - \nu_\lambda)(B) \le (\nu - \nu_\lambda)(A) < \varepsilon/2$$

for every $B \in \mathfrak{R}$ with $B \subset A$. There is a $\delta > 0$ such that $\nu_\lambda(B) < \varepsilon/2$ for every $B \in \mathfrak{R}$ with $B \subset A$ and $|\mu|(B) < \delta$. Then for all such B

$$\nu(B) = (\nu - \nu_\lambda)(B) + \nu_\lambda(B) < \varepsilon.$$

Thus $\nu \in B(\mu)$, and so $B(\mu)$ is a band in $M^\sigma(\mathfrak{R})$ (Proposition 7.7).

In order to show that $B(\mu)$ is the band generated by μ, we verify that $B(\mu) = (\{\mu\}^\perp)^\perp$. (Recall Theorem 7.11.)

Let λ be orthogonal to μ. Then $|\lambda| \wedge |\mu| = 0$. Take $\nu \in B(\mu)$, $A \in \mathfrak{R}$ and $\varepsilon > 0$. Take $\delta > 0$ such that $|\nu|(B) < \varepsilon/2$ for every $B \in \mathfrak{R}$ with $B \subset A$ and $|\mu| < \delta$. We may assume that $\delta < \varepsilon/2$. By orthogonality, there is a $B \in \mathfrak{R}$, $B \subset A$, such that

$$|\lambda|(B) + |\mu|(A \setminus B) < \delta.$$

Then $|\lambda|(B) < \delta < \varepsilon/2$ and $|\mu|(A \setminus B) < \delta$. Thus $|\nu|(A \setminus B) < \varepsilon/2$. As a result,

$$|\lambda|(B) + |\nu|(A \setminus B) < \varepsilon.$$

But $\varepsilon > 0$ was arbitrary. So $|\lambda| \wedge |\nu| = 0$ (Theorem 8.3(b)). Thus

$$\{\mu\}^\perp \subset B(\mu)^\perp \subset \{\mu\}^\perp,$$

and finally $\{\mu\}^\perp = B(\mu)^\perp$. Hence $B(\mu) = (B(\mu)^\perp)^\perp = (\{\mu\}^\perp)^\perp$. □

Where $\mathfrak{R}$ is a δ-ring, the ε–δ definition of absolute continuity reduces to an assertion about null sets. This is essentially the content of the next result, which also states that extending measures defined on a ring of sets to the δ-ring generated does not influence their behaviour with respect to absolute continuity.

Corollary 8.12 *Take* $\mu, \nu \in M^\sigma(\mathfrak{R})$. *Then the following are equivalent.*

(a) $\nu \ll \mu$.

(b) $\bar{\nu} \ll \bar{\mu}$.

(c) $|\bar{\nu}|(A) = 0$ *for every* $A \in \mathfrak{R}_\delta$ *with* $|\bar{\mu}|(A) = 0$.

Proof. (a)⟺(b) follows from Theorems 8.11 and 8.9.

(b)⟹(c) follows immediately from the definition of absolute continuity.

(c)⟹(b). Assume that there are $A \in \mathfrak{R}_\delta$ and $\varepsilon > 0$ such that for each $\delta > 0$ there is a $B \in \mathfrak{R}_\delta$, $B \subset A$, with $|\bar{\mu}|(B) < \delta$ and $|\bar{\nu}|(B) \ge \varepsilon$. Then there is a sequence $(B_n)_{n\in\mathbb{N}}$ in $\mathfrak{R}_\delta$ such that $B_n \subset A$, $|\bar{\mu}|(B_n) < 1/2^n$ and $|\bar{\nu}|(B_n) \ge \varepsilon$ for every $n \in \mathbb{N}$. Given $m \in \mathbb{N}$, define $A_m := \bigcup_{n \ge m} B_n \in \mathfrak{R}_\delta$. Then

$$|\bar{\mu}|(A_m) \le \sum_{n=m}^{\infty} \frac{1}{2^n} = \frac{1}{2^{m-1}}$$

and so $|\bar{\mu}|\left(\bigcap_{m\in\mathbb{N}} A_m\right) = 0$. On the other hand

$$|\bar{\nu}|\left(\bigcap_{m\in\mathbb{N}} A_m\right) = \inf_{m\in\mathbb{N}} |\bar{\nu}|(A_m) \geq \varepsilon,$$

which is the desired contradiction. $\qquad\qquad\qquad\qquad\qquad\qquad\Box$

As an example, let μ be counting measure on $\mathbb{N}$. Then, since $\emptyset$ is the only μ-null set, *every* measure on $\mathfrak{R}$ is absolutely continuous with respect to μ.

What is a general method for finding absolutely continuous measures with respect to a given measure μ? In order to avoid unnecessary technical difficulties, *we assume that μ is a positive measure.*

We introduced the notion of local μ-integrability in Section 4.3. Let $f \in \mathcal{L}^1_{loc}(\mu)$. We define

$$f\cdot\mu : \mathfrak{R} \longrightarrow \mathbb{R}, \qquad A \longmapsto \int_A f\,d\mu.$$

We will show in a moment that the map $f \cdot \mu$ is an absolutely continuous measure with respect to μ. This is a very pleasing result, because it offers a concrete method for producing many absolutely continuous measures with respect to μ. For instance, we obtain a measure ν on the ring $\mathfrak{J}$ of interval forms which is absolutely continuous with respect to Lebesgue measure λ by putting

$$\nu([x,y[) := \sin y - \sin x \cdot \qquad \text{for all } x, y \in \mathbb{R} \text{ with } x < y.$$

(Simply note that $\sin y - \sin x = \int_{[x,y[} \cos t\, d\lambda(t)$.) To verify the result announced above, first note that given $f, g \in \mathcal{L}^1_{loc}(\mu)$ with $f = g$ μ-a.e., we clearly have $f\cdot\mu = g\cdot\mu$. Recalling our notation for $\mathcal{L}^p$- and L^p-spaces, we let $L^1_{loc}(\mu)$ denote the set of all equivalence classes of functions in $\mathcal{L}^1_{loc}(\mu)$ with respect to the relation $f = g$ μ-a.e. Then for each $\mathcal{F} \in L^1_{loc}(\mu)$, the map

$$\mathcal{F}\cdot\mu : \mathfrak{R} \longrightarrow \mathbb{R}, \qquad A \longmapsto \int_A f\,d\mu \qquad (f \in \mathcal{F} \text{ arbitrary})$$

is well defined and, of course, $\mathcal{F}\cdot\mu = f\cdot\mu$ for every $f \in \mathcal{F}$. We have:

Theorem 8.13

(a) $L^1_{loc}(\mu)$ *is a vector lattice.*

(b) $\mathcal{F}\cdot\mu \in M^\sigma(\mathfrak{R})$ *and* $\mathcal{F}\cdot\mu \ll \mu$ *for all* $\mathcal{F} \in L^1_{loc}(\mu)$.

(c) *The map*

$$\pi : L^1_{loc}(\mu) \longrightarrow B(\mu), \qquad \mathcal{F} \longmapsto \mathcal{F}\cdot\mu$$

is an injective homomorphism of vector lattices.

(d) *Given* $\mathcal{F} \in L^1_{loc}(\mu)$,

$$\mathcal{F} \geq 0 \iff \mathcal{F}\cdot\mu \geq 0.$$

Proof. (a) The proof is similar to the proof that $L^p(\mu)$ is a vector lattice.

(b) It can easily be shown that $\mathcal{F} \cdot \mu$ is a positive measure on $\mathfrak{R}$ whenever $\mathcal{F} \in L^1_{loc}(\mu)_+$. Thus for all $\mathcal{F} \in L^1_{loc}(\mu)$,

$$\mathcal{F} \cdot \mu = (\mathcal{F}^+ \cdot \mu) - (\mathcal{F}^- \cdot \mu) \in M^\sigma(\mathfrak{R}).$$

Take $\mathcal{F} \in L^1_{loc}(\mu)_+$ and $f \in \mathcal{F}$ with $f \geq 0$. For every $n \in \mathbb{N}$,

$$f_n := f \wedge n \in \mathcal{L}^1_{loc}(\mu),$$

and $f = \uparrow f_n$. Because $0 \leq f_n \cdot \mu \leq n\mu$, we see that $f_n \cdot \mu \in B(\mu)$ for every $n \in \mathbb{N}$. But

$$\mathcal{F} \cdot \mu = f \cdot \mu = \bigvee_{n \in \mathbb{N}} (f_n \cdot \mu).$$

Then, since $B(\mu)$ is a band in $M^\sigma(\mathfrak{R})$, it follows that $\mathcal{F} \cdot \mu \in B(\mu)$. Thus $\mathcal{F} \cdot \mu \in B(\mu)$ for all $\mathcal{F} \in L^1_{loc}(\mu)$.

(c) The linearity of π is obvious. Equally trivial is the injectivity, as $\pi(\mathcal{F}) = 0$ implies $\int_A f \, d\mu = 0$ for all $f \in \mathcal{F}$ and $A \in \mathfrak{R}$. It then follows from Theorem 4.27(c) that $f \in \mathcal{N}(\mu)$ and $\mathcal{F} = 0$.

We show that $\pi(|\mathcal{F}|) = |\pi(\mathcal{F})|$ for all $\mathcal{F} \in L^1_{loc}(\mu)$. This proves (c) (see Proposition 7.12).

Take $\mathcal{F} \in L^1_{loc}(\mu)$ and $f \in \mathcal{F}$. Recalling Theorem 4.26(d), we define

$$\nu_1 : \mathfrak{R}_\delta \longrightarrow \mathbb{R}, \quad A \longmapsto \int_A f^+ d\mu$$

and

$$\nu_2 : \mathfrak{R}_\delta \longrightarrow \mathbb{R}, \quad A \longmapsto \int_A f^- d\mu.$$

Then, by (b), ν_1 and ν_2 are positive measures on $\mathfrak{R}_\delta$. Take $A \in \mathfrak{R}_\delta$. Put $A_1 := \{f^+ > 0\} \cap A$. By Theorem 4.25(a), we can find a set $B \in \mathfrak{R}_\delta$ such that $B \subset A_1$ and $A_1 \setminus B \in \mathfrak{N}(\mu)$. We obtain

$$\nu_2(B) + \nu_1(A \setminus B) = \int_B f^- d\mu + \int_{A \setminus B} f^+ d\mu = \int_{A_1 \setminus B} f^+ d\mu = 0,$$

and conclude by Theorem 8.3(b) that $\nu_1 \wedge \nu_2 = 0$. It then follows from Theorem 8.9 that $(\mathcal{F}^+ \cdot \mu) \wedge (\mathcal{F}^- \cdot \mu) = 0$. However, $\mathcal{F} \cdot \mu = (\mathcal{F}^+ \cdot \mu) - (\mathcal{F}^- \cdot \mu)$. Theorem 7.2(c) now implies that

$$(\mathcal{F} \cdot \mu)^+ = \mathcal{F}^+ \cdot \mu \quad \text{and} \quad (\mathcal{F} \cdot \mu)^- = \mathcal{F}^- \cdot \mu.$$

But then

$$\pi(|\mathcal{F}|) = \pi(\mathcal{F}^+ + \mathcal{F}^-) = (\mathcal{F}^+ \cdot \mu) + (\mathcal{F}^- \cdot \mu)$$
$$= (\mathcal{F} \cdot \mu)^+ + (\mathcal{F} \cdot \mu)^- = |\mathcal{F} \cdot \mu| = |\pi(\mathcal{F})|.$$

(d) follows from Theorem 4.27 (a). $\square$

The reader will have noticed that we did not assert that $L^1_{loc}(\mu)$ is complete. In fact there are situations where $L^1_{loc}(\mu)$ is *not* complete. However, it will be a consequence of the next theorem that in the case of σ-finite positive measure spaces, $L^1_{loc}(\mu)$ *is complete, much to our satisfaction.*

We have just seen that $\mathcal{F} \cdot \mu \ll \mu$ holds for $\mathcal{F} \in L^1_{loc}(\mu)$. Moreover, the map π which assigns to $\mathcal{F}$ the measure $\mathcal{F} \cdot \mu$ has satisfying properties. But an important question now comes to mind: under what conditions is π bijective? In other words, under what conditions is *every* μ-absolutely continuous measure of the form $\mathcal{F} \cdot \mu$, for some $\mathcal{F} \in L^1_{loc}(\mu)$? This question is not easy to answer. At any rate, π is unfortunately not bijective in every case. For the applications of integration theory, however, the following famous result – one of the most important contributions of integration theory – generally suffices.

Theorem 8.14 (Radon–Nikodym) *If* $(X, \mathfrak{R}, \mu)$ *is σ-finite, then the map*

$$\pi : L^1_{loc}(\mu) \longrightarrow B(\mu), \quad \mathcal{F} \longmapsto \mathcal{F} \cdot \mu$$

is bijective and therefore a vector lattice isomorphism. In this case, $L^1_{loc}(\mu)$ is order complete.

Proof. Because of Proposition 7.13, we need only show that π is surjective. We begin by proving the following assertion: given $A \in \mathfrak{R}$ and $\nu \in B(\mu)_+$, there is a function $h \in \mathcal{L}^1(\mu)_+$ such that $\{h \neq 0\} \subset A$ and $(h \cdot \mu)(B) = \nu(B)$ for every $B \in \mathfrak{R}$ with $B \subset A$.

Take $\gamma \in \mathbb{R}$, $\gamma > 0$. By Hahn's theorem, there is a Hahn decomposition

$$P(\gamma) \in \mathfrak{R}_\delta \cap \mathfrak{N}((\nu - \gamma\mu)^-), \quad N(\gamma) \in \mathfrak{R}_\delta \cap \mathfrak{N}((\nu - \gamma\mu)^+)$$

of A for the measure $\nu - \gamma\mu$. Define

$$A_1 := N(\gamma) \quad \text{and} \quad A_{k+1} := N((k+1)\gamma) \setminus \bigcup_{j=1}^{k} A_j$$

for $k > 1$ ($k \in \mathbb{N}$). The sets A_k are pairwise disjoint and

$$\bigcup_{j=1}^{k} N(j\gamma) = \bigcup_{j=1}^{k} A_j.$$

Then

$$A_k = N(k\gamma) \setminus \bigcup_{j=1}^{k-1} N(j\gamma) = N(k\gamma) \cap \bigcap_{j=1}^{k-1} P(j\gamma),$$

where the latter equality holds whenever $k > 1$. If $B \in \mathfrak{R}_\delta$ is a subset of A_k, we conclude that $B \subset N(k\gamma)$, and if $k > 1$, that $B \subset P((k-1)\gamma)$. Thus

$$(k-1)\gamma\bar{\mu}(B) \leq \bar{\nu}(B) \leq k\gamma\bar{\mu}(B)$$

for every $k \in \mathbb{N}$. Define

$$C := A \setminus \bigcup_{j \in \mathbb{N}} A_j = \bigcap_{j \in \mathbb{N}} P(j\gamma).$$

For any $j \in \mathbb{N}$, $C \subset P(j\gamma)$ and hence

$$0 \leq j\gamma\overline{\mu}(C) \leq \overline{\nu}(C) \leq \overline{\nu}(A) < \infty.$$

We conclude that $C \in \mathfrak{N}(\mu)$ and since $\nu \ll \mu$, it follows that $C \in \mathfrak{N}(\nu)$ (Corollary 8.12).

Define

$$g_\gamma : X \longrightarrow \mathbb{R}, \quad x \longmapsto \begin{cases} (k-1)\gamma & \text{for } x \in A_k \\ 0 & \text{for } x \in X \setminus \bigcup_{k\in\mathbb{N}} A_k. \end{cases}$$

Then g_γ is μ-integrable, and for each $B \in \mathfrak{R}_\delta$ with $B \subset A$,

$$\int_B g_\gamma d\mu = \sum_{k\in\mathbb{N}} \int_{B\cap A_k} g_\gamma d\mu = \sum_{k\in\mathbb{N}} (k-1)\gamma\overline{\mu}(B \cap A_k)$$

$$\leq \sum_{k\in\mathbb{N}} \overline{\nu}(B \cap A_k) = \overline{\nu}(B) \leq \sum_{k\in\mathbb{N}} k\gamma\overline{\mu}(B \cap A_k)$$

$$= \int_B (g_\gamma + \gamma)\, d\mu \leq \int_B g_\gamma d\mu + \gamma\mu(A).$$

(Note that for the equality $\sum_{k\in\mathbb{N}} \overline{\nu}(B \cap A_k) = \overline{\nu}(B)$ we used the fact that $C \in \mathfrak{N}(\nu)$.)

Setting $f_n := g_{2^{-n}}$ for $n \in \mathbb{N}$ defines a sequence $(f_n)_{n\in\mathbb{N}}$ of μ-integrable functions with the property that

$$\int_B f_n d\mu \leq \overline{\nu}(B) \leq \int_B f_n d\mu + 2^{-n}\mu(A) \tag{1}$$

for every $n \in \mathbb{N}$ and for every $B \in \mathfrak{R}_\delta$ with $B \subset A$. For $m \geq n$, the inequalities

$$\int_B f_n d\mu \leq \overline{\nu}(B) \leq \int_B f_m d\mu + 2^{-m}\mu(A)$$

$$\int_B f_m d\mu \leq \overline{\nu}(B) \leq \int_B f_n d\mu + 2^{-n}\mu(A)$$

imply that for every $B \in \mathfrak{R}_\delta$ with $B \subset A$

$$\left| \int_B (f_n - f_m)\, d\mu \right| \leq 2^{-n}\mu(A).$$

By Theorem 4.25(a) there is a set $D \in \mathfrak{R}_\delta$ with $D \subset \{f_n - f_m > 0\}$ such that $\{f_n - f_m > 0\} \setminus D \in \mathfrak{N}(\mu)$. First putting $B = D$ and then $B = A \setminus D$ in the above relation, we obtain

$$\int |f_n - f_m|\, d\mu = \int_A |f_n - f_m|\, d\mu \leq 2^{-n+1}\mu(A).$$

Thus $(f_n)_{n\in\mathbb{N}}$ is a Cauchy sequence in $\mathcal{L}^1(\mu)$, and we may choose an $h \in$

$\mathcal{L}^1(\mu)_+$ with $\{h \neq 0\} \subset A$ and

$$\lim_{n \to \infty} \int |h - f_n|\, d\mu = 0$$

(Theorem 6.8(b)). Given any $B \in \mathfrak{R}_\delta$ with $B \subset A$, it follows from (1) that

$$\int_B h\, d\mu = \lim_{n \to \infty} \int_B f_n d\mu = \overline{\nu}(B).$$

This proves the assertion made at the beginning of the proof.

We now make use of the σ-finiteness of $(X, \mathfrak{R}, \mu)$. It ensures the existence of a disjoint sequence $(A_n)_{n \in \mathbb{N}}$ in $\mathfrak{R}$ with $X = \bigcup_{n \in \mathbb{N}} A_n$. Given $n \in \mathbb{N}$, we construct a function h_n for A_n as described above. Put $h := \sum_{n \in \mathbb{N}} h_n$. Then $h \in \mathcal{L}^1_{loc}(\mu)$ and $h \cdot \mu = \nu$. In fact, given any $A \in \mathfrak{R}$, we have $he_A = \sum_{n \in \mathbb{N}} h_n e_A$ and

$$\sum_{n \in \mathbb{N}} \int_A h_n d\mu = \sum_{n \in \mathbb{N}} \nu(A \cap A_n) = \nu(A),$$

which implies that $he_A \in \mathcal{L}^1(\mu)$ and

$$(h \cdot \mu)(A) = \int_A h\, d\mu = \sum_{n \in \mathbb{N}} \int_A h_n d\mu = \nu(A)$$

(Theorem 3.34).

This proves that π is surjective, since for arbitrary $\nu \in B(\mu)$ we may choose $\mathcal{H}_1 \in L^1_{loc}(\mu)$ and $\mathcal{H}_2 \in L^1_{loc}(\mu)$ such that $\nu^+ = \pi(\mathcal{H}_1)$ and $\nu^- = \pi(\mathcal{H}_2)$, which implies that

$$\nu = \nu^+ - \nu^- = \pi(\mathcal{H}_1) - \pi(\mathcal{H}_2) = \pi(\mathcal{H}_1 - \mathcal{H}_2).$$

$\square$

It follows, for example, that every measure which is absolutely continuous with respect to Lebesgue measure λ, may be written in the form $f \cdot \lambda$, for some $f \in \mathcal{L}^1_{loc}(\lambda)$. We investigate this relationship more closely in Section 9.2.

As we have seen above, every measure ν on the δ-ring $\mathfrak{R}$ of finite subsets of $\mathbb{N}$ is absolutely continuous with respect to counting measure μ. Hence $\nu = f \cdot \mu$, for an appropriate $f \in \mathcal{L}^1_{loc}(\mu)$. In fact, the function $f \in \mathbb{R}^{\mathbb{N}}$, defined by $f(n) := \nu(\{n\})$ for every $n \in \mathbb{N}$, satisfies the requirements.

Exercises

1. Let λ be Lebesgue measure and δ_x Dirac measure at the point x. Do any of the following relations hold: $\delta_x \ll \lambda$, $\lambda \ll \delta_x$, $\lambda - \delta_x \ll \lambda$? Characterize $B(\delta_x)$.

2. Does $\mu = \nu$ follow from $\mu \ll \nu$ and $\nu \ll \mu$?

3. Show that for $\mu, \nu \geq 0$, the following are equivalent.

   (a) $\nu \ll \mu$.
   (b) Each μ-null set is a ν-null set.

4. Take $\mu, \nu \geq 0$ with $\nu \ll \mu$, $A \subset X$ and $f \in \overline{\mathbb{R}}^X$. Prove the following.

   (a) If A is μ-measurable, then A is ν-measurable.
   (b) If f is μ-measurable, then f is ν-measurable.

   Hint for (a): First consider the case $A \in \mathfrak{L}(\mu)$ and apply Exercise 13(b) of Section 4.1.

5. Take $\mu \in M^\sigma(\mathfrak{R})_+$.

   (a) Show that $L^1(\mu)$ and $L^\infty(\mu)$ are solid subspaces of $L^1_{loc}(\mu)$.

   Now take $\mathcal{F} \in L^1_{loc}(\mu)$ and prove the following.

   (b) $\mathcal{F} \in L^1(\mu)$ if and only if $\mathcal{F} \cdot \mu \in M^\sigma_b(\mathfrak{R})$.
   (c) $\mathcal{F} \in L^\infty(\mu)$ if and only if $\mathcal{F} \cdot \mu$ belongs to the solid subspace of $M^\sigma(\mathfrak{R})$ generated by μ.

6. Show that the map
$$L^1(\mu) \longrightarrow B(\mu) \cap M^\sigma_b(\mathfrak{R}), \quad \mathcal{F} \longmapsto \mathcal{F} \cdot \mu$$
is an isomorphism of vector lattices provided that $(X, \mathfrak{R}, \mu)$ is a σ-finite positive measure space. (This result holds even without assuming σ-finiteness.)

7. Let $(X, \mathfrak{R}, \mu)$ be a σ-finite positive measure space. Show that the following are equivalent.

   (a) $\mathcal{L}^1_{loc}(\mu) = \mathcal{L}^1(\mu)$.
   (b) If $\nu \ll \mu$, then ν is bounded.

   (This result also does not need the assumption of σ-finiteness.)

8. Prove the **Lebesgue decomposition theorem**: if $(X, \mathfrak{R}, \mu)$ is a σ-finite positive measure space and $\nu \in M^\sigma(\mathfrak{R})$, then there is a unique pair $(\mathcal{F}, \lambda) \in L^1_{loc}(\mu) \times M^\sigma(\mathfrak{R})$ such that μ and λ are orthogonal and $\nu = \mathcal{F} \cdot \mu + \lambda$.

9. The definition of 'absolutely continuous' can be carried over to $\mu, \nu \in M(\mathfrak{R})$. Prove that for $\mu \in M(\mathfrak{R})$ $\{\nu \in M(\mathfrak{R}) \,|\, \nu \ll \mu\}$ is the band in $M(\mathfrak{R})$ generated by μ.

9

Elements of the theory of real-valued functions on $\mathbb{R}$

Our investigation of positive contents on $\mathbb{R}$ has brought to light a close relationship between positive contents and positive measures on $\mathfrak{I}$ on the one hand, and increasing functions on $\mathbb{R}$ on the other. Clearly this connection can be extended to a relationship between spaces of contents or measures on $\mathfrak{I}$ and certain spaces of functions on $\mathbb{R}$. We investigate several aspects of this relationship. The theory of vector lattices is again the appropriate tool. The 'secret' to applying this theory is to introduce an order relation on $\mathbb{R}^{\mathbb{R}}$ which is different from the usual pointwise ordering.

9.1 Functions of locally finite variation

Take $f \in \mathbb{R}^{\mathbb{R}}$. For each right half-open interval $[\alpha, \beta[$ of $\mathbb{R}$, define

$$\mu_f([\alpha, \beta[) := f(\beta) - f(\alpha),$$

as we did for increasing functions. It is easy to show that μ_f admits a unique additive extension to the ring of sets $\mathfrak{I}$ of interval forms on $\mathbb{R}$. We use μ_f again to denote this extension.

From the definition of μ_f, given $f, g \in \mathbb{R}^{\mathbb{R}}$ and $\alpha \in \mathbb{R}$,

$$\mu_{f+g} = \mu_f + \mu_g \qquad \text{and} \qquad \mu_{\alpha f} = \alpha \mu_f.$$

We now introduce the order relation on $\mathbb{R}^{\mathbb{R}}$ we use for our investigations. Given $f, g \in \mathbb{R}^{\mathbb{R}}$, define

$$f \preccurlyeq g :\iff g - f \text{ is increasing and } (g - f)(0) \geq 0.$$

It is easily verified that $\preccurlyeq$ is an order relation on $\mathbb{R}^{\mathbb{R}}$. Note, however, that it is significantly different from the ordering $\leq$. The positive functions with respect to $\preccurlyeq$ are precisely the increasing functions f with $f(0) \geq 0$, and thus they may take on negative values. We write $\bigvee^{\preccurlyeq}, \vee^{\preccurlyeq}, \bigwedge_{\preccurlyeq}, \wedge_{\preccurlyeq}$ for suprema and infima with respect to $\preccurlyeq$ and $|\cdot|^{\preccurlyeq}$ for the absolute value. We retain the usual symbols $\bigvee, \vee, \bigwedge, \wedge$ and $|\cdot|$ for operations with respect to $\leq$.

$f \in \mathbb{R}^{\mathbb{R}}$ is said to be of **locally finite variation** if $\mu_f \in M(\mathfrak{I})$. (The reason for this terminology is explained below.) Let $\mathcal{V}$ denote the set of all functions of locally finite variation and put

$$\overset{\circ}{\mathcal{V}} := \{f \in \mathcal{V} \mid f(0) = 0\}.$$

We have thus defined a map which assigns to each $f \in \mathcal{V}$ a content $\mu_f \in M(\mathfrak{I})$. We now show that this map has very strong properties. In fact, when restricted to $\overset{\circ}{\mathcal{V}}$, it is a vector lattice isomorphism. In particular, $\overset{\circ}{\mathcal{V}}$ is a complete vector lattice, and as a consequence, this also holds for $\mathcal{V}$, which can be generated from $\overset{\circ}{\mathcal{V}}$ by simply adding the constant functions to the elements of $\overset{\circ}{\mathcal{V}}$. In light of this isomorphism, the theory of functions of locally finite variation is nothing but the theory of contents on $\mathfrak{I}$!

We denote by $\mathcal{P}_0$ the set of constant functions in $\mathbb{R}^{\mathbb{R}}$.

Theorem 9.1

  (a) $\mathcal{V}$ *is a complete vector lattice with respect to* $\preccurlyeq$.

  (b) *The map*

$$\pi : \mathcal{V} \longrightarrow M(\mathfrak{I}), \quad f \longmapsto \mu_f$$

  is a surjective homomorphism of vector lattices.

  (c) *Take* $f, g \in \mathcal{V}$. *Then* $\pi(f) = \pi(g)$ *if and only if* $f = g + \gamma$ *for some* $\gamma \in \mathbb{R}$.

  (d) $\overset{\circ}{\mathcal{V}}$ *is a band in* $\mathcal{V}$ *and* $\mathcal{V} = \overset{\circ}{\mathcal{V}} \oplus \mathcal{P}_0$.

  (e) $\pi|_{\overset{\circ}{\mathcal{V}}} : \overset{\circ}{\mathcal{V}} \to M(\mathfrak{I})$ *is a vector lattice isomorphism.*

Proof. We prove (e) first. Put $\varphi := \pi|_{\overset{\circ}{\mathcal{V}}}$. $\overset{\circ}{\mathcal{V}}$ is clearly an ordered vector space and φ is a linear mapping of $\overset{\circ}{\mathcal{V}}$ into $M(\mathfrak{I})$.

φ is injective: to see this, assume that $\varphi(f) = 0$ for $f \in \overset{\circ}{\mathcal{V}}$. Then

$$f(x) = f(x) - f(0) = \mu_f([0, x[) = 0 \quad \text{for } x \geq 0$$

and

$$f(x) = -(f(0) - f(x)) = -\mu_f([x, 0[) = 0 \quad \text{for } x \leq 0.$$

Thus $f = 0$.

It follows from Theorem 2.29 that φ is also surjective. Hence φ is bijective.

Take $f, g \in \overset{\circ}{\mathcal{V}}$ with $f \preccurlyeq g$. Then $g - f$ is increasing and hence $\pi(g - f)$ is positive. Thus $\varphi(f) \leq \varphi(g)$. Conversely, suppose that $\varphi(f - g) \geq 0$ for

$f, g \in \overset{\circ}{\mathcal{V}}$. Take $x, y \in \mathbb{R}$ with $x \leq y$. Then

$$(f - g)(x) = \big(\pi(f - g)\big)([0, x[) \leq \big(\pi(f - g)\big)([0, y[) = (f - g)(y)$$
$$\text{for } 0 \leq x \leq y,$$

$$(f - g)(x) = -\big(\pi(f - g)\big)([x, 0[) \leq 0 \leq \big(\pi(f - g)\big)([0, y[) = (f - g)(y)$$
$$\text{for } x \leq 0 \leq y,$$

$$(f - g)(x) = -\big(\pi(f - g)\big)([x, 0[) \leq -\big(\pi(f - g)\big)([y, 0[) = (f - g)(y)$$
$$\text{for } x \leq y \leq 0.$$

Hence $f - g$ is increasing and it follows that $g \preccurlyeq f$. Thus $\overset{\circ}{\mathcal{V}}$ is a complete vector lattice and φ is an isomorphism of $\overset{\circ}{\mathcal{V}}$ onto $M(\mathfrak{I})$ (Corollary 7.14).

To obtain the properties of $\mathcal{V}$, note that $g - g(0) \in \overset{\circ}{\mathcal{V}}$ whenever $g \in \mathcal{V}$. We conclude that

$$\mathcal{V} = \{f + \gamma \mid f \in \overset{\circ}{\mathcal{V}},\ \gamma \in \mathbb{R}\}.$$

It is not very difficult to show that for each non-empty bounded family $(f_\iota + \gamma_\iota)_{\iota \in I}$ in $\mathcal{V}$ $(f_\iota \in \overset{\circ}{\mathcal{V}},\ \gamma_\iota \in \mathbb{R})$, the relations

$$\bigvee_{\iota \in I}^{\preccurlyeq} (f_\iota + \gamma_\iota) = \bigvee_{\iota \in I}^{\preccurlyeq, 0} f_\iota + \sup_{\iota \in I} \gamma_\iota \tag{1}$$

and

$$\bigwedge_{\iota \in I}^{\preccurlyeq} (f_\iota + \gamma_\iota) = \bigwedge_{\iota \in I}^{\preccurlyeq, 0} f_\iota + \inf_{\iota \in I} \gamma_\iota$$

hold, where the symbols $\bigvee^{\preccurlyeq, 0}$ and $\bigwedge_{\preccurlyeq, 0}$ denote suprema and infima in $\overset{\circ}{\mathcal{V}}$ with respect to the restriction of $\preccurlyeq$ to $\overset{\circ}{\mathcal{V}}$. We conclude that $\mathcal{V}$ is also a complete vector lattice and that $\overset{\circ}{\mathcal{V}}$ is a band in $\mathcal{V}$.

Suppose that $|f|^{\preccurlyeq} \preccurlyeq |\alpha e_\mathbb{R}|^{\preccurlyeq}$ for some $f \in \mathcal{V}$ and some $\alpha \in \mathbb{R}$. Clearly $|\alpha e_\mathbb{R}|^{\preccurlyeq} = |\alpha| e_\mathbb{R}$. Thus $f \preccurlyeq |\alpha| e_\mathbb{R}$ and $-f \preccurlyeq |\alpha| e_\mathbb{R}$. Hence both $-f$ and f are increasing. It follows that f must be constant, which shows that $\mathcal{P}_0$ is a solid subspace of $\mathcal{V}$. Thus

$$\mathcal{V} = \overset{\circ}{\mathcal{V}} \oplus \mathcal{P}_0.$$

(b) follows from (1) and the fact that $\pi(f + \gamma) = \pi(f)$ for all $f \in \overset{\circ}{\mathcal{V}}$ and $\gamma \in \mathbb{R}$. Finally, take $f, g \in \overset{\circ}{\mathcal{V}}$ and $\gamma, \beta \in \mathbb{R}$ such that $\pi(f + \gamma) = \pi(g + \beta)$. Then $\pi(f) = \pi(g)$ and therefore $f = g$. Thus $f + \gamma = (g + \beta) + (\gamma - \beta)$, which completes the proof of (c). $\qquad \square$

Since $E = E_+ - E_+$ for any vector lattice E, and since every increasing function clearly belongs to $\mathcal{V}$, we obtain immediately the following elegant (and very practical) characterization of the elements of $\mathcal{V}$.

Corollary 9.2 *Take* $f \in \mathbb{R}^{\mathbb{R}}$. *Then* $f \in V$ *if and only if* f *is the difference of two increasing functions.*

The corollary enables us to construct many functions of locally finite variation. On the other hand, there are continuous bounded functions *not* belonging to V (see below).

The classical definition of the class of functions V appears to be different. For arbitrary $x, y \in \mathbb{R}$ with $x \le y$, let $3(x,y)$ denote the set of all subdivisions of $[x,y]$; that is, the set of all families $(x_k)_{1 \le k \le n}$ $(n \in \mathbb{N})$ of points of $[x,y]$ with the property that

$$x = x_1 \le x_2 \le \cdots \le x_n = y.$$

For $f \in \mathbb{R}^{\mathbb{R}}$ let

$$\underset{[x,y]}{V} f := \sup \left\{ \sum_{k=1}^{n-1} |f(x_{k+1}) - f(x_k)| \, \middle| \, (x_k)_{1 \le k \le n} \in 3(x,y) \right\}.$$

$V_{[x,y]}f$ is called the **variation of** f **on** $[x,y]$.

Theorem 9.3

(a) For all $f \in \mathbb{R}^{\mathbb{R}}$ *and all* $x, y \in \mathbb{R}$ *with* $x \le y$,

$$\underset{[x,y]}{V} f = (V\mu_f)([x,y[).$$

(b) $f \in V$ *if and only if* $V_{[x,y]}f < \infty$ *for all* $x, y \in \mathbb{R}$ *with* $x \le y$.

Proof. (b) follows from (a) by Theorem 8.5. But assertion (a) is a simple consequence of the definition of variation for contents and for functions. We leave the details to the reader. □

(b) provides the explanation for the name 'functions of locally finite variation'. It also gives the original definition of this important class of functions.

An example of a continuous bounded function without locally finite variation is the function

$$f : \mathbb{R} \longrightarrow \mathbb{R}, \quad x \longmapsto \begin{cases} 0 & \text{if } x \le 0 \\ x \sin \frac{1}{x} & \text{if } 0 < x < \frac{2}{\pi} \\ \frac{2}{\pi} & \text{if } x \ge \frac{2}{\pi}. \end{cases}$$

(Consider the points $\frac{2}{n\pi}$ $(n \in \mathbb{N})$ to check that $V_{[0,1]}f = \infty$, for example.)

Theorem 9.3(a) allows us to give a concrete description of the absolute value (taken in $\overset{\circ}{V}$) of an element $f \in \overset{\circ}{V}$. Again, as in the case of contents, it is the variation which makes this possible.

Corollary 9.4 *Given* $f \in \overset{\circ}{V}$, *put* $g := |f|^{\lessgtr}$. *Then*

$$g(y) - g(x) = \underset{[x,y]}{V} f \qquad \text{for all } x, y \in \mathbb{R} \text{ with } x \le y.$$

Thus, for each $x \in \mathbb{R}$,

$$g(x) = \begin{cases} \underset{[0,x]}{V} f & \text{if } x \geq 0 \\ -\underset{[x,0]}{V} f & \text{if } x < 0. \end{cases}$$

Proof. Using Theorems 9.3(a), 9.1(e) and 8.5, we have

$$\underset{[x,y]}{V} f = |\mu_f|([x,y[) = \mu_g([x,y[) = g(y) - g(x)$$

for any $x \leq y$. Since $g(0) = 0$, the second assertion follows immediately.

□

If f is left continuous, then $|f|$ (formed with respect to pointwise order) is obviously left continuous. The corresponding result for the order relation $\preccurlyeq$ also holds, but the proof is a little more complicated.

Proposition 9.5 *Suppose that $f \in \overset{\circ}{\mathcal{V}}$ is left continuous at $x \in \mathbb{R}$. Then $f \vee^{\preccurlyeq} 0$, $(-f) \vee^{\preccurlyeq} 0$ and $|f|^{\preccurlyeq}$ are also left continuous at x.*

Proof. In view of Theorem 7.2(f) it suffices to prove the assertion for $g := |f|^{\preccurlyeq}$. Note that g is increasing. Suppose that g is not left continuous at x. Then there is an $\alpha > 0$ such that

$$\underset{[y,x]}{V} f = g(x) - g(y) \geq 3\alpha \qquad \text{for every } y < x$$

(Corollary 9.4). We use this to prove the following claim.

Given any $y < x$, there are points $y = x_1 \leq x_2 \leq \cdots \leq x_m < x$ such that

$$\sum_{k=1}^{m-1} |f(x_{k+1}) - f(x_k)| > \alpha.$$

Indeed, by hypothesis there is a $\delta > 0$ such that $|f(z) - f(x)| < \alpha$ whenever $x - \delta < z < x$. Moreover, there is a subdivision $y = x_1 \leq \cdots \leq x_{m+1} = x$ of $[y,x]$ with

$$\sum_{k=1}^{m} |f(x_{k+1}) - f(x_k)| \geq 2\alpha,$$

and we may assume that $x_m > x - \delta$, adding an extra point if necessary. The claim now follows, since $|f(x_m) - f(x)| < \alpha$.

It is now easy to construct recursively an increasing sequence $(x_n)_{n \in \mathbb{N}}$ of real numbers such that $x = \lim_{n \to \infty} x_n$ and

$$\sum_{n=1}^{m} |f(x_{n+1}) - f(x_n)| \longrightarrow \infty \qquad \text{for } m \to \infty,$$

which clearly contradicts the fact that $V_{[x_1,x]} f < \infty$. □

It is natural to ask which are the elements of $\overset{\circ}{\mathcal{V}}$ corresponding to the measures on $\mathfrak{J}$ via the isomorphism of Theorem 9.1(e). In view of Theorem 2.30, the answer is quite natural, too.

Corollary 9.6

(a) *Take* $f \in \overset{\circ}{\mathcal{V}}$. *Then* $\mu_f \in M^\sigma(\mathfrak{J})$ *if and only if* f *is left continuous.*

(b) $\{f \in \overset{\circ}{\mathcal{V}} \mid f$ *is left continuous*$\}$ *is a band in* $\overset{\circ}{\mathcal{V}}$, *and*

$$\{f \in \overset{\circ}{\mathcal{V}} \mid f \text{ is left continuous}\} \longrightarrow M^\sigma(\mathfrak{J}), \quad f \longmapsto \mu_f$$

is an isomorphism of vector lattices.

Proof. In view of Theorem 9.1(e) and Theorem 8.6, it is sufficient to prove (a). Put $f_+ := f \vee^{\triangleleft} 0$ and $f_- := (-f) \vee^{\triangleleft} 0$. Suppose that $\mu_f \in M^\sigma(\mathfrak{J})$. Then $\mu_{f_+} = \mu_f \vee 0 \in M^\sigma(\mathfrak{J})$, and by Theorem 2.30, f_+ is left continuous. Similarly, f_- is left continuous. Hence the same is true of f (Theorem 7.2(b)).

Conversely, suppose that f is left continuous. By Proposition 9.5, so are f_+ and f_-. In light of Theorem 2.30, μ_{f_+} and μ_{f_-} are measures on $\mathfrak{J}$. Then the same is true of $\mu_f = \mu_{f_+} - \mu_{f_-}$.   □

9.2 Absolutely continuous functions

In the following, λ denotes Lebesgue measure on $\mathbb{R}$ and $B(\lambda)$ denotes the band of absolutely continuous measures with respect to λ.

We call $f \in \mathbb{R}^{\mathbb{R}}$ **absolutely continuous** if $\mu_f \in B(\lambda)$. Let $\mathcal{D}$ denote the set of all absolutely continuous functions on $\mathbb{R}$ and put

$$\overset{\circ}{\mathcal{D}} := \{f \in \mathcal{D} \mid f(0) = 0\}.$$

The following theorem, describing the vector lattice properties of $\mathcal{D}$, is then a consequence of Theorem 9.1.

Theorem 9.7

(a) $\mathcal{D}$ *is a band in* $\mathcal{V}$ *and thus a complete vector lattice.*

(b) *The map*
$$\pi : \mathcal{D} \longrightarrow B(\lambda), \quad f \longmapsto \mu_f$$
is a surjective homomorphism of vector lattices.

(c) *Take* $f, g \in \mathcal{D}$. *Then* $\pi(f) = \pi(g)$ *if and only if* $f = g + \gamma$ *for some* $\gamma \in \mathbb{R}$.

(d) $\overset{\circ}{\mathcal{D}}$ *is a band in* $\mathcal{D}$ *and* $\mathcal{D} = \overset{\circ}{\mathcal{D}} \oplus \mathcal{P}_0$.

(e) $\pi|_{\overset{\circ}{\mathcal{D}}} : \overset{\circ}{\mathcal{D}} \to B(\lambda)$ *is a vector lattice isomorphism.*

Proof. By virtue of Theorem 9.1(e), $\overset{\circ}{\mathcal{D}}$ is a band in $\overset{\circ}{\mathcal{V}}$ and $\pi|_{\overset{\circ}{\mathcal{D}}}$ is an isomorphism of vector lattices. Moreover,

$$\mathcal{D} = \{f + \gamma \mid f \in \overset{\circ}{\mathcal{D}}, \ \gamma \in \mathbb{R}\}.$$

Now the pre-image of a solid subspace under a homomorphism of vector lattices is again a solid subspace. Hence, using Theorem 9.1(b), $\mathcal{D}$ is a solid subspace of $\mathcal{V}$. Let $(g_\iota)_{\iota \in I}$ be a family in $\mathcal{D}_+$ for which $g := \bigvee_{\iota \in I}^{\preccurlyeq} g_\iota$ exists in $\mathcal{V}$. Each g_ι may be written as $g_\iota = f_\iota + \gamma_\iota$, with $f_\iota \in \overset{\circ}{\mathcal{D}}$ and $\gamma_\iota \in \mathbb{R}$. Then, by (1),

$$g = \bigvee_{\iota \in I}^{\preccurlyeq,0} f_\iota \ + \sup_{\iota \in I} \gamma_\iota.$$

But $\overset{\circ}{\mathcal{D}}$ is a band in $\overset{\circ}{\mathcal{V}}$, and so $\bigvee_{\iota \in I}^{\preccurlyeq,0} f_\iota$ is in $\overset{\circ}{\mathcal{D}}$. Thus $g \in \mathcal{D}$, which proves that $\mathcal{D}$ is a band in $\mathcal{V}$.

The remaining statements are now easily verified. $\qquad\square$

Thus, the fundamental properties of the important class of functions $\mathcal{D}$ are, in fact, a by-product of the general theory of spaces of measures.

The usual definition of absolute continuity of functions is also different from the one we presented above. Assertion (b) of the next result is the classical definition.

Theorem 9.8 *Take $f \in \mathbb{R}^{\mathbb{R}}$. Then the following are equivalent.*

(a) $f \in \mathcal{D}$.

(b) *For all $x, y \in \mathbb{R}$ with $x \leq y$ and every $\varepsilon > 0$ there is a $\delta > 0$ such that for every finite disjoint family $([x_\iota, y_\iota[)_{\iota \in I}$ of intervals of $[x, y[$ with $\sum_{\iota \in I} \lambda([x_\iota, y_\iota[) < \delta$,*

$$\sum_{\iota \in I} |f(y_\iota) - f(x_\iota)| < \varepsilon.$$

Proof. The claim follows easily from the definition of absolute continuity of measures. We leave the details to the reader. $\qquad\square$

Corollary 9.9 *Each function $f \in \mathcal{D}$ is uniformly continuous.*

We shall examine $\mathcal{D}$ a little more closely. First we introduce the notion of the **indefinite integral**, which may be considered in analogy with the corresponding notion for the Riemann integral.

Take $f \in \mathcal{L}^1_{loc}(\lambda)$. $F \in \mathbb{R}^{\mathbb{R}}$ is called an **indefinite integral** of f if there is a $\gamma \in \mathbb{R}$ with

$$F(x) = \begin{cases} \int_{[0,x[} f \, d\lambda + \gamma & \text{for } x \geq 0 \\ -\int_{[x,0[} f \, d\lambda + \gamma & \text{for } x \leq 0. \end{cases}$$

We then obtain the beautiful result that the absolutely continuous functions are precisely those functions which admit a representation as an indefinite integral. This can be seen as a generalization of the well-known fact that every continuously differentiable function is an indefinite (Riemann) integral of its derivative.

Theorem 9.10 *For each* $F \in \mathbb{R}^{\mathbb{R}}$, *the following are equivalent.*

(a) $F \in \mathcal{D}$.

(b) F *is an indefinite integral of a function* $f \in \mathcal{L}^1_{loc}(\lambda)$.

Furthermore, if $F \in \mathcal{D}$, *then*

(c) f *is* λ-*a.e. determined by* F, *and* $\mu_F = f \cdot \lambda$;

(d) *the map*

$$\pi : \mathcal{D} \longrightarrow L^1_{loc}(\lambda), \quad F \longmapsto \dot{f}$$

is a surjective homomorphism of vector lattices;

(e) $\pi|_{\overset{\circ}{\mathcal{D}}} : \overset{\circ}{\mathcal{D}} \to L^1_{loc}(\lambda)$ *is a vector lattice isomorphism.*

Proof. (a)$\Rightarrow$(b). By definition, '$F \in \mathcal{D}$' is equivalent to '$\mu_F \in B(\lambda)$'. The hypotheses of the Radon–Nikodym theorem are satisfied and therefore there is an $f \in \mathcal{L}^1_{loc}(\lambda)$ such that $\mu_F = f \cdot \lambda$. This means that

$$F(x) = \begin{cases} F(0) + \mu_F([0, x[) = F(0) + \int_{[0,x[} f \, d\lambda & \text{if } x \geq 0, \\ F(0) - \mu_F([x, 0[) = F(0) - \int_{[x,0[} f \, d\lambda & \text{if } x \leq 0. \end{cases}$$

Thus F is an indefinite integral of f.

(b)$\Rightarrow$(a). If F is an indefinite integral of $f \in \mathcal{L}^1_{loc}(\lambda)$, then $\mu_F = f \cdot \lambda$ by the definition of an indefinite integral. Thus $\mu_F \in B(\lambda)$, and $F \in \mathcal{D}$.

It follows from the Radon–Nikodym theorem that f is λ-a.e. determined by F. Finally, this theorem together with Theorem 9.7 implies (d) and (e). $\square$

The fundamental theorem of calculus states that an indefinite (Riemann) integral F of a continuous function f is everywhere differentiable and has derivative f. In this strong form, this assertion certainly cannot be extended to $\mathcal{L}^1_{loc}(\lambda)$. This can be seen by noting that two functions of $\mathcal{L}^1_{loc}(\lambda)$ which are λ-a.e. equal have the same indefinite integral. On this basis, we can at most maintain that an indefinite integral F of a function $f \in \mathcal{L}^1_{loc}(\lambda)$ is λ-a.e. differentiable and that the relation $F'(x) = f(x)$ holds λ-a.e. Interestingly, this in fact holds, as we prove below. In particular, the assertion then gives a concrete meaning to the claim of the preceding theorem that f is λ-a.e. determined by F'.

Recall that $f \in \mathbb{R}^{\mathbb{R}}$ is said to be **differentiable at** $x \in \mathbb{R}$ if there is a number $f'(x)$ such that for each $\varepsilon > 0$ there is a $\delta > 0$ satisfying

$$|f(y) - f(x) - f'(x)(y - x)| \leq \varepsilon |y - x|$$

whenever $y \in]x - \delta, x + \delta[$. The uniquely determined number $f'(x)$ is called the **derivative of f at x.**

We begin by proving a proposition about the derivative of the limit of a sequence.

Proposition 9.11 *Let $(f_n)_{n \in \mathbb{N}}$ be a sequence in $\mathcal{L}^1_{loc}(\lambda)$ and $f \in \mathcal{L}^1_{loc}(\lambda)$. For each $n \in \mathbb{N}$, let F_n be an indefinite integral of f_n and let F be an indefinite integral of f. For $x \in \mathbb{R}$, let the following conditions be satisfied.*

(i) $f(x) = \lim_{n \to \infty} f_n(x) \in \mathbb{R}$.

(ii) For each number $\varepsilon > 0$ there are numbers $\delta > 0$ and $m \in \mathbb{N}$ such that

$$\left|\left(F(y) - F(x)\right) - \left(F_n(y) - F_n(x)\right)\right| \leq \varepsilon|y - x|$$

whenever $n \geq m$ and $y \in]x - \delta, x + \delta[$.

If, in addition, the functions F_n are all differentiable at x and if $F_n'(x) = f_n(x)$ for every $n \in \mathbb{N}$, then F is differentiable at x and $F'(x) = f(x)$.

Proof. Given any $y \in \mathbb{R}$ and $n \in \mathbb{N}$,

$$|F(y) - F(x) - (y - x)f(x)| \leq \left|\left(F(y) - F(x)\right) - \left(F_n(y) - F_n(x)\right)\right|$$
$$+ |F_n(y) - F_n(x) - (y - x)f_n(x)| + |y - x||f(x) - f_n(x)|.$$

For $\varepsilon > 0$, there are, by condition (ii), $\delta > 0$ and $m \in \mathbb{N}$ such that

$$\left|\left(F(y) - F(x)\right) - \left(F_n(y) - F_n(x)\right)\right| \leq \frac{\varepsilon}{3}|y - x|$$

whenever $n \geq m$ and $y \in]x - \delta, x + \delta[$. By (i), we may choose m large enough to ensure that

$$|f(x) - f_m(x)| < \frac{\varepsilon}{3}$$

also holds. Finally, we may choose δ small enough to ensure that

$$|F_m(y) - F_m(x) - (y - x)f_m(x)| \leq \frac{\varepsilon}{3}|y - x|$$

for all $y \in]x - \delta, x + \delta[$. Then

$$|F(y) - F(x) - (y - x)f(x)| \leq \varepsilon|y - x|$$

holds for all such y, which shows that F is differentiable at x and that its derivative there is $f(x)$. $\square$

By Theorem 6.15, λ-integrable functions may be approximated using continuous functions for which the differentiability of the indefinite integral is well known (fundamental theorem of calculus). We now show that we can make such an approximation in a manner which preserves the λ-a.e. differentiability of the indefinite integral of the limit function. The result we announced above is then an easy corollary.

Theorem 9.12 *Let $(f_n)_{n \in \mathbb{N}}$ be a sequence in $\mathcal{L}^1(\lambda)$ converging λ-a.e. to a function $f \in \mathbb{R}^{\mathbb{R}}$. Suppose there is a $g \in \mathcal{L}^1(\lambda)_+$ such that $|f_n| \leq g$ λ-a.e.*

for every $n \in \mathbb{N}$. *Given* $n \in \mathbb{N}$, *let* F_n *be an indefinite integral of* f_n *such that* $F_n(0) = 0$. *Then:*

(a) $(F_n)_{n \in \mathbb{N}}$ *converges pointwise to a function* $F \in \mathbb{R}^{\mathbb{R}}$, *and* F *is an indefinite integral of* f.

(b) *The set* N *of those* $x \in \mathbb{R}$ *at which condition (ii) of Proposition 9.11 is not satisfied is a* λ-*null set.*

(c) *If the functions* F_n *are all* λ-*a.e. differentiable and if* $F_n'(x) = f_n(x)$ *holds* λ-*a.e. for all* $n \in \mathbb{N}$, *then* F *is* λ-*a.e. differentiable and* $F'(x) = f(x)$ λ-*a.e.*

Proof. (a) follows immediately from the Lebesgue convergence theorem.

(b) We first assume that the sequence $(f_n)_{n \in \mathbb{N}}$ is increasing. Take $k \in \mathbb{N}$. We write C_k for the set of all $x \in \mathbb{R}$ with the property that for each $\delta > 0$ and $m \in \mathbb{N}$ there are a $y \in]x - \delta, x + \delta[$ and an $n \geq m$ such that

$$\int_I |f - f_n| d\lambda > \frac{1}{k} \lambda^{\mathbb{R}}(I),$$

where I is the closed interval with endpoints x and y. Given $x \in C_k$, $\delta > 0$ and $m \in \mathbb{N}$, choose an $n \geq m$ and an open interval $G \subset]x - \delta, x + \delta[$ such that $x \in G$ and

$$\int_G |f - f_n| d\lambda > \frac{1}{k} \lambda^{\mathbb{R}}(G).$$

(Recall that by Theorem 9.10 and Corollary 9.9 an indefinite integral is continuous.) Moreover, because the sequence $(f_l)_{l \in \mathbb{N}}$ is increasing,

$$\int_G |f - f_m| d\lambda > \frac{1}{k} \lambda^{\mathbb{R}}(G).$$

Fix $m \in \mathbb{N}$. Let α and β be real numbers with $\alpha < \beta$. For all $x \in C_k \cap]\alpha, \beta[$ and all $\delta > 0$, we choose an open interval with the above properties. In this way, we obtain a Vitali cover of $C_k \cap]\alpha, \beta[$. By the Vitali covering theorem, there is a countable disjoint family $(G_\iota^{(k,m)})_{\iota \in I}$ of such intervals such that $G_\iota^{(k,m)} \subset]\alpha, \beta[$ for all $\iota \in I$ and

$$C_k \cap]\alpha, \beta[\setminus \bigcup_{\iota \in I} G_\iota^{(k,m)} \in \mathfrak{N}(\lambda).$$

Setting $G^{(k,m)} := \bigcup_{\iota \in I} G_\iota^{(k,m)}$, we see that

$$\int |f - f_m| d\lambda \geq \int_{G^{(k,m)}} |f - f_m| d\lambda \geq \frac{1}{k} \lambda^{\mathbb{R}}(G^{(k,m)}).$$

For each $m \in \mathbb{N}$, choose such a set $G^{(k,m)}$. Setting $G^k := \bigcap_{m \in \mathbb{N}} G^{(k,m)}$, we have

$$C_k \cap]\alpha, \beta[\setminus G^k \in \mathfrak{N}(\lambda)$$

and

$$\int |f - f_m| d\lambda \geq \frac{1}{k} \lambda^{\mathbb{R}}(G^k)$$

for all $m \in \mathbb{N}$. By the monotone convergence theorem,

$$\lim_{m \to \infty} \int |f - f_m| d\lambda = 0.$$

Thus $\lambda^{\mathbb{R}}(G^k) = 0$. Therefore $G^k \in \mathfrak{N}(\lambda)$ and $C_k \cap]\alpha, \beta[\in \mathfrak{N}(\lambda)$. Since α and β are arbitrary, $C_k \in \mathfrak{N}(\lambda)$. This holds for all $k \in \mathbb{N}$. Hence

$$\bigcup_{k \in \mathbb{N}} C_k \in \mathfrak{N}(\lambda).$$

But if $x \in \mathbb{R} \setminus \bigcup_{k \in \mathbb{N}} C_k$, then for each $\varepsilon > 0$ we may choose $\delta > 0$ and $m \in \mathbb{N}$ in such a way that for all $y \in]x - \delta, x + \delta[$ and all $n \geq m$

$$\int_I |f - f_n| d\lambda \leq \frac{\varepsilon}{2} \lambda^{\mathbb{R}}(I),$$

where I denotes the closed interval with endpoints x and y. The dual statement for decreasing sequences follows similarly.

Let $(f_n)_{n \in \mathbb{N}}$ be arbitrary. By assumption,

$$f(x) = \limsup_{n \to \infty} f_n(x) = \liminf_{n \to \infty} f_n(x) \quad \lambda\text{-a.e.}$$

Since $|f_n| \leq g$ λ-a.e. for every $n \in \mathbb{N}$, we may choose a decreasing sequence $(h_n)_{n \in \mathbb{N}}$ and an increasing sequence $(g_n)_{n \in \mathbb{N}}$ in $\mathcal{L}^1(\lambda)$ such that

$$g_n(x) \leq f_n(x) \leq h_n(x) \quad \lambda\text{-a.e.}$$

and

$$f(x) = \sup_{n \in \mathbb{N}} g_n(x) = \inf_{n \in \mathbb{N}} h_n(x) \quad \lambda\text{-a.e.}$$

Let C and D be λ-null sets constructed for the sequences $(g_n)_{n \in \mathbb{N}}$ and $(h_n)_{n \in \mathbb{N}}$ in accordance with the first part of the proof. Then for all $x \in \mathbb{R} \setminus (C \cup D)$ and $\varepsilon > 0$, choose a $\delta > 0$ and an $m \in \mathbb{N}$ such that

$$\int_I |f - f_n| d\lambda \leq \int_I |h_n \overset{\bullet}{-} g_n| d\lambda \leq \int_I |f - h_n| d\lambda + \int_I |f - g_n| d\lambda \leq \varepsilon \lambda^{\mathbb{R}}(I)$$

for all $y \in]x - \delta, x + \delta[$ and all $n \geq m$, where again I is the closed interval with endpoints x and y. (b) now follows from

$$\left|(F(y) - F(x)) - (F_n(y) - F_n(x))\right| = \left|\int_I f \, d\lambda - \int_I f_n d\lambda\right| \leq \int_I |f - f_n| d\lambda.$$

(c) follows from (b) and Proposition 9.11. $\qquad\square$

We are now in a position to prove the promised theorem on indefinite integrals.

Corollary 9.13 *Take* $f \in \mathcal{L}^1_{loc}(\lambda)$. *Then each indefinite integral* F *of* f *is* λ*-a.e. differentiable and*

$$F'(x) = f(x) \quad \lambda\text{- }a.e.$$

Proof. Suppose first that $f \in \mathcal{L}^1(\lambda)$. By Theorem 6.15, there is a sequence $(f_n)_{n \in \mathbb{N}}$ of continuous functions converging to f λ-a.e. such that there is a $g \in \mathcal{L}^1(\lambda)_+$ with $|f_n| \leq g$ λ-a.e. Let F be an indefinite integral of f. If G denotes the indefinite integral of f with $G(0) = 0$, then, by Theorem 9.7(c), there is a $\gamma \in \mathbb{R}$ such that $F = G + \gamma$. By the previous theorem, G is λ-a.e. differentiable and $G'(x) = f(x)$ λ-a.e. Thus F is also λ-a.e. differentiable, and $F'(x) = f(x)$ λ-a.e.

Since differentiability is a local property, the claim follows easily for all $f \in \mathcal{L}^1_{loc}(\lambda)$. $\qquad\square$

From Corollary 9.13 and Theorem 9.10, the **generalized fundamental theorem of calculus** now follows:

Corollary 9.14 *Each absolutely continuous function* F *is* λ*-a.e. differentiable. Moreover, if* F' *denotes a function in* $\mathbb{R}^\mathbb{R}$ *which is* λ*-a.e. equal to the derivative of* F, *then* $\mu_F = F' \cdot \lambda$ *and* F *is an indefinite integral of* F'.

In this case, we thus have a pleasing interpretation of the **Radon–Nikodym derivative** F': it is λ-a.e. equal to the classical derivative of F! This explains the expression 'Radon–Nikodym *derivative*'.

One might be tempted to guess that each λ-a.e. differentiable function is automatically an indefinite integral. We now present, however, an example of a λ-a.e. differentiable increasing continuous function F (and thus, in particular, of locally finite variation), for which this is not true, or, in other words, which is not absolutely continuous.

We appeal to the construction of the Cantor set C in Section 3.3 and we use the notation from there. We define the function F first on $U := \mathbb{R} \setminus C$ by setting

$$F(x) := \begin{cases} 0 & \text{if } x < 0 \\ 1 & \text{if } x > 1 \\ \frac{2i-1}{2^n} & \text{if } x \in I_{ni} \quad (n \in \mathbb{N}, \, 1 \leq i \leq 2^{n-1}). \end{cases}$$

Thus

$$F = \tfrac{1}{2} \text{ on }]\tfrac{1}{3}, \tfrac{2}{3}[;$$

$$F = \tfrac{1}{4} \text{ on }]\tfrac{1}{9}, \tfrac{2}{9}[, \quad F = \tfrac{3}{4} \text{ on }]\tfrac{7}{9}, \tfrac{8}{9}[;$$

$$F = \tfrac{1}{8} \text{ on }]\tfrac{1}{27}, \tfrac{2}{27}[, \ F = \tfrac{3}{8} \text{ on }]\tfrac{7}{27}, \tfrac{8}{27}[, \ F = \tfrac{5}{8} \text{ on }]\tfrac{19}{27}, \tfrac{20}{27}[, \ F = \tfrac{7}{8} \text{ on }]\tfrac{25}{27}, \tfrac{26}{27}[$$

and so on. (Draw a figure!) The reader is invited to verify that the above defines an increasing function on U. Finally, put

$$F(x) := \sup\{f(u) \mid u \in U, \, u \leq x\} \qquad \text{for every } x \in \mathbb{R}.$$

We again leave it as an exercise to show that the function F, usually called the **Cantor function**, is increasing and continuous. Obviously, since F is constant on the open intervals I_{ni}, the Cantor function is differentiable at every point of U, and $F'(x) = 0$ for every $x \in U$. That is, recalling that C is a λ-null set, $F' = 0$ λ-a.e. It follows that $F' \cdot \lambda = 0$. On the other hand, $\mu_F([0, 1[) = F(1) - F(0) = 1$. Thus $\mu_F \neq F' \cdot \lambda$, and in view of Corollary 9.14, F is not absolutely continuous and thus is not an indefinite integral of any locally λ-integrable function.

In conclusion, we show – in another elegant application of Vitali's covering theorem – that any function of locally finite variation is in fact λ-a.e. differentiable. However, by the considerations above, *only* the absolutely continuous functions can be reconstructed from their derivatives.

Theorem 9.15 (Lebesgue) *Every function of locally finite variation is λ-a.e. differentiable.*

Proof. For every $f \in \mathbb{R}^{\mathbb{R}}$ and every $x \in \mathbb{R}$, put

$$D^+ f(x) := \limsup_{\substack{h \to 0 \\ h > 0}} \frac{f(x + h) - f(x)}{h},$$

$$D_+ f(x) := \liminf_{\substack{h \to 0 \\ h > 0}} \frac{f(x + h) - f(x)}{h},$$

$$D^- f(x) := \limsup_{\substack{h \to 0 \\ h > 0}} \frac{f(x) - f(x - h)}{h},$$

$$D_- f(x) := \liminf_{\substack{h \to 0 \\ h > 0}} \frac{f(x) - f(x - h)}{h}.$$

(Recall the definition of $\limsup$ and $\liminf$ from Section 3.5.) If these four numbers coincide and are real, then f is clearly differentiable at x.

Now take $f \in \mathcal{V}$. In view of Corollary 9.2, we may suppose that f is increasing. Take $a, b \in \mathbb{R}$ with $a < b$ and put

$$A := \{x \in]a, b[\mid D^+ f(x) > D_- f(x)\},$$
$$B := \{x \in]a, b[\mid D^- f(x) > D_+ f(x)\}.$$

For each $x \in]a, b[\setminus (A \cup B)$,

$$D^+ f(x) \leq D_- f(x) \leq D^- f(x) \leq D_+ f(x) \leq D^+ f(x),$$

i.e. all these four numbers coincide (but may still be $\pm\infty$).

We show that $A, B \in \mathfrak{N}(\lambda)$. For $p, q \in \mathbb{Q}$ with $p < q$, put

$$A_{pq} := \{x \in A \mid D^+ f(x) > q > p > D_- f(x)\}.$$

Since A is the (countable!) union of all the sets A_{pq}, we must show that each A_{pq} belongs to $\mathfrak{N}(\lambda)$. So, having fixed p and q, put

$$\alpha := \inf\{\lambda^{\mathbb{R}}(U) \mid U \text{ open}, A_{pq} \subset U \subset]a, b[\}.$$

Take $\varepsilon > 0$. Fix an open U such that $A_{pq} \subset U \subset]a, b[$ and $\lambda^{\mathbb{R}}(U) < \alpha + \varepsilon$, and fix $n \in \mathbb{N}$.

Since $p > D_- f(x)$, given $x \in A_{pq}$ and $\delta > 0$, there is a real $h > 0$, $h < \delta/2$, such that

$$]x - h, x + \tfrac{1}{n}h[\subset U \quad \text{and} \quad f(x) - f(x - h) < ph.$$

The set of all such intervals $]x - h, x + \tfrac{1}{n}h[$ is a Vitali cover of A_{pq}. By Vitali's covering theorem we can find $x_1, \ldots, x_r \in A_{pq}$ and numbers $h_1, \ldots, h_r > 0$ such that the intervals $I_i :=]x_i - h_i, x_i + \tfrac{1}{n}h_i[$ are pairwise disjoint, contained in U and satisfy

$$\sum_{i=1}^{r} \left(1 + \tfrac{1}{n}\right)h_i > \alpha - \varepsilon.$$

We then have that

$$\sum_{i=1}^{r} \left(f(x_i) - f(x_i - h_i)\right) < p \sum_{i=1}^{r} h_i < p\lambda^{\mathbb{R}}(U) < p(\alpha + \varepsilon),$$

and, putting $T := \bigcup_{i=1}^{r} I_i$, also that $\lambda^{\mathbb{R}}(T) > \alpha - \varepsilon$. Note that there is an open set V such that $A_{pq} \setminus T \subset V \subset U$ and $\lambda^{\mathbb{R}}(V) < 2\varepsilon$. Hence

$$\inf\{\lambda^{\mathbb{R}}(W) \mid W \text{ open}, A_{pq} \cap T \subset W \subset U\} \geq \alpha - 2\varepsilon.$$

Since $q < D^+ f(y)$, given $y \in A_{pq} \cap T$ and $\delta > 0$, there is a real number $k > 0$, $k < \delta/2$, such that

$$]y - \tfrac{1}{n}k, y + k[\subset T \quad \text{and} \quad f(y + h) - f(y) > qk.$$

The set of all such intervals $]y - \tfrac{1}{n}k, y + k[$ being a Vitali cover of $A_{pq} \cap T$, we can find, again by Vitali's covering theorem, points $y_1, \ldots, y_s \in A_{pq} \cap T$ and numbers $k_1, \ldots, k_s > 0$ such that the intervals $J_l :=]y_l - \tfrac{1}{n}k_l, y_l + k_l[$ are pairwise disjoint, contained in T and satisfy

$$\sum_{l=1}^{s} \left(1 + \tfrac{1}{n}\right)k_l > \alpha - 3\varepsilon.$$

We obtain

$$\sum_{l=1}^{s} \left(f(y_l + k_l) - f(y_l)\right) > q \sum_{l=1}^{s} k_l > q(\alpha - 3\varepsilon)\frac{n}{n+1}.$$

Making use of the fact that f is increasing and noting that $\bigcup_{l=1}^{s} J_l \subset T$, we now see that

$$q(\alpha - 3\varepsilon)\frac{n}{n+1} < \sum_{l=1}^{s} \left(f(y_l + k_l) - f(y_l)\right) \leq \sum_{i=1}^{r} \left(f(x_i) - f(x_i - h_i)\right) < p(\alpha + \varepsilon).$$

This holds for every $n \in \mathbb{N}$. We conclude that

$$q(\alpha - 3\varepsilon) \le p(\alpha + \varepsilon).$$

Since ε is arbitrary, it follows that $q\alpha \le p\alpha$. But $q > p$, so $\alpha = 0$. This implies $A_{pq} \in \mathfrak{N}(\lambda)$.

We have shown that $A \in \mathfrak{N}(\lambda)$. The proof that $B \in \mathfrak{N}(\lambda)$ is analogous. Therefore for λ-almost all $x \in]a, b[$

$$g(x) := \lim_{h \to 0} \frac{f(x+h) - f(x)}{h} \in \overline{\mathbb{R}}_+$$

exists. For all $x \in \mathbb{R} \setminus]a, b[$ and for those $x \in]a, b[$ at which the limit does not exist, put $g(x) := 0$. We want to show that this function g on $\mathbb{R}$ satisfies $g < \infty$ λ-a.e. To do so, take $n \in \mathbb{N}$ and define

$$g_n(x) := \begin{cases} \dfrac{\inf\{f(b), f(x+\frac{1}{n})\} - f(x)}{\frac{1}{n}} & \text{if } x \in]a, b[\\ 0 & \text{if } x \in \mathbb{R} \setminus]a, b[. \end{cases}$$

Then $g = \lim_{n \to \infty} g_n$ λ-a.e.

Since f is increasing, it is λ-measurable. This implies that $fe_{]a,b[} \in \mathcal{L}^1(\lambda)$. We infer from this that every g_n belongs to $\mathcal{L}^1(\lambda)$. Using the translation invariance of Lebesgue measure, we calculate:

$$\liminf_{n \to \infty} \int g_n d\lambda = \liminf_{n \to \infty} n \int_{]a,b[} \left(\inf\left\{ f(b), f(x+\tfrac{1}{n}) \right\} - f(x) \right) d\lambda(x)$$

$$= \liminf_{n \to \infty} \left(n \int_{]b,b+\frac{1}{n}[} f(b) \, d\lambda - n \int_{]a,a+\frac{1}{n}[} f \, d\lambda \right)$$

$$\le f(b) - \limsup_{n \to \infty} n \int_{]a,a+\frac{1}{n}[} f(a) \, d\lambda$$

$$= f(b) - f(a).$$

By Fatou's lemma, $g \in \mathcal{L}^1(\lambda)$. We conclude that $g < \infty$ λ-a.e. (Theorem 3.30(a)). Hence

$$\{x \in]a, b[\mid f \text{ is not differentiable at } x\} \in \mathfrak{N}(\lambda).$$

Since a, b are arbitrary, it follows easily that f is differentiable λ-a.e. $\quad\square$

Exercises

1. Determine whether the following functions from $\mathbb{R}$ to $\mathbb{R}$ are of locally

bounded variation or absolutely continuous:

$$x \longmapsto |x|;$$

$$x \longmapsto \begin{cases} 0 & \text{if } x \in \mathbb{Q} \\ 1 & \text{if } x \in \mathbb{R} \setminus \mathbb{Q}; \end{cases}$$

$$x \longmapsto \begin{cases} \sin \frac{1}{x} & \text{if } x \neq 0 \\ 0 & \text{if } x = 0; \end{cases}$$

$$x \longmapsto \begin{cases} x \sin \frac{1}{x} & \text{if } x \neq 0 \\ 0 & \text{if } x = 0; \end{cases}$$

$$x \longmapsto \begin{cases} x^2 \sin \frac{1}{x} & \text{if } x \neq 0 \\ 0 & \text{if } x = 0. \end{cases}$$

Observe however that, for example, the third map is Lebesgue-a.e. differentiable.

2. Define $f : \mathbb{R} \to \mathbb{R}$, $x \mapsto |x|$, and determine $f \vee 0$, $(-f) \vee 0$ and $|f|^{\vee}$.

3. What is the relationship of Vf to $|f|^{\vee}$ when $f \in V$?

4. Show that $f \vee 0$, $(-f) \vee 0$, $|f|^{\vee}$ are continuous whenever $f \in V$ is continuous.

5. Show that $\mathcal{V}, \overset{\circ}{\mathcal{V}}, \mathcal{D}, \overset{\circ}{\mathcal{D}}$ are vector sublattices of $\mathbb{R}^{\mathbb{R}}$ with respect to the usual order.

6. Prove the following for $f \in \mathcal{V}$.

   (a) For each $x \in \mathbb{R}$

   $$f(x_-) := \lim_{\substack{y \to x \\ y < x}} f(y) \qquad \text{and} \qquad f(x_+) := \lim_{\substack{y \to x \\ y > x}} f(y)$$

   exist.

   (b) For all $x, y \in \mathbb{R}$ with $x \le y$, the families $\left(f(\gamma) - f(\gamma_-) \right)_{\gamma \in]x,y[}$, $\left(f(\gamma_+) - f(\gamma) \right)_{\gamma \in]x,y[}$ and $\left(f(\gamma_+) - f(\gamma_-) \right)_{\gamma \in]x,y[}$ are summable and

   $$\sum_{\gamma \in]x,y[} |f(\gamma_+) - f(\gamma_-)| \le \underset{[x,y]}{V} f.$$

   (c) The set of points of discontinuity of f is countable.

   (d) There is a unique left continuous real function g on $\mathbb{R}$ which coincides with f in the points of continuity of f. Furthermore, $g \in \mathcal{V}$ and $g(x) = f(x_-)$, $g(x_+) = f(x_+)$ for every $x \in \mathbb{R}$.

   (e) There is an increasing bounded left continuous real function on $\mathbb{R}$ whose set of points of discontinuity is dense in $\mathbb{R}$.

7. Given $n \in \mathbb{N}$, define the 'sawtooth' function $f_n : \mathbb{R} \to \mathbb{R}$ by the requirements that f_n be $2 \cdot 4^{-n}$-periodic and that $f_n(x) = |x|$ for $x \in [-4^{-n}, 4^{-n}]$. For $f := \sum_{n=1}^{\infty} f_n$ prove the following.

(a) f is a bounded continuous function which is nowhere differentiable.

(b) $V_{[x,y]} f = \infty$ for all $x, y \in \mathbb{R}$ with $x < y$.

Hint for (a): Take $x \in \mathbb{R}$ and $n \in \mathbb{N}$. If $(x, f_n(x))$ and $\left(x + \frac{1}{2} \cdot 4^{-n}, f_n(x + \frac{1}{2} \cdot 4^{-n})\right)$ lie on the same straight segment of the graph of f_n, then put $\delta_n := \frac{1}{2} \cdot 4^{-n}$. Otherwise, $(x, f_n(x))$ and $\left(x - \frac{1}{2} \cdot 4^{-n}, f_n(x - \frac{1}{2} \cdot 4^{-n})\right)$ must lie on the same straight segment of the graph of f_n. In this case, put $\delta_n := -\frac{1}{2} \cdot 4^{-n}$. Now consider the difference quotients $\alpha_k := \frac{f(x+\delta_k) - f(x)}{\delta_k}$ and show that for any $k \in \mathbb{N}$,

$$\alpha_k = \sum_{n=1}^{k} \frac{f_n(x + \delta_k) - f_n(x)}{\delta_k}.$$

Conclude that the α_k are alternately odd and even integers. Hence $(\alpha_k)_{k \in \mathbb{N}}$ cannot converge in $\mathbb{R}$.

8. Let $(f_n)_{n \in \mathbb{N}}$ be a sequence in $\mathcal{V}$ such that $\sum_{n \in \mathbb{N}} V_{[x,y]} f_n < \infty$ for all $x, y \in \mathbb{R}$, $x < y$. Assume that $\sum_{n \in \mathbb{N}} f_n(0)$ exists in $\mathbb{R}$. Prove the following.

(a) $\sum_{n \in \mathbb{N}} f_n$ converges to some $f \in \mathcal{V}$, uniformly in every interval $[x, y]$.

(b) If each $f_n \in \mathcal{D}$, then $f \in \mathcal{D}$, too.

9. Verify that each continuously differentiable real function on $\mathbb{R}$ is absolutely continuous.

10. Define the function f by $f(x) := 0$ for every $x \in]-\infty, 0] \cup ([0, 1] \setminus \mathbb{Q}) \cup]1, \infty[$; $f(x) := 1/n$ if $x = m/n \in [0, 1] \cap \mathbb{Q}$ such that $m, n \in \mathbb{N}$ have no common divisor. Prove the following.

(a) f is continuous at every $x \in [0, 1] \setminus \mathbb{Q}$ and discontinuous at every $x \in [0, 1] \cap \mathbb{Q}$.

(b) f is Riemann integrable and hence Lebesgue integrable.

(c) If F is an indefinite integral of f, then $F' = 0$. Hence $F'(x) = f(x)$ for $x \in [0, 1]$ if and only if $x \in]0, 1] \setminus \mathbb{Q}$.

11. In this exercise we discuss a generalization of the Riemann integral, namely the Riemann–Stieltjes integral. We fix a non-empty interval $A = [a, b]$ of $\mathbb{R}$. Let $f, g \in \mathbb{R}^A$. Then f is called **Riemann–Stieltjes integrable (on A) with respect to g** if there is a $\xi \in \mathbb{R}$ with the following property. For every $\varepsilon > 0$ there is a $\delta > 0$ such that for every subdivision $a = x_1 \leq \cdots \leq x_n = b$ of A with $\sup_{k<n} |x_{k+1} - x_k| < \delta$ and

for arbitrary $y_k \in [x_k, x_{k+1}]$, $(k < n)$,

$$\left| \sum_{k=1}^{n-1} f(y_k)\big(g(x_{k+1}) - g(x_k)\big) - \xi \right| < \varepsilon.$$

The number ξ, obviously uniquely determined, is called the **Riemann–Stieltjes integral of f with respect to g**, and we write

$$\int_a^b f(x)\, dg(x) := \xi.$$

Prove the following.

(a) If f_1 and f_2 are Riemann–Stieltjes integrable with respect to g and if $\alpha_1, \alpha_2 \in \mathbb{R}$, then $\alpha_1 f_1 + \alpha_2 f_2$ is Riemann–Stieltjes integrable with respect to g and

$$\int_a^b (\alpha_1 f_1 + \alpha_2 f_2)(x)\, dg(x) = \alpha_1 \int_a^b f_1(x)\, dg(x) + \alpha_2 \int_a^b f_2(x)\, dg(x).$$

(b) If f is Riemann–Stieltjes integrable with respect to g_1 and g_2 and if $\alpha_1, \alpha_2 \in \mathbb{R}$, then f is Riemann–Stieltjes integrable with respect to $\alpha_1 g_1 + \alpha_2 g_2$ and

$$\int_a^b f(x)\, d(\alpha_1 g_1 + \alpha_2 g_2)(x) = \alpha_1 \int_a^b f(x)\, dg_1(x) + \alpha_2 \int_a^b f(x)\, dg_2(x).$$

(c) If $a < c < b$ and if all three of the following integrals exist, then

$$\int_a^b f(x)\, dg(x) = \int_a^c f(x)\, dg(x) + \int_c^b f(x)\, dg(x).$$

(In fact, the existence of the right-hand integrals can be derived from the existence of the left-hand integral.) Find an example where the two right-hand integrals exist, but *not* the left-hand one.

(d) (Integration by parts.) If one of the following integrals exists, then so does the other and

$$\int_a^b f(x)\, dg(x) + \int_a^b g(x)\, df(x) = f(b)g(b) - f(a)g(a).$$

Hint: Suppose that $\int_a^b f(x)\, dg(x)$ exists. Take $\varepsilon > 0$. Choose $\delta > 0$ according to the definition of Riemann–Stieltjes integrability. Let $a = x_1' \le \cdots \le x_m' = b$ be a subdivision of A with $\sup_{l<m} |x_{l+1}' - x_l'| < \delta/2$, and for every $l < m$ choose $y_l' \in [x_l', x_{l+1}']$. Put

$$x_1'' := a, \qquad x_k'' := y_{k-1}' \text{ for } 1 < k \le m, \qquad x_{m+1}'' := b$$

and $y_k'' := x_k'$ for every $k \le m$. Compute $\sum_{k=1}^m f(y_k'')\big(g(x_{k+1}'') - g(x_k'')\big)$

and conclude that

$$\left| \sum_{l=1}^{m-1} g(y_l')\big(f(x_{l+1}') - f(x_l')\big) - f(b)g(b) + f(a)g(a) + \int_a^b f(x)\,dg(x) \right| < \varepsilon.$$

(e) If f is Riemann integrable on A and g is continuously differentiable on A, then f is Riemann–Stieltjes integrable with respect to g and

$$\int_a^b f(x)\,dg(x) = \int_a^b f(x)g'(x)\,dx.$$

Hint: To compute $\sum_{k=1}^{n-1} f(y_k)\big(g(x_{k+1}) - g(x_k)\big)$, use the Lagrange formula $g(x_{k+1}) - g(x_k) = g'(z_k)(x_{k+1} - x_k)$ and the uniform continuity of g'.

12. We continue the previous exercise. Let $A = [a, b]$ be a non-empty interval of $\mathbb{R}$. Take $f, g \in \mathbb{R}^A$ and suppose throughout this exercise that f is bounded and g is increasing. For each subdivision $\mathfrak{Z} : a = x_1 \leq \cdots \leq x_n = b$ of A, put

$$\ell_*(f, g, \mathfrak{Z}) := \sum_{k=1}^{n-1} \left(\inf_{y \in [x_k, x_{k+1}]} f(y) \right)\big(g(x_{k+1}) - g(x_k)\big),$$

$$\ell^*(f, g, \mathfrak{Z}) := \sum_{k=1}^{n-1} \left(\sup_{y \in [x_k, x_{k+1}]} f(y) \right)\big(g(x_{k+1}) - g(x_k)\big).$$

Call $\mathfrak{Z}'$ finer than $\mathfrak{Z}$ if every subdivision point of $\mathfrak{Z}$ is also a subdivision point of $\mathfrak{Z}'$. Prove the following.

(a) If $\mathfrak{Z}'$ is finer than $\mathfrak{Z}$, then $\ell_*(f, g, \mathfrak{Z}') \geq \ell_*(f, g, \mathfrak{Z})$ and $\ell^*(f, g, \mathfrak{Z}') \leq \ell^*(f, g, \mathfrak{Z})$.

(b) $\ell_*(f, g, \mathfrak{Z}) \leq \ell^*(f, g, \mathfrak{Z}')$ for arbitrary subdivisions $\mathfrak{Z}, \mathfrak{Z}'$. Moreover, if $\xi = \int_a^b f(x)\,dg(x)$ exists, then $\ell_*(f, g, \mathfrak{Z}) \leq \xi \leq \ell^*(f, g, \mathfrak{Z}')$.

Put

$$\xi_*(f, g) := \sup\{\ell_*(f, g, \mathfrak{Z}) \,|\, \mathfrak{Z} \text{ is a subdivision of } A\},$$
$$\xi^*(f, g) := \inf\{\ell^*(f, g, \mathfrak{Z}) \,|\, \mathfrak{Z} \text{ is a subdivision of } A\}.$$

(c) If f is Riemann–Stieltjes integrable with respect to g, then

$$\xi_*(f, g) = \int_a^b f(x)\,dg(x) = \xi^*(f, g).$$

(d) If g is continuous, then f is Riemann–Stieltjes integrable with respect to g if and only if $\xi_*(f, g) = \xi^*(f, g)$. In particular, f is Riemann–Stieltjes integrable with respect to id_A if and only if f is Riemann integrable.

(e) If f is continuous and $h \in \mathcal{V}$, then f is Riemann–Stieltjes integrable with respect to $h|_A$. (First consider the case where h is increasing.)

(f) Find functions f, g such that $\xi_*(f,g) = \xi^*(f,g)$, but f is not Riemann–Stieltjes integrable with respect to g.

(g) Let $h \in \overset{\circ}{\mathcal{V}}$, h left continuous. If f is Riemann–Stieltjes integrable with respect to $h|_A$, then f is integrable with respect to the Stieltjes measure $\mu_h^X|_A$ and $\int_a^b f(x)\, dh|_A(x) = \int f\, d(\mu_h^X|_A)$. (Again, first consider the case that h is increasing. Apply (c) and the sandwich principle.)

Symbol index

Subject index

T - #0090 - 160425 - C0 - 234/156/17 - PB - 9780412576805 - Gloss Lamination